T0305797

Stochastic versus Deterministic Systems of Iterative Processes

TRENDS IN ABSTRACT AND APPLIED ANALYSIS

ISSN: 2424-8746

This series will provide state of the art results and applications on current topics in the broad area of Mathematical Analysis. Of a more focused nature than what is usually found in standard textbooks, these volumes will provide researchers and graduate students a path to the research frontiers in an easily accessible manner. In addition to being useful for individual study, they will also be appropriate for use in graduate and advanced undergraduate courses and research seminars. The volumes in this series will not only be of interest to mathematicians but also to scientists in other areas. For more information, please go to http://www.worldscientific.com/series/taaa

Published

Vol. 11 *Stochastic versus Deterministic Systems of Iterative Processes*
by G. S. Ladde & M. Sambandham

Vol. 10 *Nonlinear Higher Order Differential and Integral Coupled Systems: Impulsive and Integral Equations on Bounded and Unbounded Domains*
by Feliz Manuel Minhós & Robert de Sousa

Vol. 9 *Boundary Value Problems for Fractional Differential Equations and Systems*
by Bashir Ahmad, Johnny Henderson & Rodica Luca

Vol. 8 *Ordinary Differential Equations and Boundary Value Problems Volume II: Boundary Value Problems*
by John R. Graef, Johnny Henderson, Lingju Kong & Xueyan Sherry Liu

Vol. 7 *Ordinary Differential Equations and Boundary Value Problems Volume I: Advanced Ordinary Differential Equations*
by John R. Graef, Johnny Henderson, Lingju Kong & Xueyan Sherry Liu

Vol. 6 *The Strong Nonlinear Limit-Point/Limit-Circle Problem*
by Miroslav Bartušek & John R. Graef

More information on this series can be found at https://www.worldscientific.com/series/taaa

Trends in Abstract and Applied Analysis
Volume **11**

Stochastic versus Deterministic Systems of Iterative Processes

G S Ladde
University of South Florida, Tampa, USA

M Sambandham
Morehouse College, USA

World Scientific

NEW JERSEY • LONDON • SINGAPORE • BEIJING • SHANGHAI • HONG KONG • TAIPEI • CHENNAI • TOKYO

Published by

World Scientific Publishing Co. Pte. Ltd.
5 Toh Tuck Link, Singapore 596224
USA office: 27 Warren Street, Suite 401-402, Hackensack, NJ 07601
UK office: 57 Shelton Street, Covent Garden, London WC2H 9HE

Library of Congress Cataloging-in-Publication Data
Names: Ladde, G. S., author. | Sambandham, M., author.
Title: Stochastic versus deterministic systems of iterative processes /
G S Ladde, University of South Florida, USA, M Sambandham, Morehouse College, USA.
Description: New Jersey : World Scientific, [2024] | Series: Trends in abstract and applied analysis, 2424-8746 ; vol. 11 | Includes bibliographical references.
Identifiers: LCCN 2024002474 | ISBN 9789811287473 (hardcover) |
ISBN 9789811287480 (ebook for institutions) | ISBN 9789811287497 (ebook for individuals)
Subjects: LCSH: Iterative methods (Mathematics) | Stochastic analysis. | Deterministic chaos.
Classification: LCC QA297.8 .L33 2024 | DDC 518/.26--dc23/eng/20240310
LC record available at https://lccn.loc.gov/2024002474

British Library Cataloguing-in-Publication Data
A catalogue record for this book is available from the British Library.

For any available supplementary material, please visit
https://www.worldscientific.com/worldscibooks/10.1142/13707#t=suppl

Desk Editors: Nambirajan Karuppiah/Rosie Williamson/Tan Rok Ting

Typeset by Stallion Press
Email: enquiries@stallionpress.com

Printed in Singapore

Preface

The classical random flow, the Newtonian mechanics approaches, and computational schemes are the most extensively studied stochastic discrete- or continuous-state dynamic modeling methods in discrete- or continuous-state dynamic processes in biological, engineering, physical and social sciences. Three of these approaches lead to discrete- or continuous-state differential or difference equations.

In the classical stochastic modeling approach, a state of a dynamic process is considered to be a random flow or process in a discrete or continuous state satisfying a certain probabilistic law, such as Markov or diffusion. From these types of probabilistic assumptions, one then needs to determine the state transition probability distribution and density functions (STPDFs). The determination of the unknown STPDF leads to the study of deterministic problems in the theory of ordinary, partial or integro-differential equations in state variables. These types of equations are referred to as master equations in the literature. The solution processes of such systems of differential or difference equations are used to find the higher moments and other statistical properties of state dynamic processes described by random flows.

On the other hand, the classical Newtonian mechanics type of stochastic modeling approach deals with a stochastic calculus to formulate stochastic mathematical models of dynamic processes. It is obvious that classical continuous-state elementary calculus almost includes discrete-state calculus as a special case with obvious differences. In fact, state measurement systems coupled with conceptual computational schemes are discrete state or time processes.

This approach leads directly to a system of stochastic differential or difference equations, and its solution processes provide a description of the states of the dynamic processes as stochastic or random processes. Moreover, continuous-state dynamic models can be reformulated into discrete-state processes by introducing numerical schemes that lead to theoretical iterative schemes via numerical analysis techniques. This method of stochastic modeling generates the following three basic problems:

(i) Concepts of solution processes depending on continuous or discrete states, modes of convergence, and the fundamental properties of solutions: existence, uniqueness, measurability and continuous dependence on system parameters.
(ii) Probabilistic and statistical properties of solution processes: probability distribution and density function, variance, and moments of solution processes and the qualitative/quantitative behavior of solutions.
(iii) Deterministic versus stochastic modeling of dynamic processes: If the deterministic mathematical model is available, then why do we need a stochastic mathematical model? If a stochastic mathematical model provides a better description of a dynamic process than the deterministic model, then the second question is to what extent the stochastic mathematical model differs from the corresponding deterministic model in the absence of random disturbances or fluctuations and uncertainties.

Recently, as of 2004, most of the work on the theory of systems of continuous state stochastic differential equations centered around problems (i), (ii) and (iii) has been reported. This is because the theory of continuous-state deterministic and stochastic systems of differential equations has provided many mathematical tools and ideas since the development of deterministic and stochastic calculi in the 1950s and 1970s. However, since the 1980s, problems (i) and (ii) have been addressed with regard to discrete-state deterministic systems of difference equations; therefore, problems of discrete state stochastic systems of difference equations deserve more attention in the context of (i), (ii) and (iii). Since 1980, serious efforts have been made to address this issue in the context of discrete-state stochastic modeling of dynamic processes by means of systems of stochastic difference equations. In light of this interest, now is an appropriate time to present an account of stochastic versus deterministic issues in discrete-state dynamic systems in a systematic and unified way.

Two of the most powerful methods for studying systems of nonlinear difference equations are nonlinear variation of parameters and Lyapunov's Second Method. About a half century ago, a hybrid of these two methods evolved. This hybrid method is called the variational comparison method. In addition, a generalized variation of constants method was also developed in the same period of time. These new techniques are very sustainable and effective tools to investigate problems concerning stochastic systems of iterative processes, in particular stochastic versus deterministic issues.

This book offers a systematic and unified treatment of systems of stochastic iterative processes in the framework of three methods: the variational comparison method, the generalized variation of constants method and the probability distribution method.

The book is divided into seven chapters. The first chapter deals with methods concerning stochastic systems of difference equations with random parameters. Chapter 2 is devoted to methods involving stochastic systems of difference equations of the Ito–Doob type. The third chapter highlights a qualitative analysis of stochastic systems of difference equations. The fourth chapter highlights random algebraic polynomials. Chapters 5 and 6 cover numerical schemes for iterative processes with random parameters and Ito-type systems, respectively. Finally, Chapter 7 is devoted to discrete-time probabilistic, stochastic dynamic modeling, and statistical conceptual data analysis.

A few important features of the monograph are as follows:

(i) This is the first book that offers a systematic study of the well-known problem of stochastic mathematical modeling in the context of discrete-state systems of stochastic iterative processes, namely, "stochastic versus deterministic."
(ii) It complements the existing books on stochastic difference equations.
(iii) It provides a unified treatment of stability, relative stability and error estimate analysis.
(iv) It exhibits the role of randomness as well as rate functions in explicit form.

(v) It provides several illustrative analytic examples to demonstrate the scope of methods in stochastic analysis.
(vi) The methods developed in the book are applied to the existing stochastic mathematical models described by stochastic difference equations in population dynamics, hydrodynamics and physics.
(vii) Last but not least, it provides several numerical examples and figures to illustrate and compare the analytic techniques that are outlined in the book.

The monograph can be used as a textbook for graduate students. It can also be used as a reference book for both experimental and applied scientists working on the mathematical modeling of continuous- and discrete-state dynamic processes.

Contents

Chapter 1

Stochastic Systems of Difference Equations with Random Parameters Methods

1.0 Introduction

The concept of a system of difference equations with random parameters generates a very general and difficult problem of "stochastic versus deterministic" (or "randomness versus non-randomness"). In broad terms, this refers to the extent to which the solution processes of systems of difference equations with random parameters deviate from those of corresponding systems of difference equations with deterministic parameters. In this chapter, two major techniques for studying nonlinear initial value problems are developed, and a solution to the above problem is addressed. A method involving a generalized variation of constants formula for stochastic discrete-time processes is discussed in Section 1.1. By employing the concept of Lyapunov-like functions and the theory of systems of random difference inequalities, several variational comparison theorems and examples are presented in Section 1.2. The examples given help illustrate the widely general results and theorems.

1.1 Variation of Constants Method

In this section, we present a generalized variation of constants formula for stochastic discrete time processes. For this purpose, we develop a new basic result. The presented results provide a mathematical tool to study discrete-time stochastic processes.

Let us present a mathematical description of discrete-time dynamic processes in chemical, engineering, medical, physical and social sciences. It is described by the following nonlinear and non-stationary iterative process under parametric random perturbations:

$$\Delta y(k) = f(k, y(k), \omega), \quad y(k_0, \omega) = y_0, \tag{1.1.1}$$

where for fixed $(k, y) \in \mathbb{I}(k_0) \times \mathbb{R}^n$, $y(k) \in \mathbb{R}^n$, $\Delta y(k) = y(k+1) - y(k)$; $f(k, y, \omega)$ and $y_0 \in \mathbb{R}^n$ are random vectors defined on a complete probability space $(\Omega, \mathcal{F}, P)$; and for each (k, y), $f(k, y, \omega)$ describes a dynamic system under random perturbations.

In the absence of random perturbations, mathematical description (1.1.1) reduces to

$$\Delta m = F(k, m), \quad m(k_0) = m_0 = E[y_0]. \tag{1.1.2}$$

In our presentation, we also utilize the following initial value problem:

$$\Delta x = F(k, x, \omega), \quad x(k_0) = x_0. \tag{1.1.3}$$

For the sake of easy reference, we list the following assumptions with regard to (1.1.1)–(1.1.3).

Hypothesis $\mathbf{H}_{(1.1.1)}$: Assume that the initial state y_0 is $\mathcal{F}_{k_0}$-measurable. $f(k, y, \omega)$ is a sequence of random vectors. We designate by $y(k, k_0, y_0) = y(k)$ a solution process of (1.1.1), for each $k \in \mathbb{I}(k_0)$ and $y(k_0) = y_0$.

Hypothesis $\mathbf{H}_{(1.1.2)}$: F is a sequence of continuous functions defined on $\mathbb{R}^n$ into $\mathbb{R}^n$, and it is twice continuously differentiable with respect to m. A solution process of (1.1.2) is denoted by $m(k, k_0, m_0) = m(k)$ for $k \geq k_0$. It is further assumed that its second derivative $\frac{\partial^2}{\partial m_0 \partial m_0} m(k, k_0, m_0)$ is locally Lipschitzian in m_0, for each (k, k_0).

Remark 1.1.1. For each $p \in \mathbb{I}$ and $y \in \mathbb{R}^n$, we observe that $y(p+1)$ satisfies

$$y(p+1, p, y) = y(p+1) = y + f(p, y, \omega), \quad y(p) = y. \tag{1.1.4}$$

In the following, we present a few auxiliary results with respect to smooth system (1.1.2). These results will be used in this section as well as in the subsequent sections of this chapter.

Lemma 1.1.1. *Let hypothesis $H_{(1.1.2)}$ be satisfied. Then, the solution process $m(k) = m(k, k_0, m_0)$ of (1.1.2) is unique, and it satisfies*

$$m(k, p, x) = m(k, p + 1, x + F(p, x)), \tag{1.1.5}$$

where $m(k, p, x)$ is the solution process of (1.1.2) through the initial data (p, x), for $k_0 \leq p \leq k$ and $p, k \in \mathbb{I}(k_0)$.

Proof. We note that the uniqueness of the solution process $m(k, k_0, m_0)$ of (1.1.2) follows from the repeated composition of a map $G(x) = x + F(p, x)$ with itself.

To prove (1.1.5), we consider the left-hand side of (1.1.5):

$$\begin{aligned}
&m(k, p, x)\\
&\quad = (G \circ G \circ \cdots \circ G \circ G)(x)\\
&\qquad \times (\text{by } (k - p) \text{ repeated composition of } G \text{ with itself})\\
&\quad = (G \circ G \circ \cdots \circ G) \circ G(x) \text{ (by regrouping)}\\
&\quad = (G \circ G \circ \cdots \circ G)(m(p + 1, p, x)\\
&\qquad \times (\text{from the definition of } G \text{ and } (1.1.4))\\
&\quad = m(k, p + 1, m(p + 1, p, x)) \text{ (by the same reasoning as before)}\\
&\quad = m(k, p + 1, x + F(p, x)) \text{ (by the definition of } G \text{ and } (1.1.4)).
\end{aligned}$$

This completes the proof of the lemma.

Lemma 1.1.2. *Let the assumption of Lemma* 1.1.1 *be satisfied. Then:*

(a) $\frac{\partial}{\partial m_0} m(k, k_0, m_0)$ *exists and satisfies the following linear homogeneous matrix iterative process:*

$$\Delta X = \frac{\partial}{\partial m} F(k, m(k))X, \quad X(k_0) = I, \tag{1.1.6}$$

along the solution process $m(k, k_0, m_0) = m(k)$ of (1.1.2), and it is denoted by

$$\frac{\partial}{\partial m_0} m(k, k_0, m_0) = \Phi(k, k_0, m_0); \tag{1.1.7}$$

(b) $\frac{\partial^2}{\partial m_0 \partial m_0} m(k, k_0, m_0)$ *exists and satisfies the following linear nonhomogeneous matrix iterative process:*

$$\Delta X = \frac{\partial}{\partial m} F(k, m(k))X + \frac{\partial^2}{\partial m^2} F(k, m(k)) \times \cdot \Phi(k, k_0, m_0)\Phi(k, k_0, m_0), \quad X(k_0) = 0, \tag{1.1.8}$$

along the solution process $m(k, k_0, m_0) = m(k)$ *of* (1.1.2), *where* $\frac{\partial^2}{\partial m^2} F(k, m, z(k))$ *and* $\frac{\partial^2}{\partial m_0^2}, m(k, k_0, m_0)$ *are* $n \times n$ *Hessian matrices of* F *and* $m(k, k_0, m_0)$, *respectively, and their elements are* $1 \times n$ *matrices, and*

$$\frac{\partial^2}{\partial m^2} F(k, m(k)) \cdot \Phi(k, k_0, mz_0) = \left(\frac{\partial^2}{\partial m \partial m_j} F_i(k, m(k))\Phi(k, k_0, m_0) \right)_{n \times n}.$$

Proof. The proof of part (a) can be imitated by following the proof of Theorem A.1.1. The proof of part (b) can be reformulated by following the steps in the proof of Theorem A.1.1. The details are left as an exercise to the reader.

Lemma 1.1.3. *Let the hypotheses of Lemma* 1.1.1 *be satisfied. Then*

(a)

$$\Phi(k, p, m) = \Phi(k, p+1, m + F(p, m))\Phi(p+1, p, m); \tag{1.1.9}$$

(b)

$$\begin{aligned} \frac{\partial^2}{\partial m_0 \partial m_0} m(k, p, m) &= \frac{\partial}{\partial m_0} \Phi(k, p, m) \\ &= \frac{\partial}{\partial m_0} \Phi(k, p+2, m(p+2)) \cdot \Phi(p+2, p, m)\Phi(p+2, p, m) \\ &\quad + (\Phi(k, p+2, m(p+2)) \cdot \frac{\partial}{\partial m_0} \Phi(p+2, p, m)\Phi(p+1, p, m); \end{aligned} \tag{1.1.10}$$

(c)

$$\begin{aligned} m(k,p+1,u+v) &- m(k,p,u) \\ &= \Phi(k,p+1,u+F(p,u))\Delta(p,u,v) \\ &\quad + \frac{1}{2}\frac{\partial}{\partial m_0}\Phi(k,p+1,u+F(p,u))\cdot\Delta(p,u,v)\Delta(p,u,v) \\ &\quad + O(k,p+1,u+F(p,u),\Delta(p,u,v)), \end{aligned} \tag{1.1.11}$$

where

$$\begin{aligned} &O(k,p+1,u+F(p,u),\Delta(p,u,v)) \\ &= \int_0^1\int_0^1 \theta\Big[\frac{\partial}{\partial m_0}\Phi(k,p+1,u+F(p,u)+\theta\psi\Delta(p,u,v)) \\ &\quad - \frac{\partial}{\partial m_0}\Phi(k,p+1,u+F(p,u))\Big]\cdot\Delta(p,u,v)d\psi\, d\theta\,\Delta(p,u,v) \end{aligned} \tag{1.1.12}$$

and

$$\Delta(p,u,v) = [v - F(p,u)] \quad \text{for } u,v \in R^n. \tag{1.1.13}$$

Proof. By differentiating partially both sides of (1.1.5) with respect to x, we have

$$\begin{aligned} &\frac{\partial}{\partial y}m(k,p,y) \\ &= \frac{\partial}{\partial y}m(k,p+1,y+F(p,y)) \quad \text{(by (1.1.4))} \\ &= \frac{\partial}{\partial y}m(k,p+1,m(p+1,p,y)) \quad \text{(by chain rule)} \\ &= \frac{\partial}{\partial m_0}m(k,p+1,m(p+1,p,y))\frac{\partial}{\partial y}m(p+1,p,y) \quad \text{(from (1.1.7))} \\ &= \Phi(k,p+1,m(p+1,p,y))\Phi(p+1,p,y). \end{aligned}$$

This completes the proof of part (a).

To prove (1.1.10), applying (1.1.4),

$$m(k,p,y) = m(k,p+2,m(p+1,p,y) + F(p+1,m(p+1,p,y))).$$

This, together with the application of (1.1.9), results in

$$\begin{aligned}\frac{\partial}{\partial m_0} m(k,p,m) &= \Phi(k,p,m)\\ &= \Phi(k,p+2,m(p+2,p,m))\Phi(p+2,p,m).\end{aligned}$$

Again, by differentiating this expression partially with respect to m, for each (k,p), we obtain

$$\begin{aligned}&\frac{\partial^2}{\partial m_0^2} m(k,p,m) = \frac{\partial}{\partial m_0}\Phi(k,p,m)\\ &= \frac{\partial}{\partial m_0}\left(\Phi(k,p+2,m(p+2,p,m))\Phi(p+2,p,m)\right)\\ &\quad \times(\text{by product and chain rules})\\ &= \frac{\partial}{\partial m_0}\Phi(k,p+2,m(p+2,p,m))\cdot\frac{\partial}{\partial m_0} m(p+2)\Phi(p+2,p,m))\\ &\quad +\Phi(k,p+2,m(p+2,p,m))\frac{\partial}{\partial m_0}\\ &\quad \times\Phi(p+2,p+1,m(p+1,p,m)) \quad (\text{by notations})\\ &= \frac{\partial}{\partial m_0}\Phi(k,p+2,m(p+2))\cdot\Phi(p+2,p,m)\Phi(p+2,p,m)\\ &\quad +\Phi(k,p+2,m(p+2))\frac{\partial}{\partial m_0}\Phi(p+2,p,m)\cdot\Phi(p+1,p,m).\end{aligned}$$

This completes the proof of (1.1.10).

To prove (1.1.11), by applying the generalized mean value Lemma A.1.1, we have

$$\begin{aligned}&m(k,p+1,u+v) - m(k,p,u) \quad (\text{from } 1.1.5)\\ &\quad = m(k,p+1,u+F(p,u)+v-F(p,u)))\\ &\qquad - m(k,p+1,u+F(p,u))\end{aligned}$$

$$
\begin{aligned}
&= \int_0^1 \Phi(k, p+1, u+F(p,u)+\theta\Delta(p,u,v))d\theta\Delta(p,u,v) \\
&\quad \times \text{(from algebra and (1.1.5))} \\
&= \Phi(k, p+1, u+F(p,u))\Delta(p,u,v) \\
&\quad + \int_0^1 [\Phi(k, p+1, u+F(p,u)+\theta\Delta(p,u,v)) \\
&\quad - \Phi(k, p+1, u+F(p,u)]d\theta\Delta(p,u,v) \\
&= \Phi(k, p+1, u+F(p,u))\Delta(p,u,v) \\
&\quad + \Psi(k, p+1, u+F(p,u), \Delta(p,u,v))\Delta(p,u,v), \qquad (1.1.14)
\end{aligned}
$$

where

$$
\begin{aligned}
&\Psi(k, p+1, u+F(p,u), \Delta(p,u,v)) \quad \text{(from (1.1.5))} \\
&\quad = \int_0^1 [\Phi(k, p+1, u+F(p,u)+\theta\Delta(p,u,v)) \\
&\quad - \Phi(k, p+1, u+F(p,u))]d\theta. \qquad (1.1.15)
\end{aligned}
$$

Again, by applying the generalized mean value Lemma A.1.1, we get

$$
\begin{aligned}
&[\Phi(k, p+1, u+F(p,u)+\theta\Delta(p,u,v)) - \Phi(k, p+1, u+F(p,u))] \\
&= \int_0^1 \frac{\partial}{\partial m_0}\Phi(k, p+1, u+F(p,u)+\theta\psi\Delta(p,u,v)) \cdot \Delta(p,u,v)\theta\, d\psi \\
&= \frac{\partial}{\partial m_0}\Phi(k, p+1, u+F(p,u)) \cdot \Delta(p,u,v)\theta \\
&\quad + \int_0^1 \theta \left[\frac{\partial}{\partial m_0}\Phi(k, p+1, u+F(p,u)+\theta\psi\Delta(p,u,v)) \right. \\
&\qquad \left. - \frac{\partial}{\partial m_0}\Phi(k, p+1, u+F(p,u))\right] \cdot \Delta(p,u,v)d\psi. \\
&\qquad (1.1.16)
\end{aligned}
$$

By integrating both sides with respect to θ from 0 to 1 and using notation (1.1.15), we have

$$
\begin{aligned}
&\Psi(k, p+1, u+F(p,u), \Delta(p,u,v)) \\
&= \frac{1}{2}\frac{\partial}{\partial m_0}\Phi(k, p+1, u+F(p,u)) \cdot \Delta(p,u,v)
\end{aligned}
$$

$$
\begin{aligned}
&+\int_0^1\int_0^1 \theta\left[\left[\frac{\partial}{\partial m_0}\Phi(k,p+1,u+F(p,u)+\theta\psi\Delta(p,u,v))\right.\right.\\
&\qquad\left.\left.-\frac{\partial}{\partial m_0}\Phi(k,p+1,u+F(p,u))\right]\cdot\Delta(p,u,v)d\psi\right]d\theta.
\end{aligned}
\tag{1.1.17}
$$

From (1.1.17), (1.1.14) reduces to

$$
\begin{aligned}
&m(k,p+1,u+v)-m(k,p,u)\\
&\quad=\Phi(k,p+1,u+F(p,u))\Delta(p,u,v)\\
&\quad\quad+\frac{1}{2}\frac{\partial}{\partial m_0}\Phi(k,p+1,u+F(p,u))\cdot\Delta(p,u,v)\Delta(p,u,v)\\
&\quad\quad+\int_0^1\int_0^1 \theta\left[\left[\frac{\partial}{\partial m_0}\Phi(k,p+1,u+F(p,u)+\theta\psi\Delta(p,u,v))\right.\right.\\
&\quad\quad\left.\left.-\frac{\partial}{\partial m_0}\Phi(k,p+1,u+F(p,u))\right]\cdot\Delta(p,u,v)d\psi\,d\theta\right]\Delta(p,u,v).
\end{aligned}
\tag{1.1.18}
$$

With this, together with the notation (1.1.12), the proof of (1.1.11) follows immediately. This completes the proof of the lemma.

By utilizing a vector Lyapunov-like function $V \in C[\mathbb{R}_+ \times \mathbb{R}^n \times \Omega, \mathbb{R}^m]$, we define an operator as follows:

$$
\begin{aligned}
\Delta V(p,m(k,p,u,v),\omega) &= V(p+1,m(k,p+1,u+v),\omega)\\
&\quad - V(p,m(k,p,u),\omega),
\end{aligned}
\tag{1.1.19}
$$

where $m(k,p,u)$ and $m(k,p+1,u+v)$ are the solution processes of (1.1.2) through (p,u) and $(p+1,u+v)$, respectively, for v and $u \in \mathbb{R}^n$.

Remark 1.1.2. For $v = f(p,y,\omega)$ and $u = y$, $\Delta V(p,m(k,p,y),\omega)$ in (1.1.19) is denoted by $\Delta_{(1.1.1)}V(p,m(k,p,y),\omega)$. Furthermore, the continuity of f guarantees the measurability of the difference operator $\Delta_{(1.1.1)}V(p,m(k,p,y),\omega)$.

In the following, we present a result that provides a basis for the definition of the generating operator L associated with a flow v. For this purpose, we need an additional condition on $V(t,y,\omega)$.

Theorem 1.1.1. *Let hypotheses $H_{(1.1.1)}$ and $H_{(1.1.2)}$ be satisfied. Let $V \in C[\mathbb{R}_+ \times \mathbb{R}^n \times \Omega, \mathbb{R}^N]$, and further assume that $\frac{\partial}{\partial t}V(t,m,\omega)$, $\frac{\partial}{\partial x}V(t,m,\omega)$ and $\frac{\partial^2}{\partial z \partial z}V(t,m,\omega)$ exist and are continuous in all (t,m) in $\mathbb{R}_+ \times \mathbb{R}^n$; moreover, $\frac{\partial^2}{\partial z \partial z}V(t,m,\omega)$ is locally Lipschitzian in m for each t. Then, an operator L is defined by*

$$\begin{aligned} &LV(p,m(k,p,u),v,\omega)\\ &\quad = \mathcal{L}_a V(p,m(k,p,u),v,\omega) + \mathcal{L}_e V(p,m(k,p,u),v,\omega)\\ &\qquad + \mathcal{L}_0 V(p,m(k,p,u),v,\omega), \end{aligned} \tag{1.1.20}$$

where

$$\begin{aligned} &\mathcal{L}_a V(p,m(k,p,u),v,\omega)\\ &\quad = V_t(p,m(k,p,u),\omega) + V_m(p,m(k,p,u),\omega)A(k,p+1,\Delta(p,u,v)), \end{aligned} \tag{1.1.21}$$

$$\begin{aligned} &\mathcal{L}_e V(p,m(k,p,u),v,\omega)\\ &\quad = \frac{1}{2}\Bigg[V_m(p,m(k,p,u),\omega)E(k,p+1,\Delta(p,u,v))\\ &\qquad + \frac{1}{2}\frac{\partial^2}{\partial m \partial m}V(p,m(k,p,u),\omega)\cdot\Theta(k,p+1,\Delta(p,u,v))\\ &\qquad \times \Theta(k,p+1,\Delta(p,u,v)\Bigg) \end{aligned} \tag{1.1.22}$$

and

$$\begin{aligned} &\mathcal{L}_0 V(p,m(k,p,u),v,\omega)\\ &\quad = \int_0^1 [V_t(p+\theta,m(k,p,u)+\theta\Delta m,\omega) - V_t(p,m(k,p,u),\omega)]d\theta\\ &\qquad + \int_0^1\int_0^1 \theta\Bigg[\frac{\partial^2}{\partial z^2}[V(p+\theta,m(k,p,u)+\psi\theta\Delta m,\omega)\\ &\qquad\qquad - V(p,m(k,p,u),\omega)]\cdot\Delta m\Delta m\Bigg]d\psi\, d\theta\\ &\qquad + \frac{1}{2}\Bigg[\frac{\partial^2}{\partial m \partial m}V(p,m(k,p,u,\omega))\cdot\Theta(k,p+1,\Delta(p,u,v))\\ &\qquad\quad \times O(k,p+1,u+F(p,u),\Delta(p,u,v)) \end{aligned}$$

$$
\begin{aligned}
&+ \frac{\partial^2}{\partial m \partial m} V(p, m(k,p,u), \omega) \cdot O(k, p+1, u \\
&+ F(p,u), \Delta(p,u,v))\Theta(k, p+1, \Delta(p,u,v)), \\
&+ \frac{\partial^2}{\partial m \partial m} V(p, m(k,p,u), \omega) \cdot O(k, p+1, u \\
&+ F(u,p), \Delta(p,u,v))O(k, p+1, u + F(p,u), \Delta(p,u,v)) \\
&+ V_m(p, m(k,p,u), \omega) O(k, p+1, u + F(p,u), \Delta(p,u,v)) \Bigg], \qquad (1.1.23)
\end{aligned}
$$

where $O(k, p+1, m(p+1), \Delta(p,u,v))$ *and* $\Delta(p,u,v)$ *are as defined in* (1.1.12) *and* (1.1.13), *respectively, and*

$$\Delta m = m(k, p+1, u+v) - m(k,p,u), \qquad (1.1.24)$$

$$A(k, p+1, \Delta(p,u,v)) = \Phi(k, p+1, u + F(p,u))\Delta(p,u,v), \qquad (1.1.25)$$

$$
\begin{aligned}
E(k, p+1, \Delta(p,u,v)) &= \frac{1}{2} \frac{\partial}{\partial m_0} \Phi(k, p+1, u + F(p,u)) \\
&\quad \times \cdot \Delta(p,u,v)\Delta(p,u,v) \qquad (1.1.26)
\end{aligned}
$$

and

$$
\begin{aligned}
\Theta(k, p+1, \Delta(p,u,v)) &= A(k, p+1, \Delta(p,u,v)) \\
&\quad + E(k, p+1, \Delta(p,u,v)). \qquad (1.1.27)
\end{aligned}
$$

Proof. Let $m(k, p+1, u+v)$ and $m(k,p,u)$ be the solution processes of (1.1.2) through $(p+1, u+v)$ and (p,u), respectively. From (1.1.5) and (1.1.13), we recall that $m(k, p+1, u + F(p,u) + \Delta(p,u,v)) = m(k, p+1, u+v)$ and $m(k, p+1, u + F(p,u)) = m(k,p,u)$. Under the assumption of the lemma, the application of Lemmas 1.1.1, 1.1.2 and 1.1.3 and the generalized mean value theorem with notation (1.1.5), followed by imitating the argument used in the proof of Lemma 1.1.3(c) and algebraic simplifications, we have

$$
\begin{aligned}
&\Delta_v V(p, m(k,p,u,v,\omega)) \\
&\quad = V(p+1, m(k, p+1, u+v), \omega) - V(p, m(k,p,u), \omega)
\end{aligned}
$$

$$
\begin{aligned}
&= \int_0^1 [V_t(p+\theta, m(k,p,u) + \theta\Delta m, \omega) \\
&\qquad + V_m(p+\theta, m(k,p,u)+\theta\Delta m, \omega)\Delta m]d\theta \\
&= V_t(p, m(k,p,u), \omega) + V_m(p, m(k,p,u), \omega)\Delta m \\
&\quad + \int_0^1 [V_t(p+\theta, m(k,p,u)+\theta\Delta m, \omega) - V_t(p, m(k,p,u), \omega) \\
&\quad + [V_m(p+\theta, m(k,p,u)+\theta\Delta m, \omega) - V_m(p, m(k,p,u), \omega)]\Delta m]d\theta \\
&= V_t(p, m(k,p,u), \omega) + V_m(p, m(k,p,u), \omega)\Delta m \\
&\quad + \frac{1}{2}\left[\frac{\partial^2}{\partial m^2} V(p, m(k,p,u), \omega)\cdot \Delta m \Delta m\right] \\
&\quad + \left[\int_0^1 [V_t(p+\theta, m(k,p,u)+\theta\Delta m, \omega) - V_t(p, m(k,p,u), \omega)]d\theta\right. \\
&\qquad + \int_0^1\int_0^1 \theta\left[\frac{\partial^2}{\partial m^2} V(p+\theta, m(k,p,u)+\psi\theta\Delta m, \omega)\right. \\
&\qquad \left.\left. - \frac{\partial^2}{\partial m^2} V(p, m(k,p,u), \omega)\right]\cdot \Delta m \Delta m\right] d\psi\, d\theta. \qquad (1.1.28)
\end{aligned}
$$

From (1.1.4), (1.1.5), (1.1.11) and (1.1.12), the definitions of operators $\mathcal{L}_a$, $\mathcal{L}_e$ and $\mathcal{L}_o$, the notations and definitions introduced in the theorem and the increments $\Delta(p,u,v)$ and Δm, with algebraic simplifications and regrouping, (1.1.28) reduces to the desired relation in (1.1.20). The details are left to the reader.

Remark 1.1.3. Please note that for $v = f(p,y,\omega)$ and $u = y$, the definition of $LV(p, m(k,p,y,v), \omega) \equiv LV(p, m(k,p,y), \omega)$ in (1.1.20) does not depend on the knowledge of the solution process of (1.1.1). It only depends on $y \in \mathbb{R}^n$, the rate function f and $\Delta(p,u,v) = f(p,y,\omega) - F(p,y)$. Moreover, in order to characterize the effects of random perturbations (both internal and external), L is decomposed into three operators $\mathcal{L}_a$, $\mathcal{L}_e$ and $\mathcal{L}_o$, representing the absence of randomness, the presence of external disturbances, and round-off errors, respectively.

The following results illustrate the computational feasibility of L-operator.

Corollary 1.1.1. *Let* $V(p, m, \omega) = \alpha(\omega)m^T m = \alpha(\omega)\|m\|^2$, $\alpha(p,\omega) > 0$, *for any* $k \in \mathbb{I}(k_0)$, $k_0 \le p \le k$. *Then,*

$$V_t(p, m, \omega) = 0, \quad V_m(p, m, \omega) = 2\alpha(p, \omega)m^T,$$

$$\frac{\partial^2}{\partial m \partial m} V(p, m(k, p, y), \omega) = 2\alpha(p, \omega)I.$$

Then, $LV(p, m(k, p, y), \omega)$ *reduces to*

$$LV(p, m(k, p), \omega) = 2\alpha(p, \omega)m^T(k, p)\Delta m + \alpha(p, \omega)(\Delta m)^T \Delta m,$$

where $\Delta m(k, p, u) = m(k, p+1, u + f(p, u, \omega)) - m(k, p, u)$ *and* $m(k, p, \omega)$ *is a solution process of* (1.1.2) *through* (p, u).

Proof. From (1.1.11) and (1.1.12), we have

$$\begin{aligned}
&V_m(p, m(k, p, u), \omega)\Delta m \\
&\quad = V_m(p, m(k, p, y), \omega)\Big[\Phi(k, p+1, m(p+1))\Delta(p, y) \\
&\qquad + \frac{1}{2}\frac{\partial}{\partial m_0}\Phi(k, p+1, u + F(p, u)) \cdot \Delta(p, u, v)\Delta(p, u, v) \\
&\qquad + O(k, p+1, u + F(p, y), \Delta(p, u, v)\Big]
\end{aligned}$$

and

$$\begin{aligned}
&\frac{1}{2}\frac{\partial^2}{\partial z^2} V(p, m(k, p, u), \omega)\Delta m^T \Delta m \\
&\quad = \alpha(p, \omega)[\Theta^T(k, p+1, m(p+1))\Theta(k, p+1, m(p+1)) \\
&\qquad + \Theta^T(k, p+1, m(p+1))O(k, p+1, y + F(p, u), \Delta(p, u)) \\
&\qquad + O^T(k, p+1, u + F(p, u), \Delta(p, u, v))\Theta(k, p+1, m(p+1)) \\
&\qquad + O^T(k, p+1, u + F(p, u), \Delta(p, u, v)) \\
&\qquad \times O(k, p+1, u + F(p, u), \Delta(p, u, v))].
\end{aligned}$$

After elementary computations and simplifications, (1.1.20) reduces to

$$\begin{aligned}
&LV(p, m(k, p, u), \omega) \\
&\quad = \mathcal{L}_a V(p, m(k, p, u, v), \omega) + \mathcal{L}_e V(p, m(k, p, u, v), \omega) \\
&\qquad + \mathcal{L}_0 V(p, m(k, p, u, v), \omega),
\end{aligned} \tag{1.1.29}$$

where

$$
\begin{aligned}
&\mathcal{L}_a V(p, m(k,p,u), \omega) \\
&\quad = 2\alpha(\omega) m(k,p,u)^T \Phi(k, p+1, m(p+1)) \Delta(p,u,v) \\
&\quad = 2\alpha(\omega) m(k,p,u)^T A(k, p+1, \Delta(p+1)) \\
&\mathcal{L}_e V_i(p, m(k,p,u), \omega) \\
&\quad = \Theta^T(k, p+1, m(p+1)) \Theta(k, p+1, m(p+1)) \\
&\qquad + \alpha(\omega) m(k,p,u)^T E(k, p+1, \Delta(p,u,v))
\end{aligned}
$$

and

$$
\begin{aligned}
&\mathcal{L}_o V(p, z(k,p,u), \omega) \\
&\quad = 2\alpha p, \omega) m(k,p,u)^T O(k, p+1, m(p+1), \Delta(p,u,v)) \\
&\qquad + O^T(k, p+1, m(p+1,p,u), \Delta(p,u,v)) \Theta(k, p+1, m(p+1)) \\
&\qquad + \Theta^T(k, p+1, m(p+1,p,u)) O(k, p+1, m(p+1), \Delta(p,u,v)) \\
&\qquad + O^T(k, p+1, m(p+1,p,u) O(k, p+1, m(p+1), \Delta(p,u,v)).
\end{aligned}
$$

We note that the scope of the L-operator defined in (1.1.20) in the context of auxiliary system (1.1.2) is illustrated by the following well-known special cases.

Corollary 1.1.2. *If $F(p, y)$ in (1.1.2) is $F(p, y) \equiv 0$, then $m(k,p,u) = u$, $\Phi(k,p,u) = I$ and $\frac{\partial^2}{\partial m_0^2} m(k,p,u) \equiv 0$. Then, the L-operator defined in (1.1.20) in Theorem 1.1.1 reduces to the L-operator $LV(p,u,v,\omega) = \mathcal{L}_a V(p,u,v,\omega) + \mathcal{L}_e V(p,u,v,\omega) + \mathcal{L}_O V(p,u,v,\omega)$, where*

$$
\mathcal{L}_a V(p,u,v,\omega) = V_t(p,u,\omega) + V_m(p,u,\omega) v,
$$

$$
\mathcal{L}_e V(p,u,v,\omega) = \frac{1}{2} \frac{\partial^2}{\partial m^2} V(p,u,\omega) \cdot v^T v,
$$

$$
\mathcal{L}_O V(p,u,v,\omega) = \int_0^1 \Big\{ V_t(p+\theta, u+\theta v, \omega) - V_t(p,u,\omega)
$$

$$+\int_0^1 \theta\left[\frac{\partial}{\partial m} V_m(p+\theta, u+\theta\psi v, \omega) - \frac{\partial}{\partial m} V_m(k, y, \omega)\right] \cdot v^T v \Big\} d\psi\, d\theta.$$

Corollary 1.1.3. *If* $F(p, m) = A(p)m$, *then in this case,* $m(k, p, u) = \prod_{i=p}^{k-1}(I + A(i))u$, $\frac{\partial}{\partial m}F(p, m) = A(p)$, $\frac{\partial^2}{\partial m^2}F(p, m) \equiv 0$, $\Phi(k, p, u) = \prod_{i=p}^{k-1}(I + A(i)) \equiv \Phi_a(k, p)$, $\frac{\partial^2}{\partial m_0^2}m(k, p, u) \equiv 0$. *Then, the L-operator defined in* (1.1.20) *in Theorem* 1.1.1 *reduces to*

$$\begin{aligned} LV&(p, m(k, p, u, v), \omega) \\ &= \mathcal{L}_a V(p, m(k, p, u), v, \omega) + \mathcal{L}_e V(p, m(k, p, u, v), \omega) \\ &\quad + \mathcal{L}_o V(p, m(k, p, u, v), \omega), \end{aligned} \tag{1.1.30}$$

where

$$\begin{aligned} \mathcal{L}_a V(p, m(k, p, u, v), \omega) &= V_t(p, m(k, p, u), \omega) + V_m(p, m(k, p, u), \omega) \\ &\quad \times [\Phi(k, p+1)\Delta(p, u, v)], \\ \mathcal{L}_e V(p, m(k, p, u, v), \omega) &= \frac{1}{2}\frac{\partial^2}{\partial m \partial m} V(p, m(k, p, u), \omega) \\ &\quad \times (\Phi(k, p+1)\Delta(p, u, v))^T \Phi(k, p+1) \\ &\quad \times \Delta(p, u, v) \end{aligned}$$

and

$$\begin{aligned} \mathcal{L}_o V(p, m(k, p, u, v), \omega) &= \int_0^1 [V_t(p+\theta, m(k, p, u) + \theta\Delta m, \omega) \\ &\quad - V_t(p, m(k, p, u), \omega)]d\theta + \int_0^1\int_0^1 \theta \\ &\quad \times \left[\frac{\partial^2}{\partial m^2} V(p+\theta, m(k, p, u) + \psi\theta\Delta m, \omega)\right. \\ &\quad \left. - \frac{\partial^2}{\partial m^2} V(p, m(k, p, u), \omega)\Delta m(\Delta m)^T\right] d\psi d\theta, \end{aligned}$$

where

$$\Delta(p, u, v) = [v - A(p)u]$$

$$\Theta(k, p+1, m(p+1)) = \Phi(k, p+1)\Delta(p, u, v)$$

and

$$\begin{aligned}\Delta m &= m(k, p+1, u+v) - m(k, p, u) \\ &= m(k, p+1, v) - m(k, p+1, A(p)u) \\ &= \Phi(k, p+1)[v - A(p)u].\end{aligned}$$

By utilizing a vector Lyapunov-like function $V(t, x, \omega)$, the following result is proved. This result lays down a foundation for the presentation of a very general variation of constants formula.

Theorem 1.1.2. *Assume that the hypotheses of Theorem* 1.1.1 *are satisfied. Then,*

$$\begin{aligned}V(k, y(k, k_0, y_0), \omega) &= V(k_0, m(k, k_0, y_0), \omega) \\ &\quad + \sum_{p=k_0}^{k-1} LV(p, m(k, p, y(p, k_0, y_0)), \omega),\end{aligned} \tag{1.1.31}$$

where L is defined as in (1.1.20).

Proof. For $k_0 \le p \le k$, let $y(p) = y(p, k_0, y_0)$ and $m(k, p, y(p))$ be the solution processes of (1.1.1) and (1.1.2) through (k_0, y_0) and $(p, y(p))$, respectively. From (1.1.5), we recall that $m(k, p+1, y(p+1)) = m(k, p+1, y(p) + F(p, y(p))) = m(k, p, y(p))$. We set $\Delta v(p) = V(p+1, m(k, p+1, y(p+1)), \omega) - V(p, m(k, p, y(p)), \omega)$ and consider

$$\begin{aligned}\Delta v(p) &= V(p+1, m(k, p+1, y(p) + f(p, y(p), \omega)), \omega)) \\ &\quad - V(p, m(k, p, y(p)), \omega) \\ &= V(p+1, m(k, p+1, y(p) + F(p, y(p)) + [f(p, y(p), \omega)) \\ &\quad - F(p, y(p))], \omega) - V(p, m(k, p, y(p)), \omega) \\ &= V(p+1, m(k, p+1, y(p) + F(p, y(p)) + \Delta(p, y(p))), \omega) \\ &\quad - V(p, m(k, p, y(p)), \omega).\end{aligned}$$

By using (1.1.20), we get

$$\Delta v(p) = LV(p, m(k, p, y(p, k_0, y_0)), \omega),$$

which implies

$$\sum_{p=k_0}^{k-1} \Delta v(p) = \sum_{p=k_0}^{k-1} LV(p, m(k, p, y(p, k_0, y_0)), \omega). \tag{1.1.32}$$

This, together with the telescopic sum, gives

$$\begin{aligned} &V(k, m(k, k, y(k, k_0, y_0)), \omega) - V(k_0, m(k, k_0, y_0), \omega) \\ &\quad = \sum_{p=k_0}^{k-1} LV(p, m(k, p, y(p, k_0, y_0)), \omega), \end{aligned}$$

which implies

$$\begin{aligned} &V(k, y(k, k_0, y_0), \omega) - V(k_0, m(k, k_0, y_0), \omega) \\ &\quad = \sum_{p=k_0}^{k-1} LV(p, m(k, p, y(p, k_0, y_0)), \omega). \end{aligned}$$

This establishes the validity of (1.1.31).

Corollary 1.1.4. *Let us assume that hypotheses $H_{(1.1.1)}$ and $H_{(1.1.2)}$ are valid.*

(i) *Let us assume that $V(t, y, \omega) = y$. Then,*

$$y(k, k_0, y_0) = m(k, k_0, y_0) + \sum_{p=k_0}^{k-1} LV(p, m(k, p, y(p), \omega), \tag{1.1.33}$$

where $LV = \mathcal{L}_a V + \mathcal{L}_e V + \mathcal{L}_0 V$,

$$\mathcal{L}_a V(p, m(k, p, x), \omega) = \Phi(k, p+1, y + F(p, y))\Delta(p, y),$$

$$\begin{aligned} \mathcal{L}_e V(p, m(k, p, x), \omega) &= \frac{1}{2}\frac{\partial}{\partial m_0}\Phi(k, p+1, y + F(p, y)) \\ &\quad \times \cdot \Delta(p, y)\Delta(p, y), \end{aligned}$$

$$\begin{aligned} \mathcal{L}_0 V(p, m(k, p, y), \omega) &= \int_0^1 \int_0^1 \theta \frac{\partial}{\partial m_0} \Big[\Phi(k, p+1, y + F(p, y) \\ &\quad + \theta\psi\Delta(p, y)) - \frac{\partial}{\partial m_0}\Phi(k, p+1, y \\ &\quad + F(p, y))\Big] \, d\psi \, d\theta \, \Delta(p, y) \end{aligned}$$

and $\Delta(p, y(p)) = f(p, y(p), \omega) - F(p, y(p))$.

(ii) *Let* $V(p, m, \omega) = \alpha m^T m$, *for constant* $\alpha > 0$. *Then,*

$$\|y(k, k_0, y_0, \omega)\|^2 = \|m(k, k_0, y_0)\|^2 + \frac{1}{\alpha} \sum_{p=k_0}^{k-1} LV(p, m(k, p, y(p)), \omega), \tag{1.1.34}$$

where LV *is as defined in* (1.1.29).

(iii) *If* $V(p, y, \omega)$ *and* $F(p, y)$ *satisfy the conditions of Corollary* 1.1.2, *then*

$$V(k, y(k, k_0, y_0), \omega) = V(k_0, m(k, k_0, y_0, \omega) + \sum_{p=k_0}^{k-1} LV(p, m(k, p, y(p)), \omega), \tag{1.1.35}$$

where LV *is as defined in Corollary* 1.1.2.

(iv) *If* $V(p, y, \omega)$ *and* $F(p, y)$ *satisfy the conditions of Corollary* 1.1.3,

$$V(k, y(k, k_0, y_0), \omega) = V(k_0, m(k, k_0, y_0), \omega) + \sum_{p=k_0}^{k-1} LV(p, m(k, p, y(p)), \omega), \tag{1.1.36}$$

where LV *is as defined in Corollary* 1.1.3.

(v) *If the assumptions in* (ii) *and* (iii) *are satisfied, then*

$$\|y(k, k_0, y_0)\|^2 = \|m(k, k_0, y_0\|^2 + \frac{1}{\alpha} \sum_{p=k_0}^{k-1} LV(p, m(k, p, y(p)), \omega), \tag{1.1.37}$$

where LV *is as defined by*

$$\mathcal{L}_a V(p, m(k, p, y), \omega) = 2\alpha m^T(k, p, y(p))(f(p, y(p), \omega) - F(p, y(p))$$

$$\mathcal{L}_e V(p, m(k, p, y), \omega) = [f(p, y(p), \omega) - F(p, y(p))]^T \times [f(p, y(p), \omega) - F(p, y(p))].$$

(vi) *If conditions in* (ii) *and* (iv) *are fulfilled, then*

$$\|y(k,k_0,y_0)\|^2 = \|\Phi(k,k_0)y_0\|^2 + \frac{1}{\alpha}\sum_{p=k_0}^{k-1} LV(p, m(k,p,y(p)),\omega), \tag{1.1.38}$$

where LV *is as defined by*

$$\begin{aligned}
&\mathcal{L}_a V(p, m(k,p,y(p),\omega)) \\
&\quad = 2\alpha m^T(k,p,y(p))\Phi(k,p)[f(p,y(p),\omega) - F(p,y(p))]. \\
&\mathcal{L}_e V(p, m(k,p,y(p)),\omega) \\
&\quad = \alpha(\Phi(k,p)[f(p,y(p),\omega) - F(p,y(p))])^T \\
&\qquad \times(\Phi(k,p)[f(p,y(p),\omega) - F(p,y(p))]).
\end{aligned}$$

(vii) *If* $V(p,y,\omega)$ *is continuous on* $[0,\infty)\times\mathbb{R}^n$ *and* $F(k,y)\equiv 0$, *then*

$$V(k, y(k,k_0,y_0),\omega) = V(k_0,y_0,\omega) + \sum_{p=k_0}^{k-1} LV(p,y(p),\omega), \tag{1.1.39}$$

where $LV(p,y(p),\omega)$ *is defined by*

$$LV(p,y(p),\omega) = V(p+1, y(p)+f(p,y(p),\omega),\omega) - V(p,y(p),\omega).$$

We observe that assumption $H_{(1.1.1)}$ implies that $\bar{m}(k) = m(k,k_0,z_0)$ is a unique solution process of (1.1.2) or (1.1.3) depending on the choice of z_0. Then, the mean value of the fundamental matrix Φ in (1.1.7) between x_0 and y_0 is given by

$$\psi(k,k_0,x_0,y_0) = \int_0^1 \Phi(k,k_0,sx_0+(1-s)y_0)ds. \tag{1.1.40}$$

We also note that if $x_0 = y_0$, then

$$\psi(k,k_0,x_0,x_0) = \Phi(k,k_0,y_0).$$

Furthermore, $\psi(k,k_0,y_0,y_0)$ is invertible, and the inverse is jointly measurable.

We now formulate a few basic results that relate the solution process of (1.1.1) with the solution process of (1.1.2) or (1.1.3).

Theorem 1.1.3. *Let the hypotheses of Theorem* 1.1.1 *be satisfied, and let* $y(k,\omega) = y(k,k_0,y_0(\omega),\omega)$ *and* $x(t,\omega) = x(t,t_0,y_0(\omega))$ *be the sample solution processes of* (1.1.1) *and* (1.1.3), *respectively, existing for all* $k \in I(k_0)$ *with* $x_0(\omega) = y_0(\omega)$. *Then,*

$$\begin{aligned} &V(k,y(k,\omega),\omega) \\ &\quad = V(k,x(k,\omega),\omega) + \sum_{s=k_0+1}^{k} W(k,k_0,y(k,\omega),\omega) \\ &\quad \times \psi^{-1}(s,k_0,p(s,\omega),p(s-1,\omega))R(s,y(s-1,\omega),\omega), \end{aligned} \tag{1.1.41}$$

where ψ *is defined in* (1.1.40),

$$\begin{aligned} R(k,y,\omega) &= f(k,y,\omega) - F(k,y), \\ W(k,k_0,y(k,\omega),\omega) &= \int_0^1 V_x(k,x(k,k_0,sp(k,\omega),\omega) + (1-s)y_0(\omega)),\omega) \\ &\quad \times \Phi(k,k_0,sp(k,\omega) + (1-s)y_0(\omega))ds, \end{aligned} \tag{1.1.42}$$

$$\begin{aligned} p(k,\omega) = y_0(\omega) &+ \sum_{s=k_0+1}^{k} \psi^{-1}(s,k_0,p(s,\omega),p(s-1,\omega)) \\ &\times R(s,x_p(s-1,\omega),\omega) \end{aligned} \tag{1.1.43}$$

and

$$x_p(k,\omega) = x(k,k_0,p(k,\omega)).$$

Proof. Let $y(k,\omega) = y(k,k_0,y_0(\omega),\omega)$ be a solution of (1.1.1). First, we show that $p(k,\omega)$ is a sample solution of (1.1.43) if and only if any sample solution $y(k,\omega)$ of (1.1.1) satisfies

$$y(k,k_0,y_0(\omega),\omega) = x(k,k_0,p(k,\omega)).$$

Let $x(k,k_0,y_0(\omega))$ be the sample solution of (1.1.3). The method of variation of constants requires determining a function $p(k,\omega)$ so that the solution $x_p(k) = x(k,k_0,p(k,\omega))$ of (1.1.3) with $p(k_0,\omega) = y_0(\omega)$ is a sample solution of (1.1.1). Therefore, to determine $p(k,\omega)$, we write (1.1.1) as

$$\begin{aligned} \Delta y(k,k_0,y_0(\omega),\omega) = f(k,y(k-1,k_0,y_0(\omega)),\omega) \\ + R((k,y(k-1,\omega),\omega), \end{aligned}$$

where

$$x(k,k_0,p(k,\omega)) - x(k,k_0,p(k-1,\omega)) = R(k,x_p(k-1,\omega),\omega). \tag{1.1.44}$$

We consider

$$\begin{aligned}&\frac{dx}{ds}(k,k_0,sp(k,\omega)+(1-s)p(k-1,\omega))\\&\quad = \Phi(k,k_0,sp(k,\omega)+(1-s)p(k-1,\omega))(p(k,\omega)-p(k-1,\omega)).\end{aligned}$$

Integrating from $s=0$ to $s=1$, we obtain

$$\begin{aligned}&x(k,k_0,p(k,\omega)) - x(k,k_0,p(k-1,\omega))\\&\quad = \psi(k,k_0,p(k,\omega),p(k-1,\omega))(p(k,\omega)-p(k-1,\omega)).\end{aligned} \tag{1.1.45}$$

This, together with (1.1.44), implies that

$$\begin{aligned}&p(k,\omega)-p(k-1,\omega)\\&\quad = \psi^{-1}(k,k_0,p(k,\omega),p(k-1,\omega))R(k,x_p(k-1,\omega),\omega).\end{aligned}$$

By iterating this equation, we obtain the proof of (1.1.43).

Now, we prove that if $p(k,\omega)$ is the solution of (1.1.43) with $p(k_0,\omega) = y_0(\omega)$, then $y(k,k_0,y_0(\omega),\omega) = x(k,k_0,p(k,\omega))$. From (1.1.43), we have

$$\begin{aligned}p(k-1,\omega) = y_0 + \sum_{s=k_0+1}^{k-1} &\psi^{-1}(s,k_0,p(s,\omega),p(s-1,\omega))\\ &\times R(s,x_p(s-1,\omega),\omega).\end{aligned}$$

Therefore,

$$\begin{aligned}&p(k,\omega)-p(k-1,\omega)\\&\quad = \psi^{-1}(k,k_0,p(k,\omega),p(k-1,\omega))R(k,x_p(k-1,\omega),\omega).\end{aligned}$$

This gives

$$\begin{aligned}&R(k,x_p(k-1,\omega),\omega)\\&\quad = \psi(k,k_0,p(k,\omega),p(k-1,\omega))(p(k,\omega)-p(k-1,\omega))\\&\quad = \int_0^1 \Phi(k,k_0,sp(k,\omega)+(1-s)p(k-1,\omega))ds(p(k,\omega)\\&\qquad -p(k-1,\omega))\end{aligned}$$

$$= \int_0^1 \frac{\partial x}{\partial x_0}(k, k_0, sp(k,\omega) + (1-s)p(k-1,\omega))ds(p(k,\omega) - p(k-1,\omega))$$

$$= \int_0^1 \frac{d}{ds} x(k, k_0, sp(k,\omega) + (1-s)p(k-1,\omega))ds$$

$$= x(k, k_0, p(k,\omega)) - x(k, k_0, p(k-1,\omega)$$

$$= x(k, k_0, p(k,\omega)) - x(k-1, k_0, p(k-1,\omega)) - [x(k, k_0, p(k-1,\omega)) - x(k-1, k_0, p(k-1,\omega))].$$

Therefore,

$$R(k, x_p(k-1,\omega),\omega) + [x(k, k_0, p(k-1,\omega)) - x(k-1, k_0, p(k-1,\omega))] = \Delta x(k, k_0, p(k,\omega)), \tag{1.1.46}$$

and hence

$$\Delta x(k, k_0, p(k,\omega)) = F(k-1, x(k-1, k_0, p(k-1,\omega)), \quad x(k_0) = y_0. \tag{1.1.47}$$

From (1.1.47), (1.1.1) and the uniqueness of the solution of (1.1.1), we have

$$y(k, k_0, y_0(\omega), \omega) = x(k, k_0, p(k,\omega)).$$

Now, to prove (1.1.41), we differentiate V, that is,

$$\frac{dV}{ds}(k, x(k, k_0, sp(k,\omega) + (1-s)y_0(\omega)), \omega)$$
$$= V_x(k, x(k, k_0, sp(k,\omega) + (1-s)y_0(\omega)), \omega)\Phi(k, k_0, sp(k,\omega) + (1-s)y_0(\omega))(p(k,\omega) - y_0(\omega)).$$

Integrating this from 0 to 1, we obtain

$$V(k, x(k, k_0, p(k,\omega)), \omega)$$
$$= V(k, x(k, k_0, x_0(\omega)), \omega) + \int_0^1 V_x(k, x(k, k_0, sp(k,\omega)$$

$$+ (1-s)y_0(\omega)), \omega)\Phi(k, k_0, sp(k,\omega)$$
$$+ (1-s)y_0(\omega))ds(p(k,\omega) - y_0(\omega)). \quad (1.1.48)$$

This equation, together with (1.1.42) and (1.1.43), proves the theorem.

In the following, we demonstrate the scope and significance of the preceding theorem.

Corollary 1.1.5. *Let the assumptions of* Theorem 1.1.3 *be satisfied, and let* $V(t, x, \omega) = x$. *Then,*

$$y(k,\omega) = x(k,\omega) + \sum_{s=k_0+1}^{k} \psi(k, k_0, p(k,\omega), x_0(\omega))$$
$$\times \psi^{-1}(s, k_0, p(s,\omega), p(s-1,\omega)R(s, y(s-1,\omega), \omega). \quad (1.1.49)$$

This corollary is a nonlinear variation of the constants formula for systems of stochastic difference equations.

Problem 1.1.1. If $V(k, x, \omega) = \|x\|^2$, then (1.1.41) in Theorem 1.1.3 reduces to

$$\|y(k,\omega)\|^2 = \|x(k,\omega)\|^2 + \sum_{s=k_0+1}^{k} W(k, k_0, x(k, k_0, y(k,\omega)))$$
$$\times \psi^{-1}(s, k_0, p(s,\omega), p(s-1,\omega))R(s, y(s-1,\omega), \omega),$$

where

$$W(k, k_0, y(k,\omega)) = 2\int_0^1 [x(k, k_0, sp(k,\omega) + (1-s)x_0(\omega))]^T$$
$$\times \Phi(k, k_0, sp(k,\omega) + (1-s)x_0(\omega))ds.$$

The following result provides an expression for the difference between the solution processes of (1.1.1) with (1.1.2) or (1.1.3).

Theorem 1.1.4. *Suppose all the hypotheses of* Theorem 1.1.3 *hold. Then,*

$$\begin{aligned} V(k, y(k,\omega) - \bar{x}(k), \omega) &= V(k, x(k,\omega) - \bar{x}(k), \omega) \\ &\quad + \sum_{s=k_0+1}^{k} W(k, k_0, y(k,\omega) - x(k, k_0, \bar{z}_0), \omega) \\ &\quad \times \psi^{-1}(s, k_0, p(s,\omega), p(s-1,\omega)) \\ &\quad \times R(s, y(s-1,\omega), \omega), \qquad (1.1.50) \end{aligned}$$

where $\bar{x}(k) = x(k, k_0, z_0)$ *is the solution process of either* (1.1.2) *or* (1.1.3) *depending on the choice of* z_0.

Proof. By following the proof of Theorem 1.1.3, we have the relation

$$\begin{aligned} &\frac{dV}{ds}(k, x(k, k_0, sp(k,\omega) + (1-s)x_0(\omega)) - \bar{x}(k), \omega) \\ &\quad = V_x(k, x(k, k_0, sp(k,\omega) + (1-s)x_0(\omega)) - \bar{x}(k)), \omega) \\ &\qquad \times \Phi(k, k_0, sp(k,\omega) + (1-s)x_0(\omega))(p(k,\omega) - x_0(\omega)). \end{aligned}$$

By integrating the above relation from 0 to 1, noting the fact that $x(k, k_0, p(k,\omega)) = y(k,\omega)$ and using (1.1.42), we complete the proof of the theorem.

Problem 1.1.2. If $V(k, x, \omega) = \|x\|^2$, then (1.1.50) reduces to

$$\begin{aligned} &\|y(k,\omega) - x(k, k_0, z_0)\|^2 = \|x(k,\omega) - x(k, k_0, z_0)\|^2 \\ &\quad + \sum_{s=k_0+1}^{k} W(k, k_0, y(k,\omega) - x(k, k_0, z_0)), \omega) \\ &\quad \times \psi^{-1}(s, k_0, p(s,\omega), p(s-1,\omega)) R(s, y(s-1,\omega), \omega), \end{aligned} \qquad (1.1.51)$$

where

$$\begin{aligned} &W(k, k_0, y(k,\omega) - x(k, k_0, z_0)), \omega) \\ &\quad = 2\int_0^1 [x(k, k_0, sp(k,\omega) + (1-s)x_0(\omega)) - x(k, k_0, z_0))]^T \\ &\qquad \times \Phi(k, k_0, sp(k,\omega) + (1-s)x_0(\omega)) ds. \end{aligned}$$

We recall that $\bar{x}(k) = x(k, k_0, z_0)$ is the solution process of either (1.1.2) or (1.1.3) depending on the choice of z_0. In other words, $\bar{x}(k)$ is either $m(k) = m(k, k_0, m_0) = x(k, k_0, m_0)$ or $x(k, k_0, x_0(\omega))$ depending on the choice of z_0.

To further illustrate the scope and usefulness of Theorems 1.1.3 and 1.1.4, let us assume some regularity conditions which will translate (1.1.1), (1.1.2) and (1.1.3) into a suitable form for our discussions. We suppose that $f(k, 0, \omega) = 0$ and the sample derivative $\frac{\partial}{\partial x} f(k, x, \omega)$ of $f(k, x, \omega)$ exists. From this and Lemma A.1.1, (1.1.1) can be rewritten as

$$y(k, \omega) = A(k, y(k-1, \omega), \omega)y(k-1, \omega), y(k_0, \omega) = y_0(\omega), \tag{1.1.52}$$

where

$$A(k, y, \omega) = \int_0^1 \frac{\partial}{\partial y} f(t, sy, \omega) ds.$$

Similarly, one can assume that $F(k, 0) \equiv 0$. This, together with the continuous differentiability of $F(k, x)$ in x, one can rewrite (1.1.2) and (1.1.3) as

$$m(k) = \hat{A}(k, m(k-1))m(k-1), \qquad m(k_0) = m_0, \tag{1.1.53}$$

and

$$x(k) = \hat{A}(k, x(k-1))x(k-1), \qquad x(k_0, \omega) = x_0(\omega), \tag{1.1.54}$$

where

$$\hat{A}(k, x) = \int_0^1 \frac{\partial \hat{F}}{\partial x}(t, sx) ds.$$

From Lemma A.1.1, we note that

$$x(k, s, y(s, \omega)) - x(k, s, \bar{x}(s)) = \psi(k, s, y(s, \omega), \bar{x}(s))(y(s, \omega) - \bar{x}(s)),$$

where

$$\psi(k, s, y(s, \omega), \bar{x}(s)) = \int_0^1 \Phi(k, s, \bar{x}(s) + u(y(s, \omega) - \bar{x}(s))) du.$$

Remark 1.1.4. We remark that relation (1.1.50), in light of $V(k,x,\omega) = \|x\|^2$, reduces to

$$\|y(k,\omega) - x(k,k_0,z_0)\|^2$$
$$= \|x(k,\omega) - x(k,k_0,z_0)\|^2 + \sum_{s=k_0+1}^{k} \left[2\int_0^1 \ell_p(\gamma) - z_0\right]^T$$
$$\times \psi^T(k,k_0,\ell_p(\gamma),\bar{x}(k))\Phi(k,k_0,\ell_p(\gamma)d\gamma\psi^{-1}(s,k_0,p(s,\omega),$$
$$\times p(s-1,\omega))R(s,y(s-1,\omega),\omega),$$

where $\ell_p(\gamma) = \gamma p(k,\omega) + (1-\gamma)x_0$, for $\gamma \in [0,1]$.

Example 1.1.1. Let us consider

$$y(k,\omega) = A(k,\omega)y(k-1,\omega), \quad y(k_0,p,\omega) = y_0(\omega), \tag{1.1.55}$$
$$m(k) = \hat{A}(k)m(k-1), \quad m(k_0) = m_0 = E[y_0(\omega)] \tag{1.1.56}$$

and

$$x(k) = \hat{A}(k)x(k-1), \quad x(k_0,\omega) = x_0(\omega), \tag{1.1.57}$$

where $A(k,\omega)$ is an $n \times n$ random matrix function and $\hat{A}(k)$ is a rate matrix which is obtained by neglecting the randomness in the system. In particular, $\hat{A}(k) = E[A(k,\omega)]$ if $E[A(k,\omega)]$ exists. In this case, $x(k,k_0,x_0(\omega)) = \Phi(k,k_0)x_0(\omega)$, where $\Phi(k,k_0)$ is the fundamental matrix solution process of either (1.1.55) or (1.1.56). Note that in the linear case, $\psi(k,s,y(s,\omega)) = \Phi(k,s)$. With regard to (1.1.55), (1.1.56) and (1.1.57) in the context of $V(k,x,\omega) = \|x\|^2$ and Remark 1.1.4, (1.1.41) and (1.1.51) reduce to

$$\|y(k,\omega)\|^2$$
$$= \|x(k,\omega)\|^2 + \sum_{s=k_0+1}^{k} (p(k,\omega) + x_0(\omega))^T\Phi^T(k,\omega)\Phi(k,\omega)$$
$$\times \Phi^{-1}(s,\omega)(A(s,\omega) - \hat{A}(s))y(s-1,\omega), \tag{1.1.58}$$

$$\|y(k,\omega) - \bar{x}(k)\|^2$$
$$= \|x(k,\omega) - \bar{x}(k))\|^2 + \sum_{s=k_0+1}^{k} (p(k,\omega) + x_0(\omega) - 2z_0)^T\Phi^T(k,\omega)$$
$$\times \Phi(k,\omega)\Phi^{-1}(s,\omega)(A(s,\omega) - \hat{A}(s))y(s-1,\omega). \tag{1.1.59}$$

In the following, we formulate an error estimation problem and develop a generalized variation of constants formula. For this purpose, let us consider the following discrete-time dynamic process:

$$\Delta w(k) = h(k, w(k), \omega), \quad w(k_0, w) = w_0, \tag{1.1.60}$$

where h satisfies hypothesis $\mathrm{H}_{(1.1.1)}$, $w, w_0, h \in \mathbb{R}^n$.

We note that by setting $u = y - w$ and $v = f - h$, the difference operator in (1.1.19) reduces to

$$\begin{aligned}
&\Delta_{(f,h)} V(p, m(k, p, y - w, f(p, y, \omega) - h(p, w, \omega), \omega) \\
&\quad = V(p + 1, m(k, p + 1, y - w + f(p, y, w) - h(p, w, \omega), \omega) \\
&\quad -V(p, m(k, p, y - w), \omega),
\end{aligned} \tag{1.1.61}$$

where $V(p, x, \omega)$ and $m(k, p, u)$ are as defined in (1.1.19).

Remark 1.1.5. By utilizing the above formulation, the results corresponding to Lemma 1.1.3 and Theorem 1.1.1 can be reestablished. We further observe that Δm, $A(k, p + 1, \Delta(p, u, v))$, $E(k, p + 1, \Delta(p, u, v))$ and $\Theta(k, p + 1, \Delta(p, u, v))$ can be expressed as follows:

$$\begin{aligned}
\Delta m &= m(k, p + 1, y - w + f(p, y, \omega) - h(p, w, \omega)) \\
&\quad - m(k, p, y - w),
\end{aligned} \tag{1.1.62}$$

$$\begin{aligned}
A(k, p + 1, \Delta(p, u, v)) &= \Phi(k, p + 1, y - w + f(p, y, \omega) \\
&\quad - h(p, w, w))\Delta(p, u, v),
\end{aligned} \tag{1.1.63}$$

$$\begin{aligned}
E(k, p + 1, \Delta(p, u, v)) &= \frac{1}{2} \frac{\partial}{\partial m} \Phi(k, p + 1, y - w + f(p, y, \omega) \\
&\quad - h(p, w, w))\Delta(p, u, v)\Delta(p, u, v),
\end{aligned} \tag{1.1.64}$$

$$\begin{aligned}
\Theta(k, p + 1, \Delta(p, u, v)) &= A(k, p + 1, \Delta(p, u, v)) \\
&\quad + E(k, p + 1, \Delta(p, u, v))
\end{aligned} \tag{1.1.65}$$

and

$$\Delta(p, u, v) = [f(p, y, \omega) - h(p, w, \omega) - F(p, y - w)]. \tag{1.1.66}$$

We further remark that the L-operator defined in (1.1.20) can be recast in the context of (1.1.60), (1.1.61), (1.1.2), (1.1.3), (1.1.64), (1.1.65) and (1.1.66).

We are ready to present a generalized variation of constants results for the error estimation of the solution process of (1.1.1) with respect to (1.1.60).

Theorem 1.1.5. *Assume that all the hypotheses of Theorem* 1.1.1 *are satisfied. Further, assume that for* $k_0 \leq p \leq k$, $y(p) = y(p, k_0, y_0)$ *and* $w(p) = w(p, k_0, w_0)$ *are the solution processes of* (1.1.1) *and* (1.1.60), *respectively. Then,*

$$\begin{aligned}&V(k, y(k, k_0.y_0) - w(k, k_0, w_0), \omega)\\&\quad = V(k_0, m(k, k_0, y_0 - w_0), \omega) + \sum_{p=k_0}^{k-1} LV(p, m(k, p, y(p) - w(p), \omega),\end{aligned} \tag{1.1.67}$$

where L *is as defined in Remark* 1.1.5.

Proof. For $k_0 \leq p \leq k$, let $y(p) = y(p, k_0, y_0)$, $w(p) = w(p, k_0, w_0)$ and $m(k, p, y(p) - w(p))$ be the solution processes of (1.1.1), (1.1.60) and (1.1.2) through (k_0, y_0), (k_0, w_0) and $(p, y(p) - w(p))$, respectively. By imitating the proof of Theorem 1.1.3, we arrive at

$$\begin{aligned}\Delta v(p) &= V(p+1, m(k, p+1, y(p+1) - w(p+1)), \omega)\\&\quad - V(p, m(k, p, y(p) - w(p)), \omega)\\&= V(p+1, m(k, p+1, y(p) - w(p) + f(p, y(p), \omega)\\&\quad - h(p, w(p), \omega)), \omega) - V(p, m(k, p, y(p) - w(p)), \omega)\\&= V(p+1, m(k, p+1, y(p) - w(p) + F(p, y(p) - w(p))\\&\quad + \Delta(p, y(p) - w(p)), \omega) - V(p, m(k, p, y(p) - w(p)), \omega),\end{aligned}$$

where $\Delta(p, y(p) - w(p)) = [f(p, y(p), \omega) - h(p, w(p), \omega) - F(p, y(p) - w(p))]$. By using (1.1.20) in the context of Remark 1.1.5, we have

$$\Delta v(p) = LV(p, m(k, p, y(p, k_0, y_0) - w(p, k_0, w_0), \omega),$$

which yields the desired result (1.1.67). The details are left to the reader.

To illustrate the scope of Theorem 1.1.5, we state a result that is parallel to Corollary 1.1.5.

Corollary 1.1.6. *Let us assume that hypotheses* $H_{(1.1.1)}$ *and* $H_{(1.1.2)}$ *are satisfied.*

(i) *Let us assume that the hypotheses of Corollary* 1.1.4(i) *are satisfied in the context of Remark* 1.1.5. *Then,*

$$\begin{aligned} y(k, k_0, y_0) - w(k, k_0, w_0) \\ = m(k, k_0, y_0 - w_0) + \sum_{p=k_0}^{k-1} LV(p, m(k, p, y(p) - w(p)), \omega), \end{aligned} \tag{1.1.68}$$

where LV *is defined in the context of Remark* 1.1.5 *and* (1.1.33), *and* $\Delta(p, y(p)-w(p)) = [f(p, y(p), \omega) - h(p, w(p), \omega) - F(p, y(p) - w(p))]$.

(ii) *Let* $v(p, m) = \alpha m^T m$, *for any constant* $\alpha > 0$. *Then,*

$$\begin{aligned} \|y(k, k_0, y_0) - w(k, k_0, w_0)\|^2 \\ = \|m(k, k_0, y_0 - w_0)\|)^2 + \frac{1}{\alpha} \sum_{p=k_0}^{k_0=1} LV \\ \times (p, m(k, k_0, y(p) - w(p), \omega), \end{aligned} \tag{1.1.69}$$

where LV *is as defined in* (1.1.34).

(iii) *If* $V(p, y, \omega)$ *and* $F(p, y)$ *satisfy the conditions of Corollary* 1.1.2, *then*

$$\begin{aligned} V(k, y(k, k_0, y_0) - w(k, k_0, w_0), \omega) \\ = V(k_0, m(k, k_0, y_0 - w_0), \omega) + \sum_{k=k_0}^{k-1} LV \\ \times (p, m(k, p, y(p) - w(p), \omega), \end{aligned} \tag{1.1.70}$$

where LV *is as defined in Corollary* 1.1.2.

(iv) *If* $V(p, y, \omega)$ *and* $F(p, y)$ *satisfy the conditions in Corollary* 1.1.3, *then*

$$\begin{aligned} &V(k, y(k, k_0, y_0) - m(k, k_0, m_0), \omega) \\ &\quad = V(k_0, m(k, k_0, y_0 - m_0), \omega) + \sum LV \\ &\quad \times (p, m(k, p, y(p) - m(p)), \omega), \end{aligned} \tag{1.1.71}$$

where LV is as defined in Corollary 1.1.3.

(v) *If the assumptions in (ii) and (iii) are satisfied, then*

$$\begin{aligned} &\|y(k, k_0, y_0) - w(k, k_0, w_0)\|^2 \\ &\quad = \|y_0 - w_0\|^2 + \frac{1}{\alpha} \sum_{p=k_0}^{k_0-1} LV(p, m(k, p, y(p) - w(p)), \omega), \end{aligned} \tag{1.1.72}$$

where LV is defined by

$$\begin{aligned} &L_a V(p, m(k, p, y - w), \omega) \\ &\quad = 2\alpha m^T(k, p, y - w)(f(p, y(p), \omega) - h(p, w(p), \omega) \\ &\qquad - F(p, y(p) - w(p))), \end{aligned} \tag{1.1.73}$$

$$\begin{aligned} &L_e V(p, m(k, p, y - w), \omega) \\ &\quad = [f(p, y(p), \omega) - h(p, w(p), \omega) - F(p, y(p) - w(p))]^T \\ &\qquad \times [f(p, y(p), \omega) - h(p, w(p), \omega) - F(p, y(p) - w(p))]. \end{aligned}$$

(vi) *If the conditions in* (ii) *and* (iv) *are fulfilled, then*

$$\begin{aligned} &\|y(k, k_0, y_0) - w(k, k_0, w_0)\|^2 \\ &\quad = \|\Phi(k, k_0)(y_0 - w_0)\|^2 + \frac{1}{\alpha} \sum_{p=k_0}^{k-1} LV \\ &\quad \times (p, m(k, p, y(p) - w(p)), \omega), \end{aligned} \tag{1.1.74}$$

where LV is defined by

$$\begin{aligned} &L_a V(p, m(k, p, y(p) - w(p)), \omega) \\ &\quad = 2\alpha m^T(k, p, y(p) - w(p))\Phi(k, p) \\ &\qquad \times [f(p, y(p), \omega) - h(p, w(p), \omega) - F(p, y(p) - w(p))], \end{aligned}$$

$$\begin{aligned}&L_eV(p,m(k,p,y(p)-w(p)),\omega)\\&\quad=\alpha(\Phi(k,p)[f(p,y(p),\omega)-h(p,w(p),\omega)\\&\qquad-F(p,y(p)-w(p))])^T\\&\qquad\times(\Phi(k,p)[f(p,y(p),\omega)-h(p,w(p),\omega)\\&\qquad-F(p,y(p)-w(p))]).\end{aligned}$$

(vii) *If* $V(p,y,\omega)$ *is continuous on* $[0,\infty)\times\mathbb{R}^n$ *and* $F(k,y)\equiv 0$, *then*

$$\begin{aligned}&V(k,y(k,k_0,y_0)-w(k,k_0,w_0),\omega)\\&\quad=V(k_0,y_0-w_0,\omega)+\sum_{p=k_0}^{k-1}LV(p,y(p)-w(p),\omega),\end{aligned}\tag{1.1.75}$$

where $LV(p,y(p)-w(p),\omega)$ *is defined by*

$$\begin{aligned}&LV(p,y(p)-w(p),\omega)\\&\quad=V(p+1,y(p)+f(p,y(p),\omega)-(w(p)+h(p,w(p),\omega)),\omega)\\&\qquad-V(p,y(p)-w(p),\omega).\end{aligned}$$

(viii) *If* $V(p,y,\omega)$ *is continuous on* $[0,\infty)+\mathbb{R}^n$ *and* $h(p,w,\omega)=F(p,w)$, *then*

$$\begin{aligned}&V(k,y(k,k_0,y_0)-m(k,k_0,m_0),\omega)\\&\quad=V(k_0,m(k,k_0,y_0-m_0),\omega)+\sum_{p=k_0}^{k-1}LV\\&\qquad\times(p,m(k,k,y(p)-m(p)),\omega),\end{aligned}$$

where

$$\begin{aligned}&LV(p,m(k,p,y-m),\omega)\\&\quad=V(p+1,m(k,p+1,y-m+f(p,y,\omega)-F(p,m)\\&\qquad-F(p,y-m)),\omega)-V(p,m(k,p,y-m),\omega).\end{aligned}\tag{1.1.76}$$

1.2 Comparison Method

In this section, by developing results concerning random difference inequalities, we present several comparison theorems which are useful to study the error estimates and stability properties of stochastic difference systems. Again, in the following, we assume that all the inequalities and relations involving random quantities are true with a probability of one (w.p. 1).

Now, we prove a fundamental result concerning difference inequalities. This result plays an important role in the further development of comparison theorems for random difference systems.

Theorem 1.2.1. *Let $\{\alpha_k(\omega)\}, \{\beta_k(\omega)\}$ be sequences of random variables defined on a complete probability space (Ω, F, P) into R^m and $k \in I(k_0) = \{k_0, k_0+1, \dots\}$. Let $G(k, x, \omega) \in R^m$ be a sequence of Borel measurable functions in Ω and satisfying the following:*

$$G(k, u, \omega) - G(k, v, \omega) \geq 0 \quad \text{w.p. 1, for } v \leq u. \tag{1.2.1}$$

Assume that

$$\begin{aligned} &\alpha_{k+1}(\omega) \leq G(k, \alpha_k(\omega), \omega), \\ &\beta_{k+1}(\omega) \geq G(k, \beta_k(\omega), \omega) \quad \text{w.p. 1.} \end{aligned} \tag{1.2.2}$$

Then,

$$\alpha_k(\omega) \leq \beta_k(\omega) \quad \text{w.p. 1,}$$

provided

$$\alpha_{k_0}(\omega) \leq \beta_{k_0}(\omega) \quad \text{w.p. 1.}$$

Proof. Suppose that

$$m_k(\omega) = \beta_k(\omega) - \alpha_k(\omega).$$

Then,

$$m_{k_0}(\omega) = \beta_{k_0}(\omega) - \alpha_{k_0}(\omega) \geq 0.$$

This, together with (1.2.1) and (1.2.2), yields

$$\begin{aligned} m_{k_0+1}(\omega) = \beta_{k_0+1}(\omega) - \alpha_{k_0+1}(\omega) &\geq G(k_0, \beta_{k_0}(\omega), \omega) \\ &\quad - G(k_0, \alpha_{k_0}(\omega), \omega) \geq 0. \end{aligned}$$

Now, let $m_i(\omega) \geq 0$, for $i = k_0, k_0 + 1, \ldots, n$. Then, for $i = n + 1$ and using (1.2.1) and (1.2.2), we get

$$\begin{aligned} m_{n+1}(\omega) = \beta_{n+1}(\omega) - \alpha_{n+1}(\omega) = G(n, \beta_n(\omega), \omega) \\ - G(n, \alpha_n(\omega), \omega) \geq 0. \end{aligned}$$

From the principle of mathematical induction, we conclude that

$$m_k(\omega) \geq 0, \qquad \text{for } k \in I(k_0).$$

Therefore,

$$\alpha_k(\omega) \leq \beta_k(\omega), \qquad \text{w.p. } 1,$$

provided

$$\alpha_0(\omega) \leq \beta_0(\omega), \qquad \text{w.p. } 1.$$

Corollary 1.2.1. *Assume that the hypotheses of Theorem* 1.2.1 *are satisfied except that* (1.2.1) *and* (1.2.2) *are replaced by*

$$G(k, u, \omega) - G(k, v, \omega) \geq -(u - v) \quad \text{w.p. } 1, \quad u \geq v, \tag{1.2.3}$$

and

$$\begin{aligned} &\alpha_k \leq G(k, \alpha_k(\omega), \omega), \\ &\beta_k \geq G(k, \beta_k(\omega), \omega) \quad \text{w.p. } 1, \end{aligned} \tag{1.2.4}$$

respectively. Then, the conclusion of Theorem 1.2.1 *remains valid.*

Now, we present a comparison theorem. This relates the solution process of an auxiliary difference equation with a stochastic sequence equation.

Theorem 1.2.2. *Let $m_k(\omega)$ be a sequence of m-dimensional random vectors such that*

$$m_{k+1}(\omega) \leq G(k, m_k(\omega), \omega). \tag{1.2.5}$$

Let $r_k(\omega)$ be the solution process of

$$u_{k+1}(\omega) = G(k, u_k(\omega), \omega), \quad u_{k_0}(\omega) = u_0(\omega), \tag{1.2.6}$$

where $G(k, r, \omega) \in R^m$ *is a sequence of Borel measurable functions on* $\Omega \times R_+^m$ *satisfying the following relation:*

$$G(k, r, \omega) - G(k, u, \omega) \geq 0 \quad \text{w.p. 1, for } u \leq r. \tag{1.2.7}$$

Then,

$$m_k(\omega) \leq r_k(\omega) \quad \text{w.p. 1, for } k \in I(k_0 + 1),$$

provided

$$m_{k_0}(\omega) \leq u_{k_0}(\omega) \quad \text{w.p. 1.}$$

Proof. We prove this theorem using the method of induction. Set

$$p(k, \omega) = r_k(\omega) - m_k(\omega).$$

We note that for k_0, we have

$$p(k_0, \omega) = r_{k_0}(\omega) - m_{k_0}(\omega) \geq 0,$$

which also implied

$$p(k_0 + 1, \omega) = r_{k_0+1}(\omega) - m_{k_0+1}(\omega) \geq G(k_0, r_{k_0}(\omega), \omega) - G(k, m_{k_0}(\omega), \omega) \geq 0.$$

We now apply the mathematical induction that to obtain

$$p(k, \omega) \geq 0, \qquad \text{for } k \in I(k_0 + 1).$$

This implies that

$$m_k(\omega) \leq r_k(\omega),$$

provided

$$m_{k_0}(\omega) \leq r_{k_0}(\omega).$$

This completes the proof of the theorem.

Remark 1.2.1. Assume that all the hypotheses of Theorem 1.2.2 are satisfied, except the inequality in (1.2.5) is reversed. Then,

$$m_k(\omega) \geq r_k(\omega) \quad \text{w.p. 1, for } k \geq k_0,$$

whenever $u_{k_0}(\omega) \leq m_{k_0}(\omega)$.

In the following, we develop another version of the comparison theorem.

Corollary 1.2.2. *Suppose that $m_k(\omega)$ is a sequence of m-dimensional random vectors and satisfies*

$$\Delta m_k(\omega) \leq G(k, m_k(\omega), \omega). \tag{1.2.8}$$

Let $r_k(\omega)$ be the solution process of

$$\Delta u_k = G(k, u_k(\omega), \omega), \tag{1.2.9}$$

where $G(k, u, \omega) \in \Omega \times R_+^m$ is a sequence of Borel measurable functions satisfying the relation

$$G(k, r, \omega) - G(k, u, \omega) \geq -(r - u), \quad \text{for } u \leq r. \tag{1.2.10}$$

Then,

$$m_k(\omega) \leq r_k(\omega) \quad w.p.\ 1,\ k \in I(k_0 + 1),$$

provided

$$m_{k_0}(\omega) \leq u_{k_0}(\omega) \quad w.p.\ 1.$$

Proof. The proof of this corollary is left as an exercise.

As a special case of the comparison Theorem 1.2.2, we present a result which has a wide range of applications.

Corollary 1.2.3. *If $m_k(\omega), k_k(\omega), p_k(\omega) \in R[\Omega, R_+]$ and $m_k(\omega)$ satisfies the inequality*

$$m_k(\omega) \leq k_k(\omega) m_{k-1}(\omega) + p_k(\omega), \quad \text{w.p. } 1, \tag{1.2.11}$$

then

$$m_k(\omega) \leq u_{k_0}(\omega) \prod_{s=k_0+1}^{k} k_s(\omega) + \sum_{s=k_0+1}^{k} p_s(\omega) \prod_{\tau=s+1}^{k} k_\tau(\omega), \quad \text{w.p. } 1; \tag{1.2.12}$$

moreover,

$$m_k(\omega) \leq u_{k_0}(\omega) \exp\left[\sum_{s=k_0+1}^{k} (k_s(\omega) - 1)\right] + \sum_{s=k_0+1}^{k} p_s(\omega) \exp\left[\sum_{\tau=s+1}^{k} (k_\tau(\omega) - 1)\right] \quad \text{w.p. } 1. \tag{1.2.13}$$

Proof. Consider

$$u_k(\omega) = k_k(\omega)u_{k-1}(\omega) + p_k(\omega), \quad u_{k_0}(\omega) = u_0(\omega). \tag{1.2.14}$$

We note that the solution of (1.2.14) is given by

$$r_k(\omega) = \begin{cases} u_0(\omega), & \text{for } k = k_0, \\ u_0(\omega)\prod_{s=k_0+1}^{k} k_s(\omega) + \sum_{s=k_0+1}^{k} p_s(\omega) & \\ \qquad \prod_{\tau=s+1}^{k} k_\tau(\omega), & k \geq k_0 + 1. \end{cases} \tag{1.2.15}$$

From (1.2.11), (1.2.14) and an application of Theorem 1.2.2, we obtain (1.2.12). The verification of (1.2.13) follows from (1.2.12) since for any x, $1 + x \leq e^x$.

Corollary 1.2.4. *Let $v_k \in R[\Omega, R_+]$. Let $v_{k-m+j} = m_k^{(j)}$, for $j = 1, 2, \ldots, m$ with some $m \in I(1) = \{1, 2, 3, \ldots, n, \ldots\}$, $m_k(\omega) \in R[\Omega, \mathcal{R}_+^m]$, satisfy the inequality*

$$m_{k+1}(\omega) \leq G(k, m_k, \omega),$$

where

$$G_i(k, m_k, \omega) = \begin{cases} m_k^{(i+1)}, & \text{for } 1 \leq i \leq m-1, \\ \alpha m_k^{(m)} m_k^{(m-1)} \cdots m_k^{(1)}, & \text{for } i = m, \end{cases} \tag{1.2.16}$$

and $m_k = [m_k^{(1)}, m_k^{(2)}, \ldots, m_k^{(m)}]^T M \in R[\Omega, R_+]$, for all $k \in I(k_0 + 1)$. Then,

$$m_k(\omega) \leq r_k(\omega) \quad \textit{w.p. } 1, \ \textit{for all } k \in I(k_0 + 1);$$

moreover,

$$v_k(\omega) \leq r_k^{(m)}(\omega), \quad \textit{w.p. } 1, \ \textit{for } k \in I(k_0 + 1),$$

where $r_k^{(m)}(\omega)$ is the solution process of

$$u_{k+1}^{(i)}(\omega) = \begin{cases} u_k^{(i+1)}(\omega), & \textit{for } 1 \leq i \leq m-1, \\ \alpha u_k^{(1)}(\omega) u_k^{(2)}(\omega) \cdots u_k^{(m)}(\omega), & \textit{for } i = m, \end{cases} \tag{1.2.17}$$

through $u_0 = z_m = [v_0, v_1, \ldots, v_m]^T$, *and it is expressed as*

$$r_k^{(j)}(\omega) = \beta_{k-m+j-1}^{q^0} u_{m,k-m+j-1}^{q^m} u_{m-1,k-m+j-1}^{q^{m-1}} \cdots u_{0,k-m+j-1}^{q^0}, \tag{1.2.18}$$

where $j = 1, 2, \ldots, m+1$ *and*

$$\begin{aligned} q_k^m &= q_{k-1}^m + q_{k-1}^{m-1}, \\ q_k^{m-1} &= q_{k-1}^{m-1} + q_{k-1}^{m-2}, \\ &\cdots \\ q_k^1 &= q_{k-1}^1 + q_{k-1}^0, \\ q_k^0 &= q_{k-1}^m, \\ [q_1^0, q_1^1, \ldots, q_1^m]^T &= [0, 1, 0, \ldots, 0]^T, \end{aligned} \tag{1.2.19}$$

with $z_{k_0}(\omega) = z_m(\omega)$.

Now, we are ready to present a very general variational comparison theorem in the context of a Lyapunov-like function and a system of difference inequalities. The presented result relates the solutions of (1.1.1) and (1.1.2) and a system of comparison difference equation. We state the result with a brief sketch of its proof.

Theorem 1.2.3 (Variational Comparison Theorem). *Let the hypotheses of Theorem* 1.1.1 *be satisfied. Further, assume that*

(i) $G \in C[\mathbb{R} \times \mathbb{R}^N \times \Omega, \mathbb{R}^N]$ *and* $G(k, a, \omega)$ *satisfies the one-sided Lipschitz condition* $G(k, b, \omega) - G(k, a, \omega) \geq -\Gamma(k, \omega)(b - a)$, *for all* $a, b \in \mathbb{R}^N$ *and* $b \geq a$, *where* $\Gamma(k, \omega) = diag\{\gamma_1, \ldots, \gamma_N\}$, *with* $0 \leq \gamma_i \leq 1$, *for* $1 \leq i \leq N$;

(ii) $LV(p, m(k, p, y), \omega) \leq G(p, V(p, m(k, p, y), \omega), \omega)$, *for all* $(p, y) \in \mathbb{I}(k_0) \times \mathbb{R}^n$, *where the generating operator* L *is defined as*

$$LV = [LV_1, LV_2, \ldots, LV_i, \ldots, LV_N]^T,$$

where $LV(k, m(t, p, x), \omega)$ *is defined analogously to* L *in* (1.1.20);

(iii) $r(k)$ *is the solution of the system of comparison difference equations*

$$u(k+1) = u(k) + G(k, u(k), \omega), \quad u(k_0, \omega) = u_0(\omega); \tag{1.2.20}$$

(iv) $E[V(p, m(k, p, y(p)), \omega)]$ *exists for any solution* $y(p) = y(p, k_0, y_0)$ *of the iterative stochastic process* (1.1.1), *for all* $p \in \mathbb{I}(k_0)$ *and* $k_0 \leq p \leq k$, *and*

$$V(k_0, m(k, k_0, y_0), \omega) \leq u_0. \tag{1.2.21}$$

Then,

$$V(p, m(k, p, y(p)), \omega) \leq r(p, \omega), \quad \text{for all } p \in \mathbb{I}(k_0). \tag{1.2.22}$$

Moreover, (1.2.22) *reduces to*

$$V(k, y(k), \omega) \leq r(k, \omega) \quad \text{for all } k \in \mathbb{I}(k_0). \tag{1.2.23}$$

Proof. For $k_0 \leq p+1 \leq k$, let $m(k, p, y) = m(k, p+1, y + F(p, y))$ and $y(p, \omega) = y(k, k_0, y_0)$ be the solution processes of (1.1.2) and (1.1.1) through $(p, y + F(p, y))$ and (k_0, y_0), respectively. Define

$$\begin{aligned} &V(p+1, m(k, p+1, y(p+1)), \omega) \\ &\quad = V(p+1, m(k, p+1, y(p) + f(p, y(p), \omega)), \omega). \end{aligned} \tag{1.2.24}$$

We also note that

$$V(p, m(k, p, y(p)), \omega) = V(p, m(k, p+1, y(p) + F(p, y(p)), \omega). \tag{1.2.25}$$

From (1.2.24), (1.2.25) and (1.1.1), we have

$$\begin{aligned} &V(p+1, m(k, p+1, y(p+1)), \omega) - V(p, m(k, p, y(p)), \omega) \\ &\quad = [\Delta_{(1.1.1)} V_1(p, m(k, p, y(p)), \omega), \ldots, \Delta_{(1.1.1)} \\ &\qquad \times V_i(p, m(k, p, y(p)), \omega), \ldots, \Delta_{(1.1.1)} V_N(p, m(k, p, y(p)), \omega)]^T. \end{aligned}$$

From the hypotheses of Theorem 1.1.1 and assumption (ii) of the theorem, we have

$$\begin{aligned} &\Delta V(p, m(k, p, y(p)), \omega) \\ &\quad = V(p+1, m(k, p+1, y(p+1)), \omega) - V(p, m(k, p, y(p)), \omega) \\ &\quad = [LV_1(p, m(k, p, y(p)), \omega), \ldots, LV_i(p, m(k, p, y(p)), \omega), \ldots, \\ &\qquad \times LV_N(p, m(k, p, y(p)), \omega)]^T \leq G(p, V(p, m(k, p, y(p))), \omega). \end{aligned} \tag{1.2.26}$$

By choosing $u_0 \geq V(k_0, m(k, k_0, y_0(\omega)), \omega)$ and applying Corollary 1.2.2, one concludes

$$V(p, m(k, p, y(p)), \omega) \leq r(p, k_0, u_0, \omega),$$
$$\text{for all } k_0 \leq p \leq k, \quad k \in \mathbb{I}(k_0). \tag{1.2.27}$$

In particular, for $p = k$, (1.2.27) reduces to

$$V(k, y(k, \omega), \omega) \leq r(k, k_0, u_0, \omega)), \quad \text{for all } k \in \mathbb{I}(k_0).$$

This completes the proof of the theorem.

The following corollary demonstrates the scope of the comparison Theorem 1.2.3. This corollary is based on Corollaries 1.1.1 and 1.1.3.

Corollary 1.2.5. *By considering Corollaries* (1.1.1) *and* (1.1.3) *and noting the fact that* $\mathcal{L}_o V_i(p, m(k, p, x), \omega) = 0$ *and* $\Delta(p, y) = f(p, y, \omega) - A(p)y$, *we assume that one can find* $\mu(p) \in \mathbb{R}$ *and* $\nu(p)$, $\beta(p) \in \mathbb{R}^+$ *such that*

$$m(k, p, y)^T \Phi(k, p)[f(p, y, \omega) - F(p, y)]$$
$$\leq \mu(p, \omega) m(k, p, y)^T m(k, p, y), \tag{1.2.28}$$
$$[f(p, y, \omega) - F(p, y)]^T \Phi^T(k, p)\Phi(k, p)[f(p, y, \omega) - F(p, y)]$$
$$\leq \nu(p, \omega) m(k, p, y)^T m(k, p, y) + \beta(p, \omega) \tag{1.2.29}$$

and

$$2\mu(p, \omega) + \nu(p, \omega) \geq -1, \quad k_0 \leq p \leq k,$$

for all $k_0 \leq p \leq k$ *and any* $k \in \mathbb{I}(k_0)$. *Then,*

$$V(k, y(k, k_0, y_0), \omega) \leq r(k, k_0, u_0, \omega),$$

provided $V(k_0, y_0, \omega) \leq u_0$.

Proof. From (1.2.21), followed by algebraic calculations and simplifications with the notation $V(p, z, \omega) = \alpha(p) z^T z$, one can obtain

$$LV(p, m(k, p, y), \omega) \leq g(p, \omega) V(p, m(k, p, y), \omega) + \beta(p, \omega),$$

where $g(p,\omega)$ is defined by $g(p,\omega) = (2\mu(p,\omega)+\nu(p,\omega)$. In this setup, the system of comparison equations then becomes

$$\Delta u(k) = G(k, u(k), \omega), \quad u_{k_0} = u_0, \tag{1.2.30}$$

where

$$G(k, u, \omega) = g(p, \omega)u + \beta(k, \omega).$$

It is obvious that the above comparison function $G(k, u, \omega)$ satisfies all the hypotheses of Corollary 1.2.2. One only needs to assume that condition (iv) holds and then pick $u_0 \geq V(k_0, x_0), \omega)$. Thus, by the application of Corollary 1.2.2, one may conclude

$$V(k, y(k), \omega) \leq r(k, k_0, u_0, \omega), \quad \forall\, k \in \mathbb{I}(k_0), \tag{1.2.31}$$

where $r(k, k_0, u_0)$ is the solution process of the comparison difference equation (1.2.30).

Corollary 1.2.6. *Let us assume that hypotheses $H_{(1.1.1)}$ and $H_{(1.1.2)}$ are satisfied.*

(a) *Assume that the conditions in Corollary* 1.2.5(iv) *are fulfilled. In addition, the operator L defined in* (1.1.30) *satisfies the following inequality:*

$$LV(p, m(k, p, y), \omega) \leq G(p, m(k, p, y), \omega). \tag{1.2.32}$$

Under the conditions of Theorem 1.2.3, *the conclusion of Theorem* 1.2.3 *remains true.*

(b) *Assume that the conditions of Corollary* 1.2.5(v) *are valid. Further, assume*

$$\begin{cases} 2\alpha y^T[f(p, y, \omega) - F(p, y)] \leq \mu(p, \omega)y^T y, \\ \alpha[f(p, y, \omega) - F(p, y)]^T[f(p, y, \omega) - F(p, y)] \leq \nu(p, \omega)y^T y, \\ 2\mu(p, \omega) + \gamma(p, \omega) \geq -1. \end{cases} \tag{1.2.33}$$

Then, the conclusion of Corollary 1.2.5 *remains valid, that is,*

$$V(k, y(k, k_0, y_0), \omega) \leq r(k, k_0, u_0, \omega), \tag{1.2.34}$$

provided

$$V(k_0, y_0, \omega) \leq u_0,$$

where $r(k, k_0, u_0)$ is the solution process of (1.2.30).

Proof. The proofs of these results can be constructed by following the arguments used in Theorem 1.2.3 and Corollary 1.2.3.

To investigate the qualitative properties of (1.1.1), we use the corresponding properties of comparison system (1.2.20). If the dimension of system (1.1.1) (dimension "n") is very high, it is difficult to obtain such information about the system. Therefore, we are forced to either reduce the system to a lower-dimension one or obtain information about such a system by comparing it with a lower-order system. Here, we choose to provide an estimate of the solution process of (1.1.1) in terms of a scalar iterative process.

Theorem 1.2.4. *Assume that conditions* (i), (ii), (iii) *and* (iv) *of Theorem* 1.2.3 *are satisfied. Further, assume that it satisfies*

$$\sum_{i=1}^{N} d_i G_i(p, m(k,p,y), \omega) \le \gamma(p,\omega) \sum_{i=1}^{m} d_i V_i(p, m(k,p,y), \omega) + \mu(p,\omega). \tag{1.2.35}$$

Let

$$\bar{v}(p, m(k,p,y), \omega) = \sum_{i=1}^{m} d_i V_i(p, m(k,p,y), \omega), \tag{1.2.36}$$

for $d_i > 0$ and $1 \le i \le N$. Then,

$$\bar{v}(p, y(p,\omega), \omega) \le r(p,\omega) \quad \text{for all } k_0 \le p \le k, \quad k \in \mathbb{I}(k_0), \tag{1.2.37}$$

whenever

$$\bar{v}(k_0, m(k,k_0,x_0), \omega) \le u_0(\omega), \tag{1.2.38}$$

where $r(p,\omega)$ is the solution of the scalar comparison functional difference equations

$$\Delta u(p) = \gamma(p,\omega) u(p) + \mu(p,\omega), \quad u_{k_0} = u_0, \tag{1.2.39}$$

with $u(p) \in \mathbb{R}_+$.

Proof. The proof of the result follows by using $\bar{v}(p, m(k,p,y), \omega)$ defined in (1.2.36) and the application of Corollary 1.2.3. The details are left to the reader.

Corollary 1.2.7. *Assume that*

(i) *the hypotheses of Corollary* 1.1.4(vii) *hold;*

(ii) $V \in M[I(k_0+1) \times R^n \times \Omega, R[\Omega, R_+^N]]$, *satisfies the relation*

$$V(k+1, y+f(k,y,\omega),\omega) \\ \leq G(k, V(k,y,\omega),\omega), \quad \text{for all } k \geq k_0, \tag{1.2.40}$$

where $V(k+1, y+f(k,y,\omega),\omega) = LV(k,y,\omega) + V(k,y,\omega)$*;*

(iii) G *satisfies the hypothesis of* Theorem 1.2.2.
Then,

$$V(k, y(k,\omega),\omega) \leq r(k,\omega), \tag{1.2.41}$$

provided

$$V(k_0, y_0(\omega),\omega) \leq u_0(\omega), \tag{1.2.42}$$

where $y(k,\omega) = y(k, k_0, y_0(\omega),\omega)$ *and* $r(k,\omega) = r(k, k_0, u_0(\omega), \omega)$ *are the solution processes of* (1.1.1) *and* (1.2.6)*, respectively.*

Proof. Let $y(k,\omega)$ be the solution process of (1.1.1). Set

$$m_k(\omega) = V(k, y(k,\omega),\omega). \tag{1.2.43}$$

From (1.2.42), we note that $m_{k_0}(\omega) \leq u_0(\omega)$. From this definition of $m_k(\omega)$ and (1.2.40) and noting the fact that $V(k+1, y(k,\omega) + f(k,y,\omega),\omega) = V(k+1, y(k+1,\omega),\omega)$, we get

$$m_{k+1}(\omega) \leq g(k, m_k(\omega),\omega).$$

This, together with the hypotheses of the theorem, verifies the hypotheses of Theorem 1.2.2. Hence, by an application of Theorem 1.2.2, we conclude that

$$m_k(\omega) \leq r(k,\omega), \qquad \text{for } k \geq k_0. \tag{1.2.44}$$

From (1.2.43) and (1.2.44), relation (1.2.41) follows immediately. Hence, the proof of the corollary is complete.

Corollary 1.2.8. *Assume that*

(i) *the hypotheses of Corollary* 1.2.4(vii) *hold;*
(ii) $LV(k,y,\omega) \leq G(k, V(k,y,\omega),\omega)$, *where* $V \in M[I(k_0) \times R^n \times \Omega, R[\Omega, R_+^N]]$;

(iii) *G satisfies relation* (1.2.2).
Then,

$$V(k, y(k, \omega), \omega) \leq r(k, \omega), \quad \text{for all } k \geq k_0, \tag{1.2.45}$$

whenever

$$V(k_0, y_0, \omega) \leq u_0(\omega),$$

where $y(k, \omega) = y(k, k_0, y_0(\omega), \omega)$ *and* $r(k, \omega) = r(k, k_0, u_0(\omega), \omega)$ *are the solution processes of* (1.1.1) *and* (1.2.20), *respectively.*

Proof. Set

$$m_{k+1}(\omega) = V(k + 1, y(k + 1, k_0, y_0, \omega), \omega). \tag{1.2.46}$$

Note that $m_{k_0}(\omega) \leq u_0(\omega)$. From (1.1.19), the definition of $m_k(\omega)$ and condition (ii), we have

$$\begin{aligned} m_k(\omega) &= V(k + 1, y(k + 1, \omega), \omega) - V(k, y(k, \omega), \omega) \\ &\leq G(k, m_k, \omega). \end{aligned} \tag{1.2.47}$$

This, together with other hypotheses of the corollary, fulfills the hypotheses of Corollary 1.2.2. Hence, one can conclude that

$$m_k(\omega) \leq r_k(\omega), \qquad \text{for } k \geq k_0.$$

This, together with the definition of $m_k(\omega)$, completes the proof of the corollary.

We demonstrate the scope of Theorem 1.2.3 in the following remark.

Remark 1.2.2. If $u_0(\omega) = V(k_0, x(k, k_0, y_0(\omega)), \omega)$, then (1.2.31) becomes

$$V(k, y(k, \omega), \omega) \leq r(k, k_0, V(k_0, x(k, \omega), \omega)), \quad k \geq k_0. \tag{1.2.48}$$

We remark that the comparison Theorems 1.2.3 and 1.2.4 are not exactly similar to the comparison results of Corollaries 1.2.7 and 1.2.8. From (1.2.48), it is obvious that Theorems 1.2.3 and 1.2.4 relate the solution processes of three kinds of difference equations, namely (1.1.1), (1.1.3) and (1.2.20)/(1.2.30). On the other hand, the usual comparison results of Corollaries 1.2.7 and 1.2.8 relate the solutions of two kinds of initial value problems, namely (1.1.1) and (1.2.20)/(1.2.30). Another important factor regarding Theorems 1.2.3 and 1.2.4 is that the initial state u_0 of the sample solution process $r(p, k_0, u_0, \omega)$ of (1.2.20)/(1.2.30) depends on k.

In the following, we give a few examples to illustrate the scope of Theorems 1.2.3 and 1.2.4.

Example 1.2.1. Consider the linear stochastic difference equation

$$\Delta y(k,\omega) = \frac{y(k,\omega)}{k+1} + H(k, y(k,\omega), \omega), \quad y(k_0,\omega) = y_0(\omega), \tag{1.2.49}$$

where $H \in M[I(k_0+1) \times R, R[\Omega, R]]$. We assume

$$\begin{aligned} yH(p,y,\omega) &\le -\lambda_1(p,\omega)y^2, \quad \lambda_1 > 0,\\ H^2(p,y,\omega) &\le \lambda_2(p,\omega)y^2, \quad \lambda_2 \ge 0. \end{aligned} \tag{1.2.50}$$

Further, we consider the difference equation

$$\Delta x(k) = \frac{1}{k+1}\, x(k), \quad x(k_0,\omega) = x_0(\omega) = y_0(\omega). \tag{1.2.51}$$

We note that

$$m(k) = \begin{cases} y_0, & k = p \ge k_0,\\ \frac{y_0(k+2)}{p+1}, & k \ge p+1 \ge k_0, \end{cases}$$

is a solution of (1.2.51). We assume that $V(k,x,\omega) = |x|^2$. From Theorem 1.2.3 and Corollary 1.1.5, we have

$$\begin{aligned} LV(p, m(k,p,y), \omega) = \mathcal{L}_a V(p, m(k,p,y), \omega) + \mathcal{L}_e V(p, m(k,p,y), \omega)\\ + \mathcal{L}_0 V(p, m(k,p,y), \omega), \end{aligned}$$

where $m(k,\ell,y)$ is the solution process of (1.2.49) through (ℓ, y),

$$\mathcal{L}_a V(p, m(k,p,y), \omega) = m(k,p,y)\left(\frac{k+2}{p+1}\right) H(p,y,w),$$

$$\mathcal{L}_e V(p, x(k,p,y), \omega) = \left(\frac{k+2}{p+1}\right)^2 H^2(p,y,w)$$

and

$$\mathcal{L}_0 V(p, x(k,p,y), \omega) = 0.$$

In the context of (1.2.49) and (1.2.49), we obtain

$$\begin{aligned}
&LV(p, x(k,p,y,\omega), \omega) \\
&\quad = \frac{(k+2)yH(p,y,\omega)}{(p+1)} + \frac{(k+2)^2 H^2(\ell,y,\omega)}{(p+1)^2} \\
&\quad \le \left(-\frac{2(p+1)\lambda_1(p,\omega)}{(k+2)} + \lambda_2(p,\omega)\right)\left(\frac{k+2}{p+1}\right)^2 y^2 \\
&\quad \le \lambda(p,\omega)V(p, m(k,p,y), \omega),
\end{aligned} \tag{1.2.52}$$

where

$$\lambda(p,\omega) = -\frac{2(p+1)\lambda_1(p,\omega)}{(k+2)} + \lambda_2(p,\omega)\left(\frac{k+2}{p+1}\right)^2.$$

In this case, the comparison equation is

$$\Delta u(p,\omega) = \lambda(p,\omega)u(p,\omega), \quad u(k_0,\omega) = u_0(\omega), \tag{1.2.53}$$

and the solution is given by

$$r(k, k_0, u_0(\omega), \omega) = \begin{cases} u_0(\omega), & k = k_0, \\ u_0(\omega)\prod_{p=k_0}^{k-1} 1 + \lambda(p,\omega), & k \ge k_0, \end{cases} \tag{1.2.54}$$

where $u_0(\omega) = |m(k, k_0, y_0(\omega))|^2$. Therefore, from (1.2.52), (1.2.53) and (1.2.54) and by applying Corollary 1.2.5, relation (1.2.23) becomes

$$|y(k,\omega)|^2 \le |m(k,\omega)|^2 \prod_{p=k_0}^{k-1}(1+\lambda(p)) \le |m(k,\omega)|^2 \exp \sum_{p=k_0}^{k-1}[\lambda(p)]. \tag{1.2.55}$$

This justifies Remark 1.2.2.

Example 1.2.2. Consider the linear stochastic difference equation

$$\Delta y(k,\omega) = y(k,\omega) + H(k, y(k,\omega), \omega), \quad y(k_0,\omega) = y_0(\omega), \tag{1.2.56}$$

where H is as defined in (1.2.49), and it satisfies

$$\begin{aligned} yH(p,y,\omega)y &\le -\lambda_3(p,w)y^2, \quad \lambda_3 > 0, \\ H^2(p,y,\omega) &\le \lambda_4(p,w)y^2, \quad \lambda_4 > 0. \end{aligned} \tag{1.2.57}$$

Further, we consider the difference equation

$$\Delta x(k) = x(k), \quad x(k_0,\omega) = y_0(\omega). \tag{1.2.58}$$

We note that $x(k) = 2^{k-k_0}y_0(\omega)$ is the solution of (1.2.58). By taking $V(k,x,\omega) = |x|^2$, we have

$$V(p, x(k,p,y),\omega) = |m(k,p,y)|^2 = 2^{2(k-p)}y^2,$$

where $m(k,p,y)$ is the solution process of (1.2.58) through (p,y). We compute $V(p,m(k,p,y),\omega)$ with respect to (1.2.56). By using (1.2.56) and following the argument used in Example 1.2.1, we have

$$\begin{aligned} LV(p,x(k,p,y),\omega) &= \mathcal{L}_a V(p,m(k,p,y),\omega) + \mathcal{L}_e V(p,m(k,p,y),\omega) \\ &\quad + \mathcal{L}_0 V(p,m(k,p,y),\omega) \\ &= 2^{2(k-p)}\left[2yH(p,y,\omega) + H^2(p,y,\omega)\right] \\ &\le [-2\lambda_3(p,\omega) + \lambda_4(p,\omega)]2^{2(k-p)}y^2 \\ &\le \lambda(p,w)V(p,m(k,p,y),\omega), \end{aligned} \tag{1.2.59}$$

where

$$\lambda(p,\omega) = [-2\lambda_3(p,\omega) + \lambda_4(p,\omega)].$$

In this case, the comparison equation is

$$\Delta u(p,\omega) = \lambda(p,\omega)u(p,\omega), \quad u(k_0,\omega) = u_0(\omega), \tag{1.2.60}$$

and its solution is given by

$$r(k,k_0,u_0(\omega),\omega) = \begin{cases} u_0(\omega), & k = k_0 \\ u_0(\omega)\prod_{p=k_0+1}^{k-1}(1+\lambda(p,\omega)), & k \ge k_0. \end{cases} \tag{1.2.61}$$

Therefore, from (1.2.59), (1.2.60) and (1.2.1) and by invoking Theorem 1.2.3, relation (1.2.23) reduces to

$$|y(k,\omega)|^2 \le |x(k,\omega)|^2 \prod_{s=k_0+1}^{k}(1+\lambda(s)) \le |x(k,\omega)|^2 \exp \sum_{s=k_0+1}^{k}[\lambda(s)]. \tag{1.2.62}$$

Remark 1.2.3. In particular, if $\lambda_3(p,\omega) = \sqrt{\lambda_4(p,\omega)} = \alpha(p,\omega)$ and $2|1-\alpha(p,\omega)| < 1$ w.p. 1, then the estimate (1.2.62) can be sharpened to the following:

$$|y(k,\omega)|^2 \leq \left(\frac{1}{4}\right)^{k-k_0} |x(k,\omega)|^2. \tag{1.2.63}$$

Example 1.2.3. We rewrite (1.2.52) as follows:

$$\Delta y(k,\omega) = \hat{A}(k,e)y(k,\omega) + R(k,y,\omega)y(k,\omega), \quad y(k_0,\omega) = y_0(\omega), \tag{1.2.64}$$

where $R(k,y,\omega) = A(k,y,\omega) - \hat{A}(k,e)$ and e is an n-dimensional parameter. Further, we consider the difference system

$$\Delta x(k) = \hat{A}(k,e)x(k), \quad x(k_0,\omega) = x_0(\omega). \tag{1.2.65}$$

We note that the solution of (1.2.65) is

$$x(k,k_0,x_0(\omega)) = \begin{cases} y_0(\omega), & k = k_0, \\ \prod_{p=k_0}^{k-1}(I + \hat{A}(p,e))y_0(\omega) = \Phi(k,k_0,e)y_0(\omega), & k \geq k_0 + 1, \end{cases} \tag{1.2.66}$$

where

$$\Phi(k,k_0,e) = \begin{cases} I, & k = k_0, \\ \prod_{p=k_0}^{k-1}(I + \hat{A}(p,e)), & k \geq k_0. \end{cases}$$

Let us assume that

$$\begin{aligned} m(k,p)\|y\|^2 &\leq y^T\Phi^T(k,p,e)\Phi(k,p,e)y \leq M(k,p)\|y\|^2, \\ y^T R^T \Phi^T(k,p,e)\Phi(k,p,e)Ry &\leq M_1(k,p)\|y\|^2, \quad M_1 > 0, \end{aligned} \tag{1.2.67}$$

where m and M are the minimal and maximal eigenvalues of $\Phi^T\Phi$, respectively, and M_2 is the maximal eigenvalue of $R^T\Phi^T(k,p,e)\Phi(k,p,e)R$. By taking $V(k,x,\omega) = \|x\|^2$, we get

$$V(p,x(k,p,y),\omega) = \|x(k,p,y)\|^2 = \|\Phi(k,p,e)y\|^2,$$

where $x(k,p,y)$ is the solution process of (1.2.64) through (p,y). Now, we compute $LV(p,x(k,p,y(p)),\omega)$ with regard to (1.2.64). By using

(1.2.67), (1.2.64), Theorem 1.2.3 and Corollaries 1.2.1 and 1.2.3, we have

$$\begin{aligned}
LV(p, x(k,p,y), \omega) &= \mathcal{L}_a V(p, x(k,p,y), \omega) + \mathcal{L}_e V(p, x(k,p,y), \omega) \\
&\quad + \mathcal{L}_0 V(p, x(k,p,y), \omega) \\
&= 2x^T(k,p,y)\Phi(k,p,e)R(p,y,\omega)y \\
&\quad + y^T R^T(p,y,\omega)\Phi^T(k,p,e)\Phi(k,p,e)R(p,y,\omega)y \\
&= 2y^T\Phi^T(k,p,e)\Phi(k,p,e)R(p,y,\omega)y \\
&\quad + y^T R^T(p,y,\omega)\Phi^T(k,p,e)\Phi(k,p,e)R(p,y,\omega)y.
\end{aligned} \tag{1.2.68}$$

By using (1.2.67), we have

$$\begin{aligned}
&\|y\|^2 \leq (1/m(k,p))y^T\Phi^T(k,p,e)\Phi(k,p,e)y \\
&\quad \leq (1/m(k,p))V(p, x(k,p,y), \omega), \\
&y^T\Phi^T(k,p,e)\Phi(k,p,e)R(p,y,\omega)y \leq -\lambda(p,\omega)\|y\|^2 \\
&\quad \leq -\frac{\lambda(p,\omega)}{M(k,p)}V(p, x(k,p,y), \omega), \quad \lambda > 0, \\
&y^T R^T(p,y,\omega)\Phi^T(k,p,e)\Phi(k,p,e)R(p,y,\omega)y \leq M_1(k,p)\|y\|^2 \\
&\quad \leq \frac{M_1(k,p)}{m(k,p)}V(p, x(k,p,y), \omega),
\end{aligned} \tag{1.2.69}$$

and using the estimate (1.2.69), (1.2.68) reduces to

$$LV(p, x(k,p,y), \omega) \leq \nu(p,\omega)V(p, x(k,p,y), \omega), \tag{1.2.70}$$

where $\nu(p,\omega) = -\frac{2\lambda(p,\omega)}{M(k,p)} + \frac{M_1(k,p)}{m(k,p)}$. In this case, the comparison equation is

$$\Delta u(p,\omega) = \nu(p,\omega)u(p,\omega), \quad u(k_0,\omega) = u_0(\omega), \tag{1.2.71}$$

and its solution is given by

$$r(p, k_0, u_0(\omega), \omega) = \begin{cases} u_0(\omega), & p = k_0, \\ u_0(\omega)\prod_{s=k_0+1}^{p}(\nu(s,\omega)+1), & p \geq s \geq k_0, \end{cases} \tag{1.2.72}$$

where $u_0(\omega) = V(k_0, x(k, k_0, y_0(\omega))) = \|x(k,\omega)\|^2$. From (1.2.70), (1.2.71) and (1.2.72) and by applying Theorem 1.2.3, relation (1.2.23) becomes

$$\begin{aligned}\|y(k,\omega)\|^2 &\leq \|x(k,\omega)\|^2 \prod_{s=k_0+1}^{k} (1+\nu(s,\omega))\\ &\leq \|x(k,\omega)\|^2 \exp \sum_{s=k_0+1}^{k} [\nu(s,\omega)]. \end{aligned} \tag{1.2.73}$$

Example 1.2.4. We consider now the linear stochastic difference system (1.2.52). In this case, $A(k,y,\omega)$ and $\hat{A}(k,m)$ play the roles of $A(k,\omega)$ and $\hat{A}(k)$, respectively, in (1.2.64) and (1.2.65) in Example 1.2.3. By proceeding with a similar technique in Example 1.2.3 with $V(k,x,\omega) = \|x\|^2$, we obtain an inequality similar to (1.2.73).

In the following example, we illustrate the usage of a comparison result (Corollary 1.2.7). The example sheds light on the differences between the comparison results, namely, Theorem 1.2.3 and Corollary 1.2.7/1.2.8.

Example 1.2.5. Consider the stochastic difference system (1.2.64). Let $V(k,x,\omega) = \|x\|^2$. Then, $V(k,y(k,\omega),\omega) = \|y(k,\omega)\|^2$. First, we assume that

$$\begin{aligned} m_1(k)\|y\|^2 \leq y^T(I+\hat{A})^T(I+\hat{A})y \leq M_2(k)\|y\|^2,\\ y^TR^TRy \leq M_3(k,\omega)\|y\|^2, \quad M_3 > 0, \end{aligned} \tag{1.2.74}$$

where m_1 and M_2 are the minimal and maximal eigenvalues of $(I+\hat{A})^T(I+\hat{A})$, respectively, and M_3 is the maximal eigenvalue of $R^T\hat{A}^TAR$. Now, we compute $V(k,y(k,\omega),\omega)$ with regard to (1.2.64):

$$\begin{aligned} LV(k,y,\omega) &= V(k+1,y+\hat{A}(k,y)+R(k,y,w)y,\omega) - V(k,y,\omega)\\ &= 2y^T\hat{A}(k,e)y + 2y^TR(k,y,\omega)y + y^T\hat{A}^T(k,e)\hat{A}(k,e)y\\ &\quad + 2y^T\hat{A}^T(k,e)R(k,y,\omega)y + y^TR^T(k,y,\omega)R(k,y,\omega)y\\ &= y^T(I+\hat{A})^T(I+\hat{A})y + 2y^T(I+\hat{A})^TR(k,y,\omega)y\\ &\quad + y^TA^T(k,y,\omega)R(k,y,\omega)y - y^Ty. \end{aligned} \tag{1.2.75}$$

By using (1.2.74), we get

$$
\begin{aligned}
\|y\|^2 &\le \frac{1}{m_1(k)} y^T(I+\hat{A}(k,e))^T(I+\hat{A}(k,e))y \\
&\le \frac{M_2(k)}{m_1(k)} V(k,y,\omega), \\
y^T\hat{A}^T(k,e)R(k,y,\omega)y &\le -\lambda_1(k,\omega)\|y\|^2 \\
&\le -\frac{\lambda_1(k,\omega)}{M_2(k)} V(k,y,\omega), \quad \lambda_1 > 0, \\
y^T R^T(k,y,\omega)R(k,y,\omega)y &\le \frac{M_3(k,\omega)}{m_1(k)} V(k,y,\omega). \qquad (1.2.76)
\end{aligned}
$$

Using the estimate (1.2.76), (1.2.75) reduces to

$$
V_{(1.2.64)}(k,y,\omega) \le \nu_1(k,\omega)V(k-1,y,\omega), \qquad (1.2.77)
$$

where $\nu_1(k,\omega) = 1 - \frac{2\lambda_1(k,\omega)}{M_2(k)} + \frac{M_3(k,\omega)}{m(k)}$. In this case, the comparison equation is

$$
u(k,\omega) = \nu_1(k,\omega)u(k-1,\omega), \quad u_{k_0}(\omega) = u_0(\omega), \qquad (1.2.78)
$$

and its solution is given by

$$
r(k,k_0,u_0(\omega),\omega) = \begin{cases} u_0(\omega), & k = k_0, \\ u_0(\omega)\prod_{s=k_0+1}^{k}(1+\nu_1(s,\omega)), & k \ge k_0+1, \end{cases} \qquad (1.2.79)
$$

where $u_0(\omega) = V(k_0,y_0,\omega) = \|y_0(\omega)\|^2$. From (1.2.77), (1.2.78) and (1.2.79) and by an application of Corollary 1.2.7, relation (1.2.23) becomes

$$
\begin{aligned}
\|y(k,\omega)\|^2 &\le \|y_0(\omega)\|^2 \prod_{s=k_0+1}^{k} (1+\nu_1(s,\omega)) \\
&\le \|y_0(\omega)\|^2 \exp \sum_{s=k_0+1}^{k} [\nu_1(s,\omega)]. \qquad (1.2.80)
\end{aligned}
$$

We now state and prove a comparison theorem which has a wide range of applications in the theory of error estimates and stability analysis of stochastic difference systems.

Theorem 1.2.5. *Let the hypotheses of Theorem* 1.2.3 *be satisfied, except* $LV(P, m(k,p,y), \omega)$, $V(p, m(k,p,y(p)), \omega)$ *and* $V(k_0, m(k, k_0, y_0), \omega)$ *are replaced by* $V(p, m(k,p,y(p) - w(p)), \omega)$, $V(t, m(k,p,y(p) - w(p)), \omega)$ *and* $V(k_0, m(k, k_0, y_0 - w_0), \omega)$, *respectively, where* $y(p) = y(p, k_0, y_0)$ *and* $w(p) = w(p, k_0, w_0)$ *are the solution processes of* (1.1.1) *and* (1.1.60), *respectively, for* $k_0 \leq p \leq k$. *Then,*

$$V(p, m(k,p,y(p) - w(p)), \omega) \leq r(p, \omega), \quad \text{for } k_0 \leq p \leq k, \tag{1.2.81}$$

provided

$$V(k_0, m(k, k_0, y_0 - w_0), \omega) \leq u_0(k_0, \omega). \tag{1.2.82}$$

Moreover, for $p = k$, (1.2.81) *reduces to*

$$V(k, y(k) - w(k), \omega) \leq r(k, \omega), \quad \text{for } k \geq k_0, \tag{1.2.83}$$

whenever (1.2.82) *remains valid.*

Proof. Let $y(p)$ and $w(p)$ be the solution processes of (1.1.1) and (1.1.60) through (k_0, y_0) and (k_0, w_0), respectively. By repeating the argument used in the proofs of Theorems 1.1.5 and 1.2.3, the proof of the theorem can be constructed analogously. The details are left to the reader.

To illustrate the significance of Theorem 1.2.5, we outline a few particular cases.

Corollary 1.2.9. *Let us suppose that hypotheses* $H_{(1.1.1)}$ *and* $H_{(1.1.2)}$ *remain true.*

(i) *Assume that all the conditions of Corollary* 1.1.6(iii) *and Theorem* 1.2.5 *remain valid. Then,*

$$V(p, y(p) - w(p), \omega) \leq r(p, \omega), \quad \text{for } k_0 \leq p \leq k,$$

and

$$V(k, y(k) - w(k), \omega) \leq r(k, \omega), \quad \text{for } k \geq k_0, \tag{1.2.84}$$

provided

$$V(k_0, y_0 - w_0, \omega) \leq u_0(k_0, \omega),$$

LV *in Theorem* 1.2.5 *is* $LV(p, y(p) - w(p), \omega)$, *and* $r(k, \omega)$ *is as defined by* (1.2.20).

(ii) *If the conditions of Corollary* 1.2.6(iv) *and Theorem* 1.2.5 *remain valid, then*

$$V(p, \Phi(k,p)(y(p) - w(p)), \omega) \leq r(p, \omega), \quad \textit{for } k_0 \leq p \leq k$$

and hence

$$V(k, y(k) - w(k), \omega) \leq r(k, \omega), \quad \textit{for } k \geq k_0, \tag{1.2.85}$$

provided

$$V(k_0, \Phi(k, k_0)(y_0 - w_0), \omega) \leq u_0(k_0, \omega),$$

and $r(k, \omega)$ *is the solution process of* (1.2.20).

(iii) *If the assumptions of Corollaries* 1.1.6(v) *and* 1.2.6(b) *remain valid, then*

$$\alpha \|y(k, k_0, y_0) - w(k, k_0, w_0)\|^2 \leq r(k, k_0, \omega), \quad \textit{for } k \geq k_0, \tag{1.2.86}$$

whenever

$$\alpha \|y_0 - w_0\|^2 \leq u_0(k_0, \omega).$$

We note that

$$\begin{cases} 2\alpha(y - w)^T [f(p, y, \omega) - h(p, w, \omega) - F(p, y - w)] \\ \qquad \leq \mu(p, \omega)\|y - w\|^2, \\ \alpha \|f(p, y, \omega) - h(p, w, \omega) - F(p, y - w)\|^2 \leq \nu(p, \omega)\|y - w\|^2, \\ 2\mu(p, \omega) + \nu(p, \omega) \geq -1, \end{cases} \tag{1.2.87}$$

and $r(k, \omega)$ *is the solution process of* (1.2.30).

(iv) *If the assumptions of Corollaries* 1.1.6(vi) *and* 1.2.5 *remain valid, then*

$$\alpha \|y(k, k_0, y_0) - w(k, k_0, w_0)\|^2 \leq r(k, \omega), \quad \textit{for } k \geq k_0, \tag{1.2.88}$$

whenever

$$\alpha \|\Phi(k, k_0)(y_0 - w_0)\|^2 \leq u_0(k_0, \omega).$$

We remark that

$$\begin{cases} m^T(k,p,y-w)\Phi(k,p)(f(p,y,\omega)-h(p,w,\omega)-F(p,y-w)) \\ \qquad\qquad \leq \mu(p,\omega)\|m(k,p,y-w)\|^2, \\ \|\Phi(k,p)(f(p,y,\omega)-h(p,w,\omega)-F(p,y-w))\|^2 \\ \qquad\qquad \leq \nu(p,\omega)\|m(k,p,y-w)\|^2+\beta(p,\omega), \\ 2\mu(p,\omega)+\nu(p,\omega)\geq -1, \end{cases}$$

and $r(p,\omega)$ *is the solution process of comparison equation* (1.2.20).

(v) *If the assumptions of Corollaries* 1.1.6(vii) *and* (1.2.7) *are fulfilled, then*

$$V(k,y(k)-w(k),\omega)\leq r(k,\omega), \quad \text{for } k\geq k_0, \tag{1.2.89}$$

provided $V(k_0,y_0-w_0,\omega)\leq u_0(\omega)$.

Theorem 1.2.6. *Assume that all the hypotheses of Corollary* 1.1.6(viii) *are satisfied with* $h=F$. *Further, assume that* L *in* (1.1.76) *satisfies*

$$\begin{aligned} &V(p,m(k,p,y-w),\omega)+LV(p,m(k,p,y-w),\omega) \\ &\quad \leq g(p,V(p,m(k,p,y-w),\omega) \end{aligned} \tag{1.2.90}$$

and

$$V(k_0,m(k,k_0,y_0-m_0,\omega),\omega)\leq u_0(\omega), \tag{1.2.91}$$

where $x(k,k_0,z_0)=\bar{x}(k)$ *is the solution process of either* (1.1.2) *or* (1.1.3) *depending on the choice of* z_0. *Then,*

$$V(k,y(k)-w(k),\omega)\leq r(k,k_0,u_0(\omega),\omega), \quad k\geq k_0. \tag{1.2.92}$$

Proof. Let $y(k,\omega)$ and $w(k,\omega)$ be the solution processes of (1.1.1) and (1.1.3), respectively. Let $m(k)=x(k,k_0,x_0)$ be a solution process of either (1.1.2) or (1.1.3) depending upon the choice of z_0. Set

$$\begin{aligned} v(p+1,\omega)&=V(p+1,x(k,p+1,y(p+1,\omega)-m(p+1)),\omega), \\ w(k_0,\omega)&=V(k_0,m(k,k_0,y_0-m_0),\omega). \end{aligned}$$

By following the proof of Corollaries 1.1.1 and 1.2.2, the proof of this theorem can be completed.

To demonstrate the scope of Theorem 1.2.5, we present an example.

Example 1.2.6. We consider the stochastic difference systems (1.1.52), (1.2.65) and

$$\Delta m(k) = \hat{A}(k,e)m(k-1), \quad m(k_0) = m_0. \tag{1.2.93}$$

Let $V(k,x,\omega) = \|x\|^2$. By following the discussion in Corollary 1.2.9(iv), we compute $LV(p, m(k,p,y(p) - m(p),\omega))$ as follows. Since $m(k) = \Phi(k,k_0,e)m_0(\omega))$, we assume that

$$\begin{aligned}
&2m^T(k,p,y-w)[f(p,y,\omega) - h(p,w,\omega) - F(p,y-w)] \\
&\quad \le \mu(p,\omega)V(p,m(k,p,y-w)) \\
&\|\Phi(k,p)[f(p,y,\omega) - h(p,w,\omega) - F(p,y-w)]\|^2 \\
&\quad \le \nu(p,\omega)V(p,m(k,p,y-w),\omega) \\
&2\mu(p,\omega) + \nu(p,\omega) \ge -\gamma,
\end{aligned}$$

and hence

$$\begin{aligned}
&LV(p,m(k,p,y-w),\omega) \\
&\quad \le [2\mu(p,\omega) + \nu(p,\omega)]V(p,m(k,p,y-w),\omega).
\end{aligned} \tag{1.2.94}$$

Therefore, the comparison equation is

$$\Delta u(k) = [2\mu(k,\omega) + \nu(k,\omega)]\, u, \quad u(k_0,\omega) = u_0(\omega), \tag{1.2.95}$$

where $u_0(\omega) = \|\Phi(k,k_0)(y_0 - w_0)\|$. By an application of Corollary 1.2.9(iv), we obtain

$$\begin{aligned}
&\|y(k,\omega) - m(k)\|^2 \\
&\quad \le \|\Phi(k,k_0)(y_0 - v_0)\|^2 \exp\left[\sum_{p=k_0}^{k-1} 2\mu(p,\omega) + \nu(p,\omega)\right].
\end{aligned} \tag{1.2.96}$$

1.3 Notes and Comments

The material adapted from Ladde and Sambandham [31, 34] includes the following: the utilization of vector Lyapunov-like functions; the generation of difference operators with three components, characterized by the absence of randomness, the presence of randomness, and discretization round-off aggregate errors, along a directional stochastic system of a discrete-time system iterative process; and several basic mathematical results presented in Section 1.1 to study qualitative and quantitative properties which are iterative under the influence of random parameters. In particular, these include: Theorem 1.1.1 and its special case, together with a highly generalized variation of constants formula; Theorem 1.1.2, for nonlinear nonstationary iterative processes, and its variants; Theorems 1.1.3, 1.1.4 and 1.1.5, with several corresponding special cases coupled with examples outlined in Section 1.1. The generalized comparison theorems presented in Section 1.2 are from Refs. [33, 34]. Theorem 1.2.1, regarding systems of differential inequalities corresponding to the comparison Theorem 1.2.2 presented with several cases, are from the works of Ladde and Siljak [37] and Ladde and Sambandham [28, 29, 33, 34]. A very general variational comparison theorem, Theorem 1.2.3, its variants, Theorems 1.2.4, 1.2.5 and 1.2.6, and its corresponding special cases and examples also appear in Section 2.2. These results are motivated by the ideas of Ladde [20]. The results in Section 1.2 are generalizations and extensions of the work of Ladde and Sambandham [28, 29, 31–34], while the deterministic discrete-time comparison results are from Lakshmikantham and Leela [38] and Lakshmikantham and Trigiante [39].

Chapter 2

Stochastic Systems of Difference Equations of Ito-Type Methods

2.0 Introduction

The concept of a stochastic system of difference equations under the influence of Gaussian processes, in particular normalized Wiener processes, generates a very general and difficult problem of, in rough terms, "stochastic versus deterministic" (or "randomness versus non-randomness"). This refers to the extent to which the solution processes of stochastic systems of difference equations of the Ito–Doob type deviate from the solution processes of corresponding average systems of difference equations with deterministic parameters. In this chapter, as in Chapter 1, two major techniques for studying nonlinear initial value problems are studied, and a solution to the above problem is addressed. A method of generalized variation of constants formula for stochastic discrete-time processes is discussed in Section 2.1. We employ the concept of Lyapunov-like functions and the theory of system of stochastic difference inequalities. We also present several variational comparison theorems along with a number of examples.

2.1 Variation of Constants Method

In this section, we present a generalized variation of constants formula for stochastic discrete-time processes under the influence of

Gaussian processes. For this purpose, we develop a new basic result. The presented results provide a mathematical tool to study such processes.

Let us present a mathematical description of discrete-time dynamic processes in the chemical, engineering, medical, physical and social sciences. It is described by the following nonlinear and nonstationary iterative process under discrete-time Gaussian perturbations:

$$\Delta y(k) = f(k, y(k), \omega) + \sigma(k, y(k, \omega), \omega) z(k, \omega), \quad y(k_0, \omega) = y_0, \tag{2.1.1}$$

where for fixed $(k, y) \in \mathbb{I}(k_0) \times \mathbb{R}^n$, $y(k) \in \mathbb{R}^n$, $\Delta y(k) = y(k+1) - y(k)$; f and $y_0 \in \mathbb{R}^n$ are defined in (1.1.1); $\sigma(k, y, \omega)$ is a sequence of $n \times q$ matrices that describes random perturbations; and $z(k, \omega)$ is a q-dimensional discrete-time Gaussian process with $E[z(k, \omega)] = 0$ and $\mathrm{var}(z(k, \omega) | \mathcal{F}_k) = I$. Moreover, f and σ are independent non-anticipatory processes with respect to $\mathcal{F}_k$, where $\mathcal{F}_k$ is a sub-σ-algebra of $\mathcal{F}$, and $(\Omega, \mathcal{F}, P)$ is a complete probability space.

In the absolute absence of random perturbations, the mathematical description (2.1.1) reduces to

$$\Delta m = F(k, m), \quad m(k_0) = m_0 = E[y_0]. \tag{2.1.2}$$

In our presentation, we also utilize the following initial value problem representing the absence of the external random perturbations:

$$\Delta x = F(k, x, \omega), \quad x(k_0) = x_0. \tag{2.1.3}$$

For the sake of easy reference, we list the following assumptions with regard to (2.1.1)–(2.1.3).

Remark 2.1.1. The admissible rate functions for the mathematical description in the absence of absolute and environmental random perturbations in (2.1.2) and (2.1.3) are (i) $F(k, y) = E[f(k, y, \omega)]$ and (ii) $F(k, y, \omega) = f(k, y, \omega)$.

Hypothesis $\mathbf{H}_{(2.1.1)}$: Assume that the initial state y_0 is $\mathcal{F}_{k_0}$-measurable. $f(k, y, \omega)$ and $\sigma(k, y, \omega)$ are sequences of random vectors and matrices, respectively. We designate by $y(k, k_0, y_0) = y(k)$ a solution process of (2.1.1), for each $k \in \mathbb{I}(k_0)$ and $y(k_0) = y_0$.

Hypothesis $\mathbf{H}_{(2.1.2)}$: $\mathcal{F}$ is a sequence of continuous random functions defined on $\mathbb{R}^n$ into $\mathbb{R}^n$, and it satisfies the assumption $\mathrm{H}_{(1.1.2)}$.

Remark 2.1.2. For each $p \in \mathbb{I}(k_0)$ and $y \in \mathbb{R}^n$, we observe that $y(p+1)$ satisfies

$$\begin{aligned} y(p+1,p,y) &= y(p+1) \\ &= y + f(p,y,\omega) + \sigma(p,y,\omega)z(p,\omega), \quad y(p) = y. \end{aligned} \tag{2.1.4}$$

Remark 2.1.3. For $v = f(p,y,\omega) + \sigma(k,y,\omega)z(k,\omega)$ and $u = y$, $\Delta V(p, m(k,p,y,v),\omega)$ in (1.1.19) is denoted by $\Delta_{(2.1.1)} V(p, m(k,p,y),\omega)$. Furthermore, the continuity of f and σ guarantees the measurability of the difference operator $\Delta_{(2.1.1)} V(p, m(k,p,y),\omega)$.

Remark 2.1.4. (a) We note that for $v = f(p,y,\omega) + \sigma(p,y,\omega) z(p,\omega)$ and $u = y$, the definition of $LV(p, m(k,p,y,v),\omega) \equiv LV(p, m(k,p,y),\omega) = E[\Delta_{(2.1.1)} V(p, m(k,p,y,\omega),\omega) \,|\, \mathcal{F}_p]$ in (1.1.20) does not depend on the knowledge of the solution process of (2.1.1). It only depends on $y \in \mathbb{R}^n$, the rate functions f, σ and $\Delta(p,u,v) = f(p,y,\omega) + \sigma(p,y,\omega)z(p,\omega) - F(p,y)$. Moreover, in order to characterize the effects of random perturbations (both internal and external), the L-operator associated with (2.1.1) in (1.1.20) is decomposed into three operators, $\mathcal{L}_a$, $\mathcal{L}_e$ and $\mathcal{L}_o$, representing the absence of randomness, the presence of external disturbances, and round-off errors, respectively. Moreover, $A(k,p+1,\Delta(p,u,v))$ and $E(k,p+1, \Delta(p,u,v))$ in (1.1.25), (1.1.26) and (1.1.27) are replaced by

$$\begin{aligned} &E[A(k,p+1,\Delta(p,u,v)) \,|\, \mathcal{F}_p] \\ &\quad = \Phi(k,p+1,u+F(p,u))(f(p,y,\omega) - F(p,y)), \end{aligned} \tag{2.1.5}$$

$$\begin{aligned} &E[E(k,p+1,\Delta(p,u,v)) \,|\, \mathcal{F}_p] \\ &\quad = \frac{1}{2} E\left[\frac{\partial}{\partial m_0}\Phi(k,p+1,u+F(p,u))\Delta(p,u,v)\Delta(p,u,v) \,|\, \mathcal{F}_p\right] \end{aligned} \tag{2.1.6}$$

and

$$\begin{aligned} &E[\Theta(k,p+1,\Delta(p,u,v)) \,|\, \mathcal{F}_p] \\ &\quad = E[A(k,p+1,\Delta(p,u,v)) \,|\, \mathcal{F}_p] \\ &\qquad + \frac{1}{2} E\left[\frac{\partial}{\partial m_0} E(k,y,u+F(p,u)\Delta) \,|\, \mathcal{F}_p\right], \end{aligned} \tag{2.1.7}$$

respectively.

(b) For $F = f$, then $E[A(k,p+1,\Delta(p,u,v)) \,|\, \mathcal{F}_k] = 0$.

The following results illustrate the computational feasibility of L-operator relative to (2.1.1). It is based on the fundamental Theorem 1.1.1.

Corollary 2.1.1. *Let* $V(p,m,\omega) = \alpha(\omega)m^T m = \alpha(\omega)\|m\|^2$, $\alpha(\omega) > 0$, *for any* $k \in \mathbb{I}(k_0)$, $k_0 \le p$, *where* α *is a random variable. Then,*

$$V_t(p,m,\omega) = 0, \quad V_m(p,m,\omega) = 2\alpha(\omega)m^T,$$

$$\frac{\partial^2}{\partial m \partial m} V(p,m(k,p,y),\omega) = 2\alpha(\omega).$$

Then, $LV(p,m(k,p,y),\omega)$ *reduces to*

$$\begin{aligned} &LV(p,m(k,p,y),\omega) \\ &\quad = 2\alpha(\omega)E[m(k,p,y)\Delta m \,|\, \mathcal{F}_p] + \alpha(\omega)E[(\Delta m)^T \Delta m \,|\, \mathcal{F}_p], \end{aligned}$$

where $m(k,p,y)$ *is the solution process of* (2.1.2) *through* (p,y) *and* $\Delta m = m(k,p+1,y+f(p,y,\omega)+\sigma(p,y,\omega)z(p,\omega)) - m(k,p,y)$.

Proof. From (1.1.11) and (1.1.12), we have

$$\begin{aligned} &V_m(p,m(k,p,u),\omega)\Delta m \\ &\quad = V_m(p,m(k,p,y),\omega)\,[\Phi(k,p+1,m(p+1))E[\Delta(p,y) \,|\, \mathcal{F}_p] \\ &\qquad + \frac{1}{2}\frac{\partial}{\partial m_0}\Phi(k,p+1,u+F(p,y))E[\Delta(p,u,v)\Delta(p,u,v) \,|\, \mathcal{F}_p] \\ &\qquad + E[O(k,p+1,u+F(p,y),\Delta(p,u,v)) \,|\, \mathcal{F}_p]]\,, \end{aligned}$$

where

$$\begin{aligned} &\frac{1}{2}\frac{\partial^2}{\partial z^2}V(p,m(k,p,u),\omega)\Delta m^T \Delta m \\ &\quad = \alpha(\omega)\Big[E[\Theta^T(k,p+1,m(p+1))\Theta(k,p+1,m(p+1)) \,|\, \mathcal{F}_p] \\ &\qquad\qquad + E[\Theta^T(k,p+1,m(p+1))O(k,p+1,y+F(p,u)\Delta(p,u)) \,|\, \mathcal{F}_p] \\ &\qquad\qquad + E[O^T(k,p+1,u+F(p,u),\Delta(p,u,v)) \\ &\qquad\qquad \times \Theta(k,p+1,m(p+1)) \,|\, \mathcal{F}_p] \\ &\qquad\qquad + E[O^T(k,p+1,u+F(p,u),\Delta(p,u,v)) \\ &\qquad\qquad \times O(k,p+1,u+F(p,u),\Delta(p,u,v)) \,|\, \mathcal{F}_p]\Big]. \end{aligned}$$

After elementary computations and simplifications, (1.1.20) reduces to

$$
\begin{aligned}
LV(p, m(k,p,u), \omega) = \mathcal{L}_a V(p, m(k,p,u,v), \omega) \\
+ \mathcal{L}_e V(p, m(k,p,u,v), \omega) \\
+ \mathcal{L}_0 V(p, m(k,p,u,v), \omega),
\end{aligned} \tag{2.1.8}
$$

where

$$
\begin{aligned}
\mathcal{L}_a V(p, m(k,p,u), \omega) &= 2\alpha(\omega) m(k,p,u)^T \\
&\quad \times \Phi(k, p+1, m(p+1)) E[\Delta(p,u,v) \mid \mathcal{F}_p] \\
&= 2\alpha(\omega) m(k,p,u)^T \\
&\quad \times E[A(k, p+1, \Delta(p+1)) \mid \mathcal{F}_p], \\
\mathcal{L}_e V(p, m(k,p,u), \omega) &= \alpha(\omega) E[\Theta^T(k, p+1, m(p+1)) \\
&\quad \times \Theta(k, p+1, m(p+1)) \mid \mathcal{F}_p] \\
&\quad + m(k,p,u)^T E[E(k, p+1, \Delta(p,u,v)) \mid \mathcal{F}_p]
\end{aligned}
$$

and

$$
\begin{aligned}
\mathcal{L}_0 V(p, z(k,p,u), \omega) &= 2\alpha(\omega) m(k,p,u)^T \\
&\quad \times E[O(k, p+1, m(p+1), \Delta(p,u,v)) \mid \mathcal{F}_p] \\
&\quad + E[O^T(k, p+1, m(p+1,p,u), \Delta(p,u,v)) \\
&\qquad \times \Theta(k, p+1, m(p+1)) \mid \mathcal{F}_p] \\
&\quad + E[\Theta^T(k, p+1, m(p+1,p,u)) \\
&\qquad \times O(k, p+1, m(p+1), \Delta(p,u,v)) \mid \mathcal{F}_p] \\
&\quad + E[O^T(k, p+1, m(p+1,p,u)) \\
&\qquad \times O(k, p+1, m(p+1), \Delta(p,u,v)) \mid \mathcal{F}_p].
\end{aligned}
$$

By utilizing a vector Lyapunov-like function $V(t, x, \omega)$, the following result is proved. This result lays down a foundation for the presentation of a very general variation of constants formula.

Theorem 2.1.1. *Assume that the hypotheses of Theorem* 1.1.1 *are satisfied. Then,*

$$E[V(k, y(k, k_0, y_0), \omega) \,|\, \mathcal{F}_{k-1}] = V(k_0, m(k, k_0, y_0), \omega) + \sum_{p=k_0}^{k-1} LV(p, m(k, p, y(p, k_0, y_0)), \omega), \tag{2.1.9}$$

where L is defined as in Remark 2.1.4, *and it is relative to* (2.1.1). *Moreover,*

$$E[V(k, y(k, k_0, y_0), \omega) \,|\, \mathcal{F}_{k_0}] = V(k_0, m(k, k_0, y_0), \omega) + \sum_{p=k_0}^{k-1} E[LV(\ldots) \,|\, \mathcal{F}_{k_0}].$$

Proof. For $k_0 \leq p \leq k$, let $y(p) = y(p, k_0, y_0)$ and $m(k, p, y(p))$ be the solution processes of (2.1.1) and (2.1.2) through (k_0, y_0) and $(p, y(p))$, respectively. From (1.1.5), we recall that $m(k, p+1, m(p+1, \omega)) = m(k, p+1, y(p) + F(p, y(p))) = m(k, p, y(p))$. We set $\Delta v(p) = V(p+1, m(k, p+1, y(p+1)), \omega) - V(p, m(k, p, y(p)), \omega)$ and consider

$$\begin{aligned}
\Delta v(p) &= V(p+1, m(k, p+1, y(p) + f(p, y(p), \omega) \\
&\quad + \sigma(p, y(p), \omega) z(p, \omega))) - V(p, m(k, p, y(p)), \omega) \\
&= V(p+1, m(k, p+1, y(p) + F(p, y(p)) \\
&\quad + [f(p, y(p), \omega) + \sigma(p, y(p), \omega) z(p, \omega)) - F(p, y(p))], \omega) \\
&\quad - V(p, m(k, p, y(p)), \omega) \\
&= V(p+1, m(k, p+1, y(p) + F(p, y(p)) + \Delta(p, y(p))), \omega) \\
&\quad - V(p, m(k, p, y(p)), \omega),
\end{aligned}$$

where $\Delta(p, y(p)) = f(p, y(p), \omega) + \sigma(p, y(p), \omega) z(p, \omega) - F(p, y(p))$. By following the proof of Theorem 1.1.1, we get

$$E[\Delta v(p) \,|\, \mathcal{F}_p] = LV(p, m(k, p, y(p, k_0, y_0)), \omega),$$

where L is defined in (1.1.20) relative to (2.1.1). For this, we have

$$\sum_{p=k_0}^{k-1} E[\Delta v(p) \mid \mathcal{F}_p] = \sum_{p=k_0}^{k-1} LV(p, m(k, p, y(p, k_0, y_0)), \omega). \quad (2.1.10)$$

This, together with the idea of a telescopic sum, gives

$$E[V(k, m(k, k, y(k, k_0, y_0)), \omega) \mid \mathcal{F}_{k-1}] - V(k, m(k, k_0, y_0), \omega)$$
$$= \sum_{p=k_0}^{k-1} LV(p, m(k, p, y(p, k_0, y_0)), \omega),$$

which implies

$$E[V(k, y(k, k_0, y_0), \omega) \mid \mathcal{F}_{k_0}] - V(k, m(k, k_0, y_0), \omega)$$
$$= \sum_{p=k_0}^{k-1} E[LV(p, m(k, p, y(p, k_0, y_0)), \omega) \mid \mathcal{F}_{k_0}].$$

This establishes the validity of (2.1.9).

Remark 2.1.5. For $F = f$, (2.1.9) gives

$$V(k, y(k, k_0, y_0), \omega) = V(k_0, x(k, k_0, y_0), \omega)$$
$$+ \sum_{i=k_0}^{k-1} LV(i, x(k, i, y(i, k_0, y_0)), \omega).$$

Corollary 2.1.2. *Let us assume that hypotheses $H_{(2.1.1)}$ and $H_{(2.1.2)}$ are valid in the context of $F(k, y)$.*

(i) *Let us assume that $V(t, y, \omega) = y$. Then,*

$$E[y(k, k_0, y_0) \mid \mathcal{F}_{k_0}] = m(k, k_0, y_0) + \sum_{p=k_0}^{k-1} LV(p, m(k, p, y(p))), \quad (2.1.11)$$

where $LV = \mathcal{L}_a V + \mathcal{L}_e V + \mathcal{L}_0 V$,

$$\mathcal{L}_a V(p, m(k, p, x), \omega) = \Phi(k, p+1, y + F(p, y)) E[\Delta(p, y) \mid \mathcal{F}_p],$$

$$\mathcal{L}_e V(p, m(k, p, x), \omega)$$
$$= \frac{1}{2} E\left[\frac{\partial}{\partial m_0} \Phi(k, p+1, y + F(p, y)) \cdot \Delta(p, y)\Delta(p, y) \mid \mathcal{F}_p\right],$$

$$\mathcal{L}_0 V(p, m(k,p,y), \omega)$$
$$= E\left[\int_0^1 \int_0^1 \theta \frac{\partial}{\partial m_0}\left[\Phi(k, p+1, y+F(p,y)+\theta\psi\Delta(p,y)) - \frac{\partial}{\partial m_0}\Phi(k, p+1, y+F(p,y))\right]\psi d\theta \Delta(p,y) \,|\, \mathcal{F}_p\right]$$

and $\Delta(p, y(p)) = f(p, y(p), \omega) + \sigma(p, y(p), \omega) z(p, \omega) - F(p, y(p))$.

(ii) *Let* $V(p, m, \omega) = \alpha m^T m$, *for constant* $\alpha > 0$. *Then,*

$$E[\|y(k, k_0, y_0, \omega)\|^2 \,|\, \mathcal{F}_{k_0}]$$
$$= \|m(k, k_0, y_0)\|^2 + \frac{1}{\alpha}\sum_{p=k_0}^{k-1} E[LV(p, m(k, p, y(p)), \omega) \,|\, \mathcal{F}_{k_0}], \tag{2.1.12}$$

where LV *is as defined in* (1.1.20) *relative to* (2.1.1).

(iii) *If* $V(p, y, \omega)$ *and* $F(p, y)$ *satisfy the conditions of Corollary* 1.1.2, *then*

$$E[V(k, y(k, k_0, y_0), \omega) \,|\, \mathcal{F}_{k_0}]$$
$$= V(k_0, m(k, k_0, y_0, \omega)) + \sum_{p=k_0}^{k-1} E[LV(p, m(k, p, y(p)), \omega) \,|\, \mathcal{F}_{k_0}], \tag{2.1.13}$$

where LV *is as defined in Corollary* 1.1.2 *whenever* $v = f(p, y, \omega) + \sigma(p, y, \omega) z(p, \omega)$.

(iv) *If* $V(p, y, \omega)$ *and* $F(p, y)$ *satisfy the conditions of Corollary* 1.1.3, *then*

$$E[V(k, y(k, k_0, y_0), \omega) \,|\, \mathcal{F}_{k_0}]$$
$$= V(k_0, m(k, k_0, y_0), \omega) + \sum_{p=k_0}^{k-1} E[LV(p, m(k, p, y(p)), \omega) \,|\, \mathcal{F}_{k_0}], \tag{2.1.14}$$

where LV *is as defined in Corollary* 1.1.3, *for* $v = f(p, y(p), \omega) + \sigma(p, y(p), \omega) z(p, \omega)$.

(v) *If the assumptions in* (ii) *and* (iii) *are satisfied, then*

$$E[\|y(k,k_0,y_0)\|^2 \mid \mathcal{F}_{k_0}] = \|m(k,k_0,y_0)\|^2 + \frac{1}{\alpha}\sum_{p=k_0}^{k-1} E[LV(p,m(k,p,y(p)),\omega) \mid \mathcal{F}_{k_0}], \tag{2.1.15}$$

where LV *is defined by*

$$\mathcal{L}_a V(p,m(k,p,y),\omega) = 2\alpha E\big[m^T(k,p,y(p))(f(p,y(p),\omega) + \sigma(p,y(p),\omega)z(p,\omega) - F(p,y(p))) \mid \mathcal{F}_p\big],$$

$$\mathcal{L}_e V(p,m(k,p,y),\omega) = \alpha E\Big[\big[f(p,y(p),\omega) + \sigma(p,y(p),\omega)z(p,\omega) - F(p,y(p))\big]^T \times [f(p,y(p),\omega) + \sigma(p,y(p),\omega)z(p,\omega) - F(p,y(p))] \mid \mathcal{F}_p\Big].$$

(vi) *If the conditions in* (ii) *and* (iv) *are fulfilled, then*

$$\begin{cases} \|y(k,k_0,y_0\|^2 = \|\Phi(k,k_0)y_0\|^2 \\ \quad + \sum_{p=k_0}^{k-1} \Delta V(p,m(k,p,y(p)),\omega) \\ E\|y(k,k_0,y_0)^2 \mid \mathcal{F}_{k_0}] = \|\Phi(k,k_0)y_0\|^2 \\ \quad + \sum_{p=k_0}^{k-1} E[LV(p,m(k,p,y(p)),\omega) \mid \mathcal{F}_{k_0}], \end{cases} \tag{2.1.16}$$

where LV *is defined by*

$$\mathcal{L}_a V(p,m(k,p,y(p),\omega)) = 2\alpha E[m^T(k,p,y(p))\Phi(k,p)[f(p,y(p),\omega) + \sigma(p,y(p),\omega)z(p,\omega) - F(p,y(p))] \mid \mathcal{F}_p],$$

$$\mathcal{L}_e V(p,m(k,p,y(p)),\omega) = \alpha E[(\Phi(k,p)[f(p,y(p),\omega) + \sigma(p,y(p),\omega)z(p,\omega) - F(p,y(p))])^T(\Phi(k,p)[f(p,y(p),\omega) + \sigma(p,y(p),\omega)z(p,\omega) - F(p,y(p))]) \mid \mathcal{F}_p]$$

$$= \alpha[f(p, y(p), \omega) - F(p, y(p))]^T \Phi^T(k, p)$$
$$\times \Phi(k, p)[f(p, y(p), \omega) - F(p, y(p))]$$
$$+ \alpha tr((\Phi(k, p)\sigma(p, y(p), \omega))^T (\Phi(k, p)\sigma(p, y(p), \omega))).$$

(vii) *If* $V(p, y, \omega)$ *is continuous on* $[0, \infty) \times \mathbb{R}^n$ *and* $F(k, y) \equiv 0$, *then*

$$V(k, y(k, k_0, y_0), \omega) = V(k_0, y_0, \omega) + \sum_{p=k_0}^{k-1} \Delta V(p, y(p), \omega), \tag{2.1.17}$$

where $\Delta V(p, y(p), \omega)$ *is defined by*

$$\Delta V(p, y(p), \omega) = V(p + 1, y(p) + f(p, y(p), \omega), \omega)$$
$$+ \sigma(p, y(p), \omega)z(p, \omega) - V(p, y(p), \omega).$$

(viii) *If* $F = f$, *then* $\Delta(p, y(p)) = f(p, y, \omega) + \sigma(p, y, \omega)z(p, \omega)$ *is reduced to* $\Delta(p, y) = \sigma(p, y, \omega)z(p, \omega)$.

We now formulate a few basic results that relate the solution process of (2.1.1) with the solution process of (2.1.2) or (2.1.3).

Theorem 2.1.2. *Let the hypotheses of Theorem* 1.1.1 *be satisfied, and let* $y(k, \omega) = y(k, k_0, y_0(\omega), \omega)$ *and* $x(t, \omega) = x(t, t_0, y_0(\omega))$ *be the sample solution processes of* (2.1.1) *and* (2.1.3), *respectively, existing for all* $k \in I(k_0)$ *with* $x_0(\omega) = y_0(\omega)$. *Then,*

$$V(k, y(k, \omega), \omega) = V(k, x(k, \omega), \omega) + \sum_{s=k_0}^{k-1} W(k, k_0, y(k, \omega), \omega)$$
$$\times \psi^{-1}(s, k_0, p(s, \omega), p(s - 1, \omega))$$
$$\times R(s - 1, y(s - 1, \omega), \omega), \tag{2.1.18}$$

where

$$R(k, y, \omega) = f(k, y, \omega) + \sigma(k, y(k, \omega), \omega)z(k, \omega) - F(k, y, \omega),$$

$\psi(k, k_0, y_0)$ *is as described in* (1.1.40),

$$W(k, k_0, y(k, \omega), \omega)$$
$$= \int_0^1 V_x(k, x(k, k_0, y_0(\omega) + \theta(p(k, \omega) - y_0(\omega))), \omega)$$
$$\times \Phi(k, k_0, y_0(\omega) + \theta(p(k, \omega) - y_0(\omega)))d\theta, \tag{2.1.19}$$

$$p(k,\omega) = y_0(\omega) + \sum_{s=k_0}^{k-1} \psi^{-1}(s, k_0, p(s,\omega), p(s-1,\omega))$$
$$\times R(s-1, x_p(s-1,\omega), \omega) \tag{2.1.20}$$

and

$$x_p(k,\omega) = x(k, k_0, p(k,\omega)).$$

Proof. Let $y(k,\omega) = y(k, k_0, y_0(\omega), \omega)$ be a solution of (2.1.1). First, we show that $p(k,\omega)$ is a sample solution of (2.1.20) if and only if any sample solution $y(k,\omega)$ of (2.1.1) satisfies

$$y(k, k_0, y_0(\omega), \omega) = x(k, k_0, p(k,\omega)).$$

Let $x(k, k_0, y_0(\omega))$ be the sample solution of (2.1.3). The method of variation of constants requires determining a function $p(k,\omega)$ so that the solution $x_p(k) = x(k, k_0, p(k,\omega))$ of (2.1.3) with $p(k_0,\omega) = y_0(\omega)$ is also a sample solution of (2.1.1). Therefore, to determine $p(k,\omega)$, we consider

$$y(k, k_0, y_0) = x(k, k_0, p(k,\omega)),$$
$$y(k+1, k_0, y_0) = x(k+1, k_0, p(k+1,\omega)),$$

and hence

$$\begin{aligned}\Delta y(k, k_0, y_0) &= f(k, y(k, k_0, y_0(\omega)), \omega) + \sigma(k, y(k,\omega), \omega) z(k,\omega)\\ &= x(k+1, k_0, p(k+1,\omega)) - x(k, k_0, p(k,\omega))\\ &= x(k+1, k_0, p(k+1,\omega)) - x(k+1, k_0, p(k,\omega))\\ &\quad + x(k+1, k_0, p(k,\omega)) - x(k, k_0, p(k,\omega))\\ &= x(k+1, k_0, p(k+1,\omega)) - x(k+1, k_0, p(k,\omega))\\ &\quad + F(k, x(k, k_0, p(k,\omega))).\end{aligned}$$

Thus,

$$\begin{aligned}&x(k+1, k_0, p(k+1,\omega)) - x(k+1, k_0, p(k,\omega))\\ &\quad = f(k, y(k, k_0, y_0(\omega)), \omega) + \sigma(k, y(k,\omega), \omega) z(k-1,\omega)\\ &\qquad - F(k, y(k,\omega)) = R(k, y(k, k_0, y_0), \omega).\end{aligned} \tag{2.1.21}$$

We observe that

$$\begin{aligned}
&\frac{dx}{ds}(k+1, k_0, sp(k+1,\omega) + (1-s)p(k,\omega)) \\
&\quad = \Phi(k+1, k_0, sp(k+1,\omega) \\
&\qquad + (1-s)p(k,\omega))(p(k+1,\omega) - p(k,\omega)).
\end{aligned}$$

Integrating from $s = 0$ to $s = 1$, we obtain

$$\begin{aligned}
&x(k+1, k_0, p(k+1,\omega)) - x(k+1, k_0, p(k,\omega)) \\
&\quad = \psi(k+1, k_0, p(k+1,\omega), p(k,\omega))(p(k+1,\omega) - p(k,\omega)).
\end{aligned} \tag{2.1.22}$$

Together with (2.1.21), it implies that

$$\begin{aligned}
&p(k+1,\omega) - p(k,\omega) \\
&\quad = \psi^{-1}(k+1, k_0, p(k+1,\omega), p(k,\omega))R(k, y(k,\omega), \omega).
\end{aligned}$$

By iterating this equation, we obtain the proof of (2.1.20).

Now, we prove that if $p(k,\omega)$ is the solution of (2.1.20) with $p(k_0,\omega) = y_0(\omega)$, then $y(k, k_0, y_0(\omega), \omega) = x(k, k_0, p(k,\omega))$. From (2.1.20), we have

$$p(k,\omega) = y_0 + \sum_{s=k_0}^{k} \psi^{-1}(s, k_0, p(s,\omega), p(s-1,\omega))R(s, y(s,\omega), \omega).$$

Therefore,

$$\begin{aligned}
&p(k+1,\omega) - p(k,\omega) \\
&\quad = \psi^{-1}(k+1, k_0, p(k+1,\omega), p(k,\omega))R(k, y(k,\omega), \omega).
\end{aligned}$$

This gives

$$\begin{aligned}
R(k, y(k,\omega), \omega) &= \psi(k+1, k_0, p(k+1,\omega), p(k,\omega)) \\
&\quad \times (p(k+1,\omega) - p(k,\omega)) \\
&= \int_0^1 \Phi(k+1, k_0, sp(k+1,\omega) + (1-s)p(k,\omega))ds \\
&\quad \times (p(k+1,\omega) - p(k,\omega)) \\
&= \int_0^1 \frac{\partial x}{\partial x_0}(k+1, k_0, sp(k+1,\omega) + (1-s)p(k,\omega))ds \\
&\quad \times (p(k+1,\omega) - p(k,\omega))
\end{aligned}$$

$$\begin{aligned}
&= \int_0^1 \frac{d}{ds} x(k+1, k_0, sp(k+1,\omega) + (1-s)p(k,\omega))ds \\
&= x(k+1, k_0, p(k+1,\omega)) - x(k+1, k_0, p(k,\omega)) \\
&= x(k+1, k_0, p(k+1,\omega)) - x(k, k_0, p(k,\omega)) \\
&\quad - [x(k+1, k_0, p(k,\omega)) - x(k, k_0, p(k,\omega))].
\end{aligned}$$

Therefore,

$$\begin{aligned}
&R(k, y(k,\omega), \omega) + [x(k+1, k_0, p(k,\omega)) - x(k, k_0, p(k,\omega))] \\
&\quad = \Delta x(k, k_0, p(k,\omega)),
\end{aligned} \tag{2.1.23}$$

and hence

$$\Delta x(k, k_0, p(k,\omega)) = F(k, x(k, k_0, p(k,\omega))), \quad x(k_0) = y_0. \tag{2.1.24}$$

From (2.1.24), (2.1.1) and the uniqueness of the solution of (2.1.3), we have

$$y(k, k_0, y_0(\omega), \omega) = x(k, k_0, p(k,\omega)).$$

Now, to prove (2.1.18),

$$\begin{aligned}
&\frac{dV}{ds}(k, x(k, k_0, sp(k,\omega) + (1-s)y_0(\omega)), \omega) \\
&\quad = V_x(k, x(k, k_0, sp(k,\omega) \\
&\qquad + (1-s)y_0(\omega)), \omega)\Phi(k, k_0, sp(k,\omega) \\
&\qquad + (1-s)y_0(\omega))(p(k,\omega) - y_0(\omega)).
\end{aligned}$$

Integrating this from 0 to 1, we obtain

$$\begin{aligned}
&V(k, x(k, k_0, p(k,\omega)), \omega) \\
&\quad = V(k, x(k, k_0, x_0(\omega)), \omega) \\
&\qquad + \int_0^1 V_x(k, x(k, k_0, y_0(\omega) + \theta(p(k,\omega) - y_0(\omega)), \omega) \\
&\qquad \times \Phi(k, k_0, y_0(\omega) + \theta(p(k,\omega) - y_0(\omega)))d\theta \\
&\qquad \times (p(k,\omega) - y_0(\omega)).
\end{aligned} \tag{2.1.25}$$

This equation, together with (2.1.19) and (2.1.20), proves the theorem.

In the following, we demonstrate the scope and significance of the preceding theorem.

Corollary 2.1.3. *Let the assumptions of Theorem 2.1.2 be satisfied, and let $V(t,x,\omega) = x$. Then,*

$$
\begin{aligned}
y(k,\omega) = x(k,\omega) + \sum_{s=k_0+1}^{k} & \psi(k,k_0,p(k,\omega),x_0(\omega)) \\
& \times \psi^{-1}(s,k_0,p(s,\omega),p(s-1,\omega)) \\
& \times R(s-1,y(s-1,\omega),\omega). \qquad (2.1.26)
\end{aligned}
$$

This corollary is a nonlinear variation of constants formula for systems of stochastic difference equations.

Problem 2.1.1. (a) If $V(k,x,\omega) = \|x\|^2$, then (2.1.18) in Theorem 2.1.2 reduces to

$$
\begin{aligned}
\|y(k,\omega)\|^2 = \|x(k,\omega)\|^2 + \sum_{s=k_0+1}^{k} & W(k,k_0,x(k,k_0,y(k,\omega))) \\
& \times \psi^{-1}(s,k_0,p(s,\omega),p(s-1,\omega)) \\
& \times R(s-1,y(s-1,\omega),\omega),
\end{aligned}
$$

where

$$
\begin{aligned}
W(k,k_0,y(k,\omega)) = 2\int_0^1 & [x(k,k_0,x_0(\omega)+\theta(p(k,\omega)-x_0(\omega)))]^T \\
& \times \Phi(k,k_0,x_0(\omega)+\theta(p(k,\omega)-x_0(\omega)))d\theta.
\end{aligned}
$$

The following result provides an expression for the difference between the solution processes of (2.1.1) with (2.1.2) or (2.1.3).

Theorem 2.1.3. *Suppose all the hypotheses of Theorem 2.1.2 hold. Then,*

$$
\begin{aligned}
V(k,y(k,\omega)-\bar{x}(k),\omega) &= V(k,x(k,\omega)-\bar{x}(k),\omega) \\
&\quad + \sum_{s=k_0}^{k} W(k,k_0,y(k,\omega)-x(k,k_0,\bar{z}_0),\omega)
\end{aligned}
$$

$$\times \psi^{-1}(s, k_0, p(s,\omega), p(s-1,\omega))$$
$$\times R(s-1, y(s-1,\omega), \omega), \qquad (2.1.27)$$

where $\bar{x}(k) = x(k, k_0, z_0)$ *is the solution process of either* (2.1.2) *or* (2.1.3) *depending on the choice of* z_0.

Proof. By following the proof of Theorem 2.1.2, we have the relation

$$\frac{dV}{d\theta}(k, x(k, k_0, x_0(\omega) + \theta(p(k,\omega) + x_0(\omega))) - \bar{x}(k), \omega)$$
$$= V_x(k, x(k, k_0, x_0(\omega) + \theta(p(k,\omega) - x_0(\omega))) - \bar{x}(k), \omega)$$
$$\times \Phi(k, k_0, x_0(\omega) + \theta(p(k,\omega) - x_0(\omega)))$$
$$\times (p(k,\omega) - x_0(\omega)).$$

By integrating the above relation from 0 to 1 and noting the fact that $x(k, k_0, p(k,\omega)) = y(k,\omega)$ and (2.1.19), we complete the proof of the theorem.

Problem 2.1.2. If $V(k, x, \omega) = \|x\|^2$, then (2.1.27) reduces to

$$\|y(k,\omega) - x(k, k_0, z_0)\|^2$$
$$= \|x(k,\omega) - x(k, k_0, z_0)\|^2$$
$$+ \sum_{s=k_0}^{k-1} W(k, k_0, y(t,\omega) - x(k, k_0, z_0)), \omega)$$
$$\times \psi^{-1}(s, k_0, p(s,\omega), p(s-1,\omega)) R(s-1, y(s-1,\omega), \omega), \qquad (2.1.28)$$

where

$$W(k, k_0, y(k,\omega) - x(k, k_0, z_0)), \omega)$$
$$= 2\int_0^1 [x(k, k_0, x_0(\omega) + \theta(p(k,\omega) - x_0(\omega))) - x(k, k_0, z_0)]^T$$
$$\times \Phi(k, k_0, x_0(\omega) + \theta(p(k,\omega) - x_0(\omega)))d\theta.$$

We recall that $\bar{x}(k) = x(k, k_0, z_0)$ is the solution process of either (2.1.2) or (2.1.3) depending on the choice of z_0. In other words, $\bar{x}(k)$ is either $m(k) = m(k, k_0, m_0) = x(k, k_0, m_0)$ or $x(k, k_0, x_0(\omega))$ depending on the choice of z_0.

To further illustrate the scope and usefulness of Theorems 2.1.2 and 2.1.3, let us assume some regularity conditions which will translate (2.1.1), (2.1.2) and (2.1.3) into a suitable form for our discussions. We suppose that $f(k, 0, \omega) = 0$ and the sample derivative $\frac{\partial}{\partial x} f(k, x, \omega)$ of $f(k, x, \omega)$ exists. From this and Lemma A.1.1, (2.1.1) can be rewritten as

$$\begin{aligned} y(k, \omega) &= A(k, y(k-1, \omega), \omega) y(k-1, \omega) \\ &\quad + \sigma(k, y(k-1, \omega), \omega) z(k-1, \omega), \quad y(k_0, \omega) = y_0(\omega), \end{aligned} \tag{2.1.29}$$

where

$$A(k, y, \omega) = \int_0^1 \frac{\partial}{\partial y} f(t, sy, \omega) ds.$$

Similarly, one can assume that $F(k, 0) \equiv 0$. This, together with the continuous differentiability of $F(k, x)$ in x, one can rewrite (2.1.2) and (2.1.3) as

$$m(k) = \hat{A}(k, m(k-1)) m(k-1), \quad m(k_0) = m_0, \tag{2.1.30}$$

and

$$x(k) = \hat{A}(k, x(k-1)) x(k-1), \quad x(k_0, \omega) = x_0(\omega), \tag{2.1.31}$$

respectively, where

$$\hat{A}(k, x) = \int_0^1 \frac{\partial F}{\partial x}(t, sx) ds.$$

From Lemma A.1.1, we note that

$$x(k, s, y(s, \omega)) - x(k, s, \bar{x}(s)) = \psi(k, s, y(s, \omega), \bar{x}(s))(y(s, \omega) - \bar{x}(s)),$$

where

$$\psi(k, s, y(s, \omega), \bar{x}(s)) = \int_0^1 \Phi(k, s, \bar{x}(s) + u(y(s, \omega) - \bar{x}(s))) du.$$

Remark 2.1.6. We remark that relation (2.1.27) in light of $V(k,x,\omega) = \|x\|^2$ reduces to

$$\begin{aligned}
&\|y(k,\omega) - x(k,k_0,z_0)\|^2 \\
&\quad = \|x(k,\omega) - x(k,k_0,z_0)\|^2 \\
&\qquad + \sum_{s=k_0+1}^{k} \left[2\int_0^1 \ell_p(\gamma) - z_0\right]^T \psi^T(k,k_0,\ell_p(\gamma),\bar{x}(k)) \\
&\qquad \times \Phi(k,k_0,\ell_p(\gamma))d\gamma\psi^{-1}(s,k_0,p(s,\omega),p(s-1,\omega)) \\
&\qquad \times R(s-1,y(s-1,\omega),\omega),
\end{aligned}$$

where $\ell_p(\gamma) = \gamma p(k,\omega) + (1-\gamma)x_0$, for $\gamma \in [0,1]$.

Example 2.1.1. Let us consider

$$\begin{aligned}
y(k,\omega) &= A(k,\omega)y(k-1,\omega) + \sigma(k,y(k-1,\omega))z(k-1,\omega), \\
y(k_0,\omega) &= y_0(\omega), \qquad (2.1.32)
\end{aligned}$$

$$m(k) = \hat{A}(k)m(k-1), \quad m(k_0) = m_0 = E[y_0(\omega)] \qquad (2.1.33)$$

and

$$x(k) = \hat{A}(k)x(k-1), \quad x(k_0,\omega) = x_0(\omega), \qquad (2.1.34)$$

where $A(k,\omega)$ is an $n \times n$ random matrix function and $\hat{A}(k)$ is a rate matrix which is obtained by neglecting the randomness in the system. In particular, $\hat{A}(k) = E[A(k,\omega)]$ if $E[A(k,\omega)]$ exists. In this case, $x(k,k_0,x_0(\omega)) = \Phi(k,k_0)x_0(\omega)$, where $\Phi(k,k_0)$ is the fundamental matrix solution process of either (2.1.33) or (2.1.34). Note that in the linear case, $\psi(k,s,y(s,\omega)) = \Phi(k,s)$. With regard to (2.1.32), (2.1.33) and (2.1.34) in the context of $V(k,x,\omega) = \|x\|^2$ and Remark 2.1.6, (2.1.18) and (2.1.28) reduce to

$$\begin{aligned}
\|y(k,\omega)\|^2 = \|x(k,\omega)\|^2 + \sum_{s=k_0+1}^{k} (p(k,\omega) + x_0(\omega))^T \Phi^T(k,\omega)\Phi(k,\omega) \\
\times \Phi^{-1}(s,\omega)R(s-1,y(s-1,\omega),\omega) \qquad (2.1.35)
\end{aligned}$$

and

$$\|y(k,\omega)-\bar{x}(k)\|^2 = \|x(k,\omega)-\bar{x}(k)\|^2 + \sum_{s=k_0+1}^{k} (p(k,\omega)+x_0(\omega)-2z_0)^T\Phi^T(k,\omega) \times \Phi(k,\omega)\Phi^{-1}(s,\omega)R(s-1,y(s-1,\omega),\omega), \tag{2.1.36}$$

respectively, where

$$R(s,y,\omega) = A(s,\omega)y + \sigma(s,y,\omega)z(s-1,\omega) - \hat{A}(s)y.$$

Remark 2.1.7. Let us consider a discrete-time perturbed system with respect to (1.1.60) as an unperturbed system:

$$\Delta w(k) = h(k,w(k),\omega) + \epsilon(k,w(k),\omega)\xi(k,\omega), \quad w(k_0) = w_0. \tag{2.1.37}$$

By setting $u = y - w$ and $v = f + \sigma - h - \epsilon$, a remark parallel to Remark 1.1.5 can be reformulated.

We are ready to present a generalized variation of constants results for the error estimation of the solution process of (2.1.1) with respect to (2.1.37).

Theorem 2.1.4. *Assume that all the hypotheses of Theorem* 1.1.1 *are satisfied. Further, assume that for* $k_0 \le p \le k$, $y(p) = y(p,k_0,y_0)$ *and* $w(p) = w(p,k_0,w_0)$ *are the solution processes of* (2.1.1) *and* (2.1.37), *respectively. Then,*

$$E[V(k,y(k,k_0,y_0)-w(k,k_0,w_0),\omega)\,|\,\mathcal{F}_{k_0}] = V(k_0,m(k,k_0,y_0-w_0),\omega) + \sum_{p=k_0}^{k-1} E[LV(p,m(k,p,y(p)-w(p),\omega))\,|\,\mathcal{F}_{k_0}], \tag{2.1.38}$$

where L is defined in Remark 1.1.5 *in the context of Remark* 2.1.4.

Proof. For $k_0 \le p \le k$, let $y(p) = y(p, k_0, y_0)$, $w(p) = w(p, k_0, w_0)$ and $m(k, p, y(p) - w(p))$ be the solution processes of (2.1.1), (2.1.37) and (2.1.2) through (k_0, y_0), (k_0, w_0) and $(p, y(p) - w(p))$, respectively. By imitating the proof of Theorem 2.1.2, we arrive at

$$\begin{aligned}\Delta v(p) &= V(p+1, m(k, p+1, y(p+1) - w(p+1)), \omega)\\ &\quad - V(p, m(k, p, y(p) - w(p)), \omega)\\ &= V(p+1, m(k, p+1, y(p) - w(p) + f(p, y(p), \omega)\\ &\quad + \sigma(p, y(p), \omega) z(p, \omega) - h(p, w(p), \omega)), \omega)\\ &\quad - V(p, m(k, p, y(p) - w(p)), \omega)\\ &= V(p+1, m(k, p+1, y(p) - w(p) + F(p, y(p) - w(p))\\ &\quad + \Delta(p, y(p) - w(p)), \omega) - V(p, m(k, p, y(p) - w(p)), \omega),\end{aligned}$$

where $\Delta(p, y(p) - w(p)) = [f(p, y(p), \omega) + \sigma(p, y(p), \omega) z(p, \omega) - h(p, w(p), \omega) - \epsilon(p, w(p), \omega) - F(p, y(p) - w(p))]$. By following the proof of Theorem 2.1.1, we have

$$E[\Delta v(p) \,|\, \mathcal{F}_p] = LV(p, m(k, p, y(p, k_0, y_0) - w(p, k_0, w_0), \omega)), \tag{2.1.39}$$

where L is defined in (1.1.20) relative to (2.1.1). From (2.1.39) and following the proof of Theorem 2.1.4, we get the desired result (2.1.38). The details are left to the reader.

To illustrate the scope of Theorem 2.1.4, we state a result that is parallel to Corollary 2.1.3.

Corollary 2.1.4. *Let us assume that hypotheses $H_{(2.1.1)}$ and $H_{(2.1.2)}$ are satisfied.*

(i) *Let us assume that the hypotheses of Corollary* 2.1.2(i) *are satisfied in the context of Remark* 1.1.5. *Then,*

$$\begin{aligned}&E[y(k, k_0, y_0) \,|\, \mathcal{F}_k] - w(k, k_0, w_0)\\ &\quad = m(k, k_0, y_0 - w_0)\\ &\qquad + \sum_{p=k_0}^{k-1} E[LV(p, m(k, p, y(p) - w(p)), \omega) \,|\, \mathcal{F}_{k_0}],\end{aligned} \tag{2.1.40}$$

where LV *is defined in the context of Remark* 1.1.5, (2.1.11) *and* $\Delta(p, y(p) - w(p)) = [f(p, y(p), \omega) + \sigma(p, y(p), \omega)z(p, \omega) - h(p, w(p), \omega) - \epsilon(p, w(p), \omega) - F(p, y(p) - w(p))]$.

(ii) *Let* $V(p, m) = \alpha m^T m$, *for any constant* $\alpha > 0$. *Then,*

$$\begin{aligned} E[\|y(k, k_0, y_0) - w(k, k_0, w_0)\|^2 \,|\, \mathcal{F}_{k_0}] &= \|m(k, k_0, y_0 - w_0)\|^2 \\ &\quad + \frac{1}{\alpha} \sum_{p=k_0}^{k-1} E[LV(p, m(k, k_0, y(p) - w(p)), \omega) \,|\, \mathcal{F}_{k_0}], \end{aligned} \tag{2.1.41}$$

where LV *is as defined in* (2.1.12).

(iii) *If* $V(p, y, \omega)$ *and* $F(p, y)$ *satisfy the conditions of Corollary* 1.1.2, *then*

$$\begin{aligned} E[V(k, y(k, k_0, y_0) - w(k, k_0, w_0), \omega) \,|\, \mathcal{F}_{k_0}] &= V(k_0, m(k, k_0, y_0 - w_0), \omega) \\ &\quad + \sum_{k=k_0}^{k-1} E[LV(p, m(k, p, y(p) - w(p)), \omega) \,|\, \mathcal{F}_{k_0}], \end{aligned} \tag{2.1.42}$$

where LV *is as defined in Corollary* 1.1.2, *whenever* $\Delta(p, y(p) - w(p)) = f(p, y(p), \omega) + \sigma(p, y(p), \omega)z(p, \omega) - h(p, y(p), \omega) - \epsilon(p, w(p), \omega)\xi(p, \omega)$.

(iv) *If* $V(p, y, \omega)$ *and* $F(p, y)$ *satisfy the conditions in Corollary* 1.1.3, *then*

$$\begin{aligned} E[V(k, y(k, k_0, y_0) - m(k, k_0, m_0), \omega) \,|\, \mathcal{F}_{k_0}] &= V(k_0, m(k, k_0, y_0 - m_0), \omega) \\ &\quad + \sum_{p=k_0}^{k-1} E[LV(p, m(k, p, y(p) - m(p)), \omega) \,|\, \mathcal{F}_{k_0}], \end{aligned} \tag{2.1.43}$$

where LV *is as defined in Corollary* 1.1.3, *for* $\Delta(p, y(p) - w(p)) = f(p, y(p), \omega) + \sigma(p, y(p), \omega)z(p, \omega) - h(p, y(p), \omega) - \epsilon(p, w(p), \omega)\xi(p, \omega)$.

(v) *If the assumptions in* (ii) *and* (iii) *are satisfied, then*

$$
\begin{aligned}
&E[\|y(k,k_0,y_0) - w(k,k_0,w_0)\|^2 \,|\, \mathcal{F}_{k_0}] \\
&\quad = \|y_0 - w_0\|^2 + \frac{1}{\alpha}\sum_{p=k_0}^{k-1} E\,[LV(p, m(k,p,y(p) \\
&\qquad - w(p)), \omega) \,|\, \mathcal{F}_{k_0}], \qquad (2.1.44)
\end{aligned}
$$

where LV is defined by

$$
\begin{aligned}
&L_a V(p, m(k,p,y-w), \omega) \\
&\quad = E[2\alpha m^T(k,p,y-w)(f(p,y(p),\omega) + \sigma(p,y(p),\omega)z(p,\omega) \\
&\qquad - h(p,w(p),\omega) - \epsilon(p,w(p),\omega)\xi(p,\omega) \\
&\qquad - F(p, y(p) - w(p))) \,|\, \mathcal{F}_p], \qquad (2.1.45) \\
&L_e V(p, m(k,p,y-w), \omega) \\
&\quad = E[\alpha[f(p,y(p),\omega) + \sigma(p,y(p),\omega)z(p,\omega) \\
&\qquad - h(p,w(p),\omega) - F(p, y(p) - w(p))]^T \\
&\qquad \times [f(p,y(p),\omega) + \sigma(p,y(p),\omega)z(p,\omega) - h(p,w(p),\omega) \\
&\qquad - \epsilon(p,w(p),\omega)\xi(p,\omega) - F(p, y(p) - w(p))] \,|\, \mathcal{F}_p].
\end{aligned}
$$

(vi) *If the conditions in* (ii) *and* (iv) *are fulfilled, then*

$$
\begin{aligned}
&E[\|y(k,k_0,y_0) - w(k,k_0,w_0)\|^2 \,|\, \mathcal{F}_{k_0}] \\
&\quad = \|\Phi(k,k_0)(y_0 - w_0)\|^2 \\
&\qquad + \frac{1}{\alpha}\sum_{p=k_0}^{k-1} E[LV(p, m(k,p,y(p) - w(p)), \omega) \,|\, \mathcal{F}_{k_0}], \\
&\qquad\qquad (2.1.46)
\end{aligned}
$$

where LV is defined by

$$
\begin{aligned}
&L_a V(p, m(k,p,y(p) - w(p)), \omega) \\
&\quad = 2\alpha E[m^T(k,p,y(p) - w(p))\Phi(k,p) \\
&\qquad \times [f(p,y(p),\omega) + \sigma(p,y(p),\omega)z(p,\omega) - h(p,w(p),\omega) \\
&\qquad - \epsilon(p,w(p),\omega)\xi(p,\omega) - F(p, y(p) - w(p))] \,|\, \mathcal{F}_p],
\end{aligned}
$$

$$
\begin{aligned}
&L_eV(p, m(k,p,y(p) - w(p)), \omega) \\
&\quad = \alpha E[(\Phi(k,p)[f(p,y(p),\omega) + \sigma(p,y(p),\omega)z(p,\omega) \\
&\qquad - h(p,w(p),\omega) - \epsilon(p,w(p),\omega)\xi(p,\omega) \\
&\qquad - F(p,y(p) - w(p))])^T \\
&\qquad \times (\Phi(k,p)[f(p,y(p),\omega) + \sigma(p,y(p),\omega)z(p,\omega) \\
&\qquad - h(p,w(p),\omega) - \epsilon(p,w(p),\omega)\xi(p,\omega) \\
&\qquad - F(p,y(p) - w(p))]) \,|\, \mathcal{F}_p].
\end{aligned}
$$

(vii) *If* $V(p,y,\omega)$ *is continuous on* $[0,\infty)\times\mathbb{R}^n$ *and* $F(k,y)\equiv 0$*, then*

$$
\begin{aligned}
&V(k, y(k,k_0,y_0) - w(k,k_0,w_0), \omega) \\
&\quad = V(k_0, y_0 - w_0, \omega) + \sum_{p=k_0}^{k-1} \Delta V(p, y(p) - w(p), \omega),
\end{aligned}
\tag{2.1.47}
$$

where $\Delta V(p,y(p) - w(p),\omega)$ *is defined by*

$$
\begin{aligned}
&\Delta V(p, y(p) - w(p), \omega) \\
&\quad = V(p+1, y(p) + f(p,y(p),\omega) \\
&\qquad + \sigma(p,y(p),\omega)z(p,\omega) - (w(p) + h(p,w(p),\omega) \\
&\qquad + \epsilon(p,w(p),\omega)\xi(p,\omega) - V(p, y(p) - w(p), \omega).
\end{aligned}
$$

(viii) *If* $V(p,y,\omega)$ *is continuous on* $[0,\infty)\times\mathbb{R}^n$ *and* $h(p,w,\omega) = F(p,w)$*, then*

$$
\begin{aligned}
&E[V(k, y(k,k_0,y_0) - m(k,k_0,m_0), \omega) \,|\, \mathcal{F}_{k_0}] \\
&\quad = V(k_0, m(k,k_0,y_0 - m_0), \omega) \\
&\qquad + \sum_{p=k_0}^{k-1} E[LV(p, m(k,p,y(p) - m(p)), \omega) \,|\, \mathcal{F}_{k_0}],
\end{aligned}
$$

where

$$\begin{aligned}
&\Delta V(p, m(k,p,y-m),\omega)) \\
&\quad = V(p+1, m(k,p+1,y-m+f(p,y,\omega) \\
&\qquad + \sigma(p,y(p),\omega)z(p,\omega) - h(p,y(p),\omega) - \epsilon(p,w(p),\omega)\xi(p,\omega) \\
&\qquad - F(p,y-m)),\omega) - V(p,m(k,p,y-m),\omega). \qquad (2.1.48)
\end{aligned}$$

2.2 Comparison Method

In this section, we present variational comparison theorems with regard to (2.1.1). Again, in the following, we assume that all the inequalities and relations involving random quantities are true with a probability of one (w.p. 1). These results are based on the basic comparison results developed in Chapter 1.

Now, we are ready to present a very general variational comparison theorem in the context of a Lyapunov-like function and the system of difference inequalities. The presented result relates the solutions of (2.1.1), (2.1.2) and a system of comparison difference equations. The result is followed by a brief sketch of its proof.

Theorem 2.2.1 (Variational Comparison Theorem). *Let the hypotheses of Theorem* 1.1.1 *be satisfied. Further, assume that*

(i) $G \in C[\mathbb{R} \times \mathbb{R}^N \times \Omega, \mathbb{R}^N]$, *and* $G(k,a,\omega)$ *satisfies the one-sided Lipschitz condition* $G(k,b,\omega) - G(k,a,\omega) \geq -\Gamma(k,\omega)(b-a)$, *for all* $a, b \in \mathbb{R}^N$ *and* $b \geq a$, *where* $\Gamma(k,\omega) = diag(\gamma_1, \ldots, \gamma_N)$, *with* $0 \leq \gamma_i \leq 1$, *for* $1 \leq i \leq N$;

(ii) $E[LV(p, m(k,p,y),\omega) \,|\, \mathcal{F}_p] \leq G(p, V(p, m(k,p,y),\omega),\omega)$, *for all* $(p,y) \in \mathbb{I}(k_0) \times \mathbb{R}^n$, *where the generating operator* L *is defined as*

$$LV = [LV_1, LV_2, \ldots, LVi, \ldots, LV_N]^T,$$

where $LV(k, m(t,p,x),\omega)$ *is defined analogously to* L *in* (1.1.20);

(iii) $r(k)$ *is the solution of the system of comparison difference equations*

$$u(k+1) = u(k) + G(k,u(k),\omega), \quad u(k_0,\omega) = u_0(\omega); \quad (2.2.1)$$

(iv) $E[V(p, m(k, p, y(p)), \omega)|\mathcal{F}_p]$ *exists for any solution* $y(p) = y(p, k_0, y_0)$ *of the stochastic iterative process* (2.1.1), *for all* $p \in \mathbb{I}(k_0)$ *and* $k_0 \le p \le k$, *and*

$$E[V(k_0, m(k, k_0, y_0), \omega) \,|\, \mathcal{F}_{k_0}] \le u_0. \tag{2.2.2}$$

Then,

$$E[V(p, m(k, p, y(p)), \omega) \,|\, \mathcal{F}_{k_0}] \le r(p, \omega), \quad \text{for all } p \in \mathbb{I}(k_0). \tag{2.2.3}$$

Moreover, (2.2.3) *reduces to*

$$E[V(k, y(k), \omega) \,|\, \mathcal{F}_{k_0}] \le r(k, \omega), \quad \text{for all } k \in \mathbb{I}(k_0). \tag{2.2.4}$$

Proof. For $k_0 \le p+1 \le k$, let $m(k, p, y) = m(k, p+1, y + F(p, y))$ and $y(p, \omega) = y(p, k_0, y_0)$ be the solution processes of (2.1.2) and (2.1.1) through $(p, y + F(p, y))$ and (k_0, y_0), respectively. Define

$$\begin{aligned} &V(p+1, m(k, p+1, y(p+1)), \omega) \\ &\quad = V(p+1, m(k, p+1, y(p)) + f(p, y(p), \omega) \\ &\qquad + \sigma(p, y(p), \omega) z(p, \omega)), \omega). \end{aligned} \tag{2.2.5}$$

We also note that

$$V(p, m(k, p, y(p)), \omega) = V(p, m(k, p{+}1, y(p){+}F(p, y)), \omega). \tag{2.2.6}$$

From (2.2.5), (2.2.6) and (2.1.1), we have

$$\begin{aligned} &E[V(p+1, m(k, p+1, y(p+1)), \omega) - V(p, m(k, p, y(p)), \omega) \,|\, \mathcal{F}_p] \\ &\quad = [\Delta^E_{(2.1.1)} V_1(p, m(k, p, y(p)), \omega), \dots, \\ &\qquad \Delta^E_{(2.1.1)} V_i(p, m(k, p, y(p)), \omega), \dots, \\ &\qquad \Delta^E_{(2.1.1)} V_N(p, m(k, p, y(p)), \omega)]^T, \end{aligned}$$

where $\Delta^E_{(2.1.1)} V_i(p, m(k, p, y(p))) = E[\Delta_{(2.1.1)} V_i(p, m(k, p, y(p))) \,|\, \mathcal{F}_p]$. From the hypotheses of Theorem 1.1.1 and assumption (ii) of the theorem, we have

$$
\begin{aligned}
&E[\Delta V(p, m(k,p,y(p)), \omega) \,|\, \mathcal{F}_p] \\
&\quad = E[V(p+1, m(k,p+1,y(p+1)), \omega) - V(p, m(k,p,y(p)), \omega) \,|\, \mathcal{F}_p] \\
&\quad = E[[LV_1(p, m(k,p,y(p)), \omega), \ldots, LV_i(p, m(k,p,y(p)), \omega), \ldots, \\
&\qquad LV_N(p, m(k,p,y(p)), \omega)]^T \,|\, \mathcal{F}_p] \\
&\quad \le G(p, V(p, m(k,p,y(p))), \omega). \qquad (2.2.7)
\end{aligned}
$$

By choosing $u_0 \ge V(k_0, m(k, k_0, y_0(\omega)), \omega)$ and applying Corollary 1.2.2, we conclude that

$$
\begin{aligned}
&E[V(p, m(k,p,y(p)), \omega) \,|\, \mathcal{F}_{k_0}] \\
&\quad \le r(p, k_0, u_0, \omega), \quad \text{for all } k_0 \le p \le k, k \in \mathbb{I}(k_0). \qquad (2.2.8)
\end{aligned}
$$

In particular, for $p = k$, (2.2.8) reduces to

$$E[V(k, y(k,\omega), \omega) \,|\, \mathcal{F}_{k_0}] \le r(k, k_0, u_0, \omega), \quad \text{for all } \; k \in \mathbb{I}(k_0).$$

This completes the proof of the theorem.

The following corollary demonstrates the scope of the comparison Theorem 2.2.1. This corollary is based on Corollaries 2.1.1 and 2.1.2.

Corollary 2.2.1. *By considering Corollaries* 2.1.1(iv) *and* 2.1.2 *and noting the fact that* $L_oV_i(p, m(k,p,x), \omega) = 0$ *and* $\Delta(p,y) = f(p,y,\omega) - A(p)y$, *we assume that one can find* $\mu(p) \in \mathbb{R}$ *and* $\nu(p), \beta(p) \in \mathbb{R}^+$ *such that*

$$
\begin{aligned}
&E[m(k,p,y)^T \Phi(k,p)[f(p,y,\omega) + \sigma(p,y,\omega)z(p,\omega) - F(p,y)] \,|\, \mathcal{F}_p] \\
&\quad \le \mu(p,\omega) m(k,p,y)^T m(k,p,y), \qquad (2.2.9)
\end{aligned}
$$

$$
\begin{aligned}
&E\Big[[(f(p,y,\omega) + \sigma(p,yp,\omega)z(p,\omega) - F(p,y)]^T \Phi^T(k,p)\Phi(k,p) \\
&\qquad \times [f(p,y,\omega) + \sigma(p,yp,\omega)z(p,\omega) - F(p,y)] \,|\, \mathcal{F}_p\Big] \\
&\quad \le \nu(p,\omega) m(k,p,y)^T m(k,p,y) + \beta(p,\omega) \qquad (2.2.10)
\end{aligned}
$$

and

$$2\mu(p,\omega) + \nu(p,\omega) \ge -1, \quad k_0 \le p \le k,$$

for all $k_0 \leq p \leq k$ *and any* $k \in \mathbb{I}(k_0)$. *Then,*

$$E[V(k, y(k, k_0, y_0), \omega) \,|\, \mathcal{F}_{k_0}] \leq r(k, k_0, u_0, \omega),$$

provided $V(k_0, y_0, \omega) \leq u_0$.

Proof. From (1.2.21), followed by algebraic calculations and simplifications with the notation $V(p, z, \omega) = \alpha(p) z^T z$, one can obtain

$$E[LV(p, m(k, p, y), \omega) \,|\, \mathcal{F}_p] \leq g(p, \omega) V(p, m(k, p, y), \omega) + \beta(p, \omega),$$

where $g(p, \omega)$ is defined by $g(p, \omega) = 2\mu(p, \omega) + \nu(p, \omega)$. In this setup, the system of comparison equations then becomes

$$\Delta u(k) = G(k, u(k), \omega), \quad u_{k_0} = u_0, \tag{2.2.11}$$

where

$$G(k, u, \omega) = g(p, \omega) u(k) + \beta(k, \omega).$$

It is obvious that the above comparison function $G(k, u, \omega)$ satisfies all the hypotheses of Corollary 1.2.2. One only needs to assume that condition (iv) holds and then pick $u_0 \geq V(k_0, x_0, \omega)$. Thus, by the application of Corollary 1.2.2, one may conclude that

$$V(k, y(k), \omega) \leq r(k, k_0, u_0, \omega), \quad \forall k \in \mathbb{I}(k_0), \tag{2.2.12}$$

where $r(k, k_0, u_0)$ is the solution process of the comparison difference equation (2.2.11).

Corollary 2.2.2. *Let us assume that hypotheses* $H_{(2.1.1)}$ *and* $H_{(2.1.2)}$ *are satisfied.*

(a) *Assume that the conditions in Corollary* 2.1.2(iv) *are fulfilled. In addition, the operator* L *defined in* (1.1.30) *satisfies the following inequality*:

$$E[LV(p, m(k, p, y), \omega) \,|\, \mathcal{F}_p] \leq G(p, m(k, p, y), \omega), \tag{2.2.13}$$

Under the conditions of Theorem 2.2.1, *the conclusion of Theorem* 2.2.1 *remains true.*

(b) *Assume that the conditions of Corollary* 2.1.2(v) *are valid. Further, assume that*

$$\begin{cases} 2\alpha E[y^T[f(p,y,\omega)+\sigma(p,y(p),\omega)z(p,\omega)-F(p,y)]\,|\,\mathcal{F}_p] \\ \quad \leq \mu(p,\omega)y^Ty, \\ \alpha E[[f(p,y,\omega)+\sigma(p,y(p),\omega)z(p,\omega)-F(p,y)]^T \\ \quad [f(p,y,\omega)+\sigma(p,y(p),\omega)z(p,\omega)-F(p,y)]\,|\,\mathcal{F}_p] \\ \quad \leq \nu(p,\omega)y^Ty, \\ 2\mu(p,\omega)+\gamma(p,\omega) \geq -1. \end{cases} \tag{2.2.14}$$

Then, the conclusion of Corollary 2.2.1 *remains valid, that is,*

$$E[V(k,y(k,k_0,y_0),\omega)\,|\,\mathcal{F}_{k_0}] \leq r(k,k_0,u_0,\omega), \tag{2.2.15}$$

provided

$$E[V(k_0,y_0,\omega)] \leq u_0,$$

where $r(k,k_0,u_0)$ *is the solution process of* (2.2.11).

Proof. The proofs of these results can be constructed by following the arguments used in Theorem 2.2.1 and Corollary 2.2.1.

To investigate the qualitative properties of (2.1.1), we use the corresponding properties of the comparison system (2.2.1). If the dimension of system (2.1.1) (dimension "n") is very high, it is difficult to obtain such information about the system. Therefore, we are forced to either reduce the system to a lower-dimension one or obtain information about such a system by comparing it with lower-order systems. Here, we choose to provide an estimate of the solution process of (2.1.1) in terms of a scalar iterative process.

Theorem 2.2.2. *Assume that conditions* (i), (ii), (iii) *and* (iv) *of Theorem* 2.2.1 *are satisfied. Further, assume that the following is satisfied:*

$$\sum_{i=1}^{N} d_i G_i(p,m(k,p,y),\omega) \\ \leq \gamma(p,\omega)\sum_{i=1}^{m} d_i V_i(p,m(k,p,y),\omega)+\mu(p,\omega). \tag{2.2.16}$$

Let

$$\bar{v}(p, m(k, p, y), \omega) = \sum_{i=1}^{n} d_i V_i(p, m(k, p, y), \omega), \tag{2.2.17}$$

for $d_i > 0$ *and* $1 \leq i \leq N$. *Then,*

$$\bar{v}(p, y(p, \omega), \omega) \leq r(p, \omega), \quad \text{for all } k_0 \leq p \leq k, k \in \mathbb{I}(k_0), \tag{2.2.18}$$

whenever

$$\bar{v}(k_0, m(k, k_0, x_0), \omega) \leq u_0(\omega), \tag{2.2.19}$$

where $r(p, \omega)$ *is the solution of the scalar comparison functional difference equation*

$$\Delta u(p) = \gamma(p, \omega) u(p) + \mu(p, \omega), \quad u_{k_0} = u_0, \tag{2.2.20}$$

with $u(p) \in \mathbb{R}_+$.

Proof. The proof of the result follows by using $\bar{v}(p, m(k, p, y), \omega)$ in (2.2.17) and the application of Corollary 1.2.3. The details are left to the reader.

Corollary 2.2.3. *Assume that*

(a) *the hypotheses of Corollary* 2.1.2*(vii) hold;*
(b) $V \in M[I(k_0 + 1) \times R^n \times \Omega, R[\Omega, R_+^N]]$ *satisfies the relation*

$$\begin{aligned} &E[V(k+1, y + f(k, y, \omega) + \sigma(k, y, \omega) z(k, \omega), \omega) \mid \mathcal{F}_k] \\ &\quad \leq G(k, V(k, y, \omega), \omega), \quad \text{for all } k \geq k_0, \end{aligned} \tag{2.2.21}$$

where $E[V(k+1, y + f(k, y, \omega) + \sigma(k, y, \omega) z(k, \omega), \omega) \mid \mathcal{F}_k] = LV(k, y, \omega) + V(k, y, \omega)$;
(c) G *satisfies the hypothesis of Theorem* 1.2.2.
Then,

$$E[V(k, y(k, \omega), \omega) \mid \mathcal{F}_{k_0}] \leq r(k, \omega), \tag{2.2.22}$$

provided

$$E[V(k_0, y_0(\omega), \omega)] \leq u_0(\omega), \tag{2.2.23}$$

where $y(k, \omega) = y(k, k_0, y_0(\omega), \omega)$ *and* $r(k, \omega) = r(k, k_0, u_0(\omega), \omega)$ *are the solution processes of* (2.1.1) *and* (1.2.6), *respectively.*

Proof. Let $y(k,\omega)$ be the solution process of (2.1.1). Set

$$m_k(\omega) = V(k, y(k,\omega), \omega). \tag{2.2.24}$$

From (2.2.23), we note that $m_{k_0}(\omega) \leq u_0(\omega)$. From this definition of $m_k(\omega)$, (2.2.21) and noting the fact that $V(k+1, y(k,\omega)+f(k,y,\omega)+\sigma(k,y,\omega)z(k,\omega),\omega) = V(k+1, y(k+1,\omega),\omega)$, we get

$$m_{k+1}(\omega) \leq g(k, m_k(\omega), \omega).$$

This, together with the hypotheses of the theorem, verifies the hypotheses of Theorem 1.2.2. Hence, by the application of Theorem 1.2.2, we conclude that

$$m_k(\omega) \leq r(k,\omega), \quad \text{for } k \geq k_0. \tag{2.2.25}$$

From (2.2.24) and (2.2.25), relation (2.2.22) follows immediately. Hence, the proof of the corollary is complete.

Corollary 2.2.4. *Assume that*

(a) *the hypotheses of Corollary* 2.1.2(vii) *hold;*
(b) $E[LV(k,y,\omega) \,|\, \mathcal{F}_k] \leq G(k, V(k,y,\omega),\omega)$, *where* $V \in M[I(k_0) \times R^n \times \Omega, R[\Omega, \mathbb{R}_+^N]]$;
(c) G *satisfies relation* (2.2.1). *Then,*

$$E[V(k, y(k,\omega), \omega)|\mathcal{F}_{k_0}] \leq r(k,\omega), \quad \text{for all } k \geq k_0, \tag{2.2.26}$$

whenever

$$E[V(k_0, y_0, \omega)] \leq u_0(\omega),$$

where $y(k,\omega) = y(k, k_0, y_0(\omega), \omega)$ *and* $r(k,\omega) = r(k, k_0, u_0(\omega), \omega)$ *are the solution processes of* (2.1.1) *and* (2.2.1), *respectively.*

Proof. Set

$$m_{k+1}(\omega) = V(k+1, y(k+1, k_0, y_0, \omega), \omega). \tag{2.2.27}$$

Note that $m_{k_0}(\omega) \leq u_0(\omega)$. From (1.1.19), the definition of $m_k(\omega)$ and condition (ii), we have

$$\begin{aligned}\Delta m(\omega) &= V(k+1, y(k+1,\omega),\omega) - V(k, y(k,\omega),\omega) \\ &\leq G(k, m_k, \omega).\end{aligned} \tag{2.2.28}$$

This, together with the other hypotheses of the corollary, fulfills the hypotheses of Corollary 1.2.2. Hence, we can conclude that

$$m_k(\omega) \leq r_k(\omega), \quad \text{for } k \geq k_0.$$

This, together with the definition of $m_k(\omega)$, completes the proof of the corollary.

We demonstrate the scope of Theorem 2.2.1 in the following remark.

Remark 2.2.1. If $u_0(\omega) = V(k_0, x(k, k_0, y_0(\omega)), \omega)$, then (1.2.31) becomes

$$\begin{aligned} &E[V(k, y(k,\omega), \omega) \,|\, \mathcal{F}_{k_0}] \\ &\quad \leq E[r(k, k_0, V(k_0, x(k,\omega), \omega)) \,|\, \mathcal{F}_{k_0}], \quad k \geq k_0. \end{aligned} \tag{2.2.29}$$

We remark that the comparison Theorems 2.2.1 and 2.2.2 are not exactly similar to the comparison results of Corollaries 2.2.3 and 2.2.4. From (2.2.29), it is obvious that Theorems 2.2.1 and 2.2.2 relate the solution processes of three kinds of difference equations, namely (2.1.1), (2.1.3) and (2.2.1)/(2.2.11). On the other hand, the usual comparison results of Corollaries 2.2.3 and 2.2.4 relate the solutions of two kinds of initial value problems, namely (1.1.1) and (2.2.1)/(2.2.11). Another important aspect regarding Theorems 2.2.1 and 2.2.2 is that the initial state u_0 of the sample solution process $r(p, k_0, u_0(\omega))$ of (2.2.1)/(2.2.11) depends on k.

In the following, we give a few examples to illustrate the scope of Theorems 2.2.1 and 2.2.2.

Example 2.2.1. Consider the stochastic difference equation

$$\Delta y(k,\omega) = \frac{y(k,\omega)}{k+1} + \sigma(k, y(k,\omega), \omega) z(k,\omega), \quad y(k_0,\omega) = y_0(\omega), \tag{2.2.30}$$

where $\sigma \in M[I(k_0+1) \times R, R[\Omega, R]]$ and z is an independent normalized discrete process with $E[z(k,\omega)] = 0$ and $\mathrm{var}(z(k,\omega)) = 1$. We assume that

$$\sigma^2(p, y, \omega) \leq \lambda(p,\omega) y^2, \quad \lambda \geq 0. \tag{2.2.31}$$

Further, we consider the difference equation

$$\Delta x(k) = \frac{1}{k+1} x(k), \quad x(k_0, \omega) = x_0(\omega) = y_0(\omega). \tag{2.2.32}$$

We note that

$$m(k,p,y) = \begin{cases} y, & k = p \geq k_0, \\ \dfrac{y(k+2)}{(p+1)}, & k \geq p+1 \geq k_0, \end{cases} \tag{2.2.33}$$

is the solution of (2.2.32). We assume that $V(k, x, \omega) = |x|^2$. From Theorem 2.2.1 and Corollary 2.1.1, we have

$$\begin{aligned} LV(p, m(k,p,y), \omega) &= \mathcal{L}_a V(p, m(k,p,y), \omega) + \mathcal{L}_e V(p, m(k,p,y), \omega) \\ &\quad + \mathcal{L}_0 V(p, m(k,p,y), \omega), \end{aligned}$$

where $m(k, p, y)$ is the solution process of (2.2.30) through (p, y),

$$\mathcal{L}_a V(p, m(k,p,y), \omega) = 0,$$

$$\mathcal{L}_e V(p, m(k,p,y), \omega) = \left(\frac{k+2}{p+1}\right)^2 \sigma^2(p, y, \omega)$$

and

$$\mathcal{L}_0 V(p, m(k,p,y), \omega) = 0.$$

In the context of (2.2.30) and (2.2.32), we obtain

$$\begin{aligned} &E[\Delta V(p, m(k,p,y), \omega) \mid \mathcal{F}_p] \\ &\quad = \frac{(k+2)^2 y \sigma(p,y,\omega) z(p,\omega)}{(p+1)} + \frac{(k+2)^2 \sigma^2(p,y,\omega)}{(p+1)^2}. \end{aligned}$$

Hence, from (2.2.31),

$$\begin{aligned} LV(p, m(k,p,y), \omega) &\leq \lambda(p,\omega) \left(\frac{k+2}{p+1}\right)^2 y^2 \\ &\leq \lambda(p,\omega) V(p, m(k,p,y), \omega). \end{aligned} \tag{2.2.34}$$

In this case, the comparison equation is

$$\Delta u(p,\omega) = \lambda(p,\omega) u(p,\omega), \quad u(k_0,\omega) = u_0(\omega), \tag{2.2.35}$$

and its solution is given by

$$r(k,k_0,u_0(\omega),\omega) = \begin{cases} u_0(\omega), & k = k_0, \\ u_0(\omega)\prod_{p=k_0}^{k-1}(1+\lambda(p,\omega)), & k \geq k_0, \end{cases} \tag{2.2.36}$$

where $u_0(\omega) = |m(k,k_0,y_0(\omega))|^2$. Therefore, from (2.2.34), (2.2.35) and (2.2.36) and applying Corollary 2.2.4, relation (2.2.4) becomes

$$\begin{aligned} E\left[|y(k,\omega)|^2 \mid \mathcal{F}_{k_0}\right] &\leq |m(k,\omega)|^2 \prod_{p=k_0}^{k-1}(1+\lambda(p,\omega)) \\ &\leq |m(k,\omega)|^2 \exp\left[\sum_{p=k_0}^{k-1}\lambda(p,\omega)\right]. \end{aligned} \tag{2.2.37}$$

Example 2.2.2. Consider the stochastic difference equation

$$\Delta y(k,\omega) = y(k,\omega) + \sigma(k,y(k,\omega),\omega)z(k,\omega), \quad y(k_0,\omega) = y_0(\omega), \tag{2.2.38}$$

where σ and $z(k,\omega)$ are as defined in (2.2.30), and it satisfies

$$\sigma^2(p,y,\omega) \leq \lambda(p,\omega)y^2, \quad \lambda > 0. \tag{2.2.39}$$

Further, we consider the difference equation

$$\Delta x(k) = x(k), \quad x(k_0,\omega) = y_0(\omega). \tag{2.2.40}$$

We note that $x(k) = 2^{k-k_0}y_0(\omega)$ is the solution of (2.2.40). By taking $V(k,x,\omega) = |x|^2$, we have

$$V(p,m(k,p,y),\omega) = |m(k,p,y)|^2 = 2^{2(k-p)}y^2, \tag{2.2.41}$$

where $m(k,p,y)$ is the solution process of (2.2.40) through (p,y). We compute $LV(p,m(k,p,y),\omega)$ with respect to (2.2.38). By using (2.2.38) and following the argument used in Example 2.2.1, we have

$$\begin{aligned} \Delta V(p,m(k,p,y),\omega) &= \mathcal{L}_a V(p,m(k,p,y),\omega) + \mathcal{L}_e V(p,m(k,p,y),\omega) \\ &\quad + \mathcal{L}_0 V(p,m(k,p,y),\omega) \\ &= 2^{2(k-p)}[2y\sigma(p,y,\omega)z(k) + \sigma^2(p,y,\omega)]. \end{aligned}$$

Hence,

$$E[LV(p,m(k,p,y),\omega) \mid \mathcal{F}_p] \le \lambda(p,\omega)2^{2(k-p)}y^2 \\ \le \lambda(p,\omega)V(p,m(k,p,y),\omega). \quad (2.2.42)$$

In this case, the comparison equation is

$$\Delta u(p,\omega) = \lambda(p,\omega)u(p,\omega), \quad u(k_0,\omega) = u_0(\omega), \quad (2.2.43)$$

and its solution is given by

$$r(k,k_0,u_0(\omega),\omega) = \begin{cases} u_0(\omega), & k = k_0, \\ u_0(\omega)\prod_{p=k_0+1}^{k-1}(1+\lambda(p,\omega)), & k \ge k_0. \end{cases} \quad (2.2.44)$$

Therefore, from (2.2.42), (2.2.43) and (2.2.44) and invoking Theorem 2.2.1, relation (2.2.4) reduces to

$$E[\|y(k,\omega)\|^2 \mid \mathcal{F}_{k_0}] \le |x(k,\omega)|^2 \prod_{s=k_0+1}^{k}(1+\lambda(s)) \\ \le |x(k,\omega)|^2 \exp \sum_{s=k_0+1}^{k}[\lambda(s)]. \quad (2.2.45)$$

Example 2.2.3. We rewrite (2.1.29) as follows:

$$\Delta y(k,\omega) = \hat{A}(k,e)y(k,\omega) + R(k,y,\omega)y(k,\omega) \\ + \sigma(k,y,\omega)z(k,\omega), \quad y(k_0,\omega) = y_0(\omega), \quad (2.2.46)$$

where $R(k,y,\omega) = A(k,y,\omega) - \hat{A}(k,e)$ and e is an n-dimensional parameter. Further, we consider the difference system

$$\Delta x(k) = \hat{A}(k,e)x(k), \quad x(k_0,\omega) = x_0(\omega). \quad (2.2.47)$$

We note that the solution of (2.2.47) is

$$x(k,k_0,x_0(\omega)) = \begin{cases} y_0(\omega), & k = k_0, \\ \prod_{p=k_0}^{k-1}(I+\hat{A}(p,e))y_0(\omega) & \\ \quad = \Phi(k,k_0,e)y_0(\omega), & k \ge k_0+1, \end{cases} \quad (2.2.48)$$

where

$$\Phi(k, k_0, e) = \begin{cases} I, & k = k_0, \\ \prod_{p=k_0}^{k-1}(I + \hat{A}(p, e)), & k \geq k_0. \end{cases}$$

Let us assume that

$$\begin{cases} \operatorname{tr}(\Phi^T(k,p,e)\sigma(p,y,\omega)\Phi^T(p,y,\omega)\Phi(k,p,e)) \\ \quad \leq \gamma(p,\omega)V(p,m(k,p,y)), \\ y^T R^T \Phi^T(k,p,e)\Phi(k,p,e)Ry \leq M(p,\omega)V(p,m(k,p,y)), \quad M > 0, \\ y^T \Phi^T(k,p,e)\Phi(k,p,\omega)R(p,y,\omega)y \leq -\lambda(p,\omega)V(p,m(k,p,y)), \end{cases} \tag{2.2.49}$$

where γ, M and λ are discrete-time processes, with $\gamma \geq 0$ and $\lambda > 0$.

By taking $V(k, x, \omega) = \|x\|^2$, we get

$$V(p, m(k,p,y), \omega) = \|m(k,p,y)\|^2 = \|\Phi(k,p,e)y\|^2,$$

where $m(k,p,y)$ is the solution process of (2.2.47) through (p,y). Now, we compute $LV(p, m(k,p,y,p), \omega)$ with regard to (2.2.46). By using (2.2.49), (2.2.46), Theorem 2.2.1 and Corollaries 1.2.2 and 1.1.3, we have

$$\begin{aligned} \Delta V(p, m(k,p,y), \omega) &= \mathcal{L}_a V(p, m(k,p,y), \omega) + \mathcal{L}_e V(p, m(k,p,y), \omega) \\ &\quad + \mathcal{L}_0 V(p, m(k,p,y), \omega) \\ &= 2m^T(k,p,y)\Phi(k,p,e) \\ &\quad \times [R(p,y,\omega)y + \sigma(p,y,\omega)z(p,\omega)] \\ &\quad + [z^T(p,\omega)\sigma^T(p,y,\omega) + y^T R^T(p,y,\omega)] \\ &\quad \times \Phi^T(k,p,e)\Phi(k,p,e) \\ &\quad \times [R(p,y,\omega)y + \sigma(p,y,\omega)z(p,\omega)]. \end{aligned}$$

Thus,

$$\begin{aligned} LV(p, m(k,p,y), \omega) &= 2y^T \Phi^T(k,p,e)\Phi(k,p,e)R(p,y,\omega)y \\ &\quad + y^T R^T(p,y,\omega)\Phi^T(k,p,e)\Phi(k,p,e)R(p,y,\omega)y \\ &\quad + \operatorname{tr}(\Phi^T(k,p,e)\sigma(p,y,\omega)\sigma^T(p,y,\omega)\Phi(k,p,e)). \end{aligned} \tag{2.2.50}$$

Using the estimates in (2.2.49), (2.2.50) reduces to

$$LV(p, x(k, p, y), \omega) \leq \nu(p, \omega)V(p, x(k, p, y), \omega), \tag{2.2.51}$$

where $\nu(p, \omega) = -2\lambda(p, \omega) + M(p, \omega) + \gamma(p, \omega) \geq 1$. In this case, the comparison equation is

$$\Delta u(p, \omega) = \nu(p, \omega)u(p, \omega), \quad u(k_0, \omega) = u_0(\omega), \tag{2.2.52}$$

and its solution is given by

$$r(p, k_0, u_0(\omega), \omega) = \begin{cases} u_0(\omega), & p = k_0, \\ u_o(\omega) \prod_{s=k_0+1}^{p} (\nu(s, \omega) + 1), & p \geq s \geq k_0, \end{cases} \tag{2.2.53}$$

where $u_0(\omega) = V(k_0, m(k, k_0, y_0(\omega))) = \|m(k, \omega)\|^2$. From (2.2.51), (2.2.52) and (2.2.53) and applying Theorem 1.2.3, relation (2.2.4) becomes

$$\begin{aligned} E[\|y(k, \omega)\|^2 \mid \mathcal{F}_{k_0}] &\leq \|x(k, \omega)\|^2 \prod_{s=k_0+1}^{k} (1 + \nu(s, \omega)) \\ &\leq \|x(k, \omega)\|^2 \exp \sum_{s=k_0+1}^{k} [\nu(s, \omega)]. \end{aligned} \tag{2.2.54}$$

Example 2.2.4. We consider now the linear stochastic difference system (1.1.52). In this case, $A(k, y, \omega)$ and $\hat{A}(k, m)$ play the roles of $A(k, \omega)$ and $\hat{A}(k)$ in (2.2.46) and in Example 2.2.3, respectively. By proceeding with a similar technique in Example 2.2.3 with $V(k, x, \omega) = \|x\|^2$, we obtain an inequality similar to (2.2.54).

In the following example, we illustrate the usage of a comparison result (Corollary 2.2.3). The example sheds light on the differences between the comparison results of Theorem 2.2.1 and Corollary 2.2.3/2.2.4.

Example 2.2.5. Consider the stochastic difference system (2.2.46). Let $V(k,x) = \|x\|^2$. Then, $V(k, y(k,\omega), \omega) = \|y(k,\omega)\|^2$. First, we assume that

$$\begin{cases} y^T(I+\hat{A}(k,e))^T(I+\hat{A}(k,e))y \le \beta(k)\|y\|^2, \\ y^T(I+\hat{A}(k,e))^T R(k,y,\omega)y \le -\lambda(k,\omega)\|y\|^2, \\ y^T R^T R y \le M_3(k,\omega)\|y\|^2, \quad M > 0, \\ \operatorname{tr}(\sigma(k,y,\omega)\sigma^T(k,y,\omega)) \le \gamma(k,\omega)\|y\|^2, \end{cases} \tag{2.2.55}$$

where β, γ, M and λ are as defined in (2.2.55) and λ and γ are discrete-time processes. Now, we compute $V(k, y(k,\omega), \omega)$ with regard to (2.2.46):

$$\begin{aligned} \Delta V(k,y,\omega) &= V(k+1, y+\hat{A}(k,e)+R(k,y,\omega)y \\ &\quad + \sigma(k,y,\omega)z(k,\omega)) - V(k,y,\omega) \\ &= 2y^T\hat{A}(k,e)y + 2y^T R(k,y,\omega)y \\ &\quad + y^T\hat{A}^T(k,e)\hat{A}(k,e)y \\ &\quad + 2z^T(k,\omega)\sigma^T(k,y,\omega)((I+\hat{A})y + Ry) \\ &\quad + z^T(k,\omega)\sigma^T(k,y,\omega)\sigma(k,y,\omega)z(k,\omega) \\ &\quad + 2y^T\hat{A}^T(k,e)R(k,y,\omega)y \\ &\quad + y^T R^T(k,y,\omega)R(k,y,\omega)y, \qquad (2.2.56) \\ E[\Delta V(k,y,\omega) \,|\, \mathcal{F}_{k-1}] &= y^T(I+\hat{A})^T(I+\hat{A})y \\ &\quad + 2y^T(I+\hat{A})^T R(k,y,\omega)y \\ &\quad + \operatorname{tr}(\sigma^T(k,y,\omega)\sigma(k,y,\omega)) \\ &\quad + y^T R^T(k,y,\omega)R(k,y,\omega)y - y^T y. \end{aligned} \tag{2.2.57}$$

By using (2.2.55) and (2.2.57), we get

$$\begin{cases} \operatorname{tr}(\sigma(k,y,\omega)\sigma^T(k,y,\omega)) \le \gamma(k,\omega)V(k,y), \\ y^T(I+\hat{A}(k,e))^T R(k,y,\omega)y \le -\lambda(k,\omega)V(k,y), \quad \lambda > 0, \\ y^T(I+\hat{A}(k,e))^T(I+\hat{A}(k,e))y \\ \quad \le \beta(k)V(k,y)y^T R^T(k,y,\omega)R(k,y,\omega)y \le M(k,\omega)V(k,y). \end{cases} \tag{2.2.58}$$

Using estimates (2.2.58), (2.2.56) reduces to

$$E[V_{(2.2.46)}(k,y,\omega) \mid \mathcal{F}_{k-1}] \le \nu_1(k,\omega)V(k-1,y,\omega), \tag{2.2.59}$$

where $\nu(k,\omega) = 1-2\lambda(k,\omega)+\beta(k)+\gamma(k,\omega)+M(k,\omega) = 1+\nu_1(k,\omega)$. In this case, the comparison equation is

$$u(k,\omega) = \nu(k,\omega)u(k-1,\omega), \quad u(k_0,\omega) = u_0(\omega), \tag{2.2.60}$$

and its solution is given by

$$r(k,k_0,u_0(\omega),\omega) = \begin{cases} u_0(\omega), & k = k_0, \\ u_0(\omega)\prod_{s=k_0+1}^{k}\nu(s,\omega), & k \ge k_0+1, \end{cases} \tag{2.2.61}$$

where $u_0(\omega) = V(k_0,y_0,\omega) = \|y_0(\omega)\|^2$. From (2.2.59), (2.2.60), (2.2.61) and an application of Corollary 2.2.3, relation (2.2.4) becomes

$$\begin{aligned} E[\|y(k,\omega)\|^2] &\le E[\|y_0(\omega)\|^2]E\left[\prod_{s=k_0+1}^{k}(1+\nu_1(s,\omega))\right] \\ &\le E[\|y_0(\omega)\|^2]E\left[\exp\sum_{s=k_0+1}^{k}[\nu_1(s,\omega)]\right]. \end{aligned} \tag{2.2.62}$$

Example 2.2.6. We consider now the following difference equation:

$$\begin{aligned} \Delta y(k,\omega) &= -\alpha(k,\omega)y(k,\omega) + \sigma(k,y(k,\omega),\omega)z(k,\omega), \\ y(k_0,\omega) &= y_0(\omega), \end{aligned} \tag{2.2.63}$$

where $\sigma \in M[I(k_0+1) \times R, R[\Omega, \mathbb{R}]]$ and z is an independent normalized discrete process with $E(z(k,\omega)) = 0$ and $\text{var}(z(k,\omega)) = 1$. We assume that

$$\sigma^2(p, y, \omega) \leq \lambda(p,\omega)y^2, \quad \lambda \geq 0. \tag{2.2.64}$$

We consider the following difference equation in the absence of external random perturbation and in the context of Remark 2.1.1 and (2.2.63),

$$\Delta x(k) = -\alpha(k,\omega)x(k), \quad x(k_0,\omega) = x_0(\omega) = y_0(\omega). \tag{2.2.65}$$

We note that

$$m(k,p,y) = \begin{cases} y, & k = p \geq k_0, \\ y\prod_{s=p}^{k-1}(1-\alpha(s,\omega)), & k \geq p+1 \geq k_0, \end{cases} \tag{2.2.66}$$

is the solution to (2.2.65). We assume that $V(k, m(k,p,y) = (m(k,p,y))^2$. From Theorem 2.2.1 and Corollary 2.1.3, we have

$$\begin{aligned} LV(p, m(k,p,y),\omega) &= L_aV(p, m(k,p,y),\omega) + L_eV(p, m(k,p,y),\omega) \\ &\quad + L_0V(p, m(k,p,y),\omega), \end{aligned}$$

where $m(k,p,y)$ is the solution process of (2.2.65) through (p,y). By the application of Corollary 2.1.2(ii) and (viii), knowing $F = f$, we have

$$\begin{aligned} &L_aV(p, m(k,p,y)) = 0, \\ &L_eV(p, m(k,p,y)) = \left(\prod_{s=p}^{k-1}(1-\alpha(s,\omega))\right)^2 \sigma^2(p,y,\omega), \\ &L_0V(p, m(k,p,y)) = 0. \end{aligned}$$

Thus,

$$\begin{aligned} \Delta V(p, m(k,p,y),\omega) = \left(\prod_{s=p}^{k-1}(1-\alpha(s,\omega))\right)^2 y\sigma(p,y,\omega)z(p,\omega) \\ + \left(\prod_{s=p}^{k-1}(1-\alpha(s,\omega))\right)^2 \end{aligned}$$

$$\times \sigma^2(p, y, \omega) z^2(p, \omega),$$

$$E[LV(p, m(k, p, y), \omega) \mid \mathcal{F}_p] \le \left(\lambda(p, \omega) \prod_{s=p}^{k-1} (1 - \alpha(s, \omega)) \right)^2 y^2$$

$$\le \lambda(p, \omega) V(p, m(k, p, y)). \tag{2.2.67}$$

In this case, the comparison equation is

$$\Delta u(p, \omega) = \lambda(p, \omega) u(p, \omega), \quad u(k_0, \omega) = u_0(\omega), \tag{2.2.68}$$

and the solution is given by

$$r(k, k_0, u_0(\omega), \omega) = \begin{cases} u_0(\omega), \\ u_0(\omega) \prod_{p=k_0}^{k-1} (1 + \lambda(p, \omega)), \end{cases} \tag{2.2.69}$$

where $u_0(\omega) = |m(k, k_0, y_0(\omega))|^2$. Therefore, from (2.2.67), (2.2.68) and (2.2.69) and applying Corollary 2.2.1, relation (2.2.4) becomes

$$E\left[|y(k, \omega)|^2 \mid \mathcal{F}_{k_0} \right] \le |m(k, \omega)|^2 \prod_{p=k_0}^{k-1} (1 + \lambda(p, \omega)). \tag{2.2.70}$$

We now state and prove a comparison theorem which has a wide range of applications in the theory of error estimation and stability analysis of stochastic difference systems.

Theorem 2.2.3. *Let the hypotheses of Theorem* 2.2.1 *be satisfied, except that* $LV(p, m(k, p, y), \omega)$, $V(p, m(k, p, y(p)), \omega)$ *and* $V(k_0, m(k, k_0, y_0), \omega)$ *are replaced by* $LV(p, m(k, p, y(p) - w(p)), \omega)$, $V(t, m(k, p, y(p) - w(p)), \omega)$ *and* $V(k_0, m(k, k_0, y_0 - w_0), \omega)$, *respectively, where* $y(p) = y(p, k_0, y_0)$ *and* $w(p) = w(p, k_0, w_0)$ *are the solution processes of* (2.1.1) *and*

$$\Delta w(k) = h(k, w(k), \omega) + \epsilon(k, w(k), \omega) \xi(k), \quad w(k_0) = w_0, \tag{2.2.71}$$

respectively, for $k_0 \le p \le k$, *where* $w(t_0)$, h *and* ϵ *are* $\mathcal{F}_k$*-measurable processes and* $\xi(k)$ *is a standard q-dimensional discrete-time processes with* $E[\xi(k)] = 0$, $E[\xi(k)\xi(j)] = 0$ *and* $var[\xi(k)] = I$. *Then,*

$$E[V(p, m(k, p, y(p) - w(p)), \omega) \mid \mathcal{F}_{k_0}] \le r(p, \omega), \quad \text{for } k_0 \le p \le k, \tag{2.2.72}$$

provided

$$V(k_0, m(k, k_0, y_0 - w_0), \omega) \le u_0. \tag{2.2.73}$$

Moreover, for $p = k$, (2.2.72) *reduces to*

$$E[V(k, y(k) - w(k), \omega) \mid \mathcal{F}_{k_0}] \le r(k, \omega), \quad \text{for } k \ge k_0, \tag{2.2.74}$$

whenever (2.2.73) *remains valid.*

Proof. Let $y(p, k_0, y_0)$ and $w(p, k_0, w_0)$ be the solution processes of (2.1.1) and (2.2.71) through (k_0, y_0) and (k_0, w_0), respectively. By repeating the argument used in the proofs of Theorems 2.2.1 and 2.1.4, the proof of the theorem can be constructed analogously. The details are left to the reader.

To illustrate the significance of Theorem 2.2.3, we outline a few particular cases.

Corollary 2.2.5. *Let us suppose that hypotheses* $H_{(1.1.1)}$ *and* $H_{(1.1.2)}$ *remain true.*

(i) *Assume that all the conditions of Corollary* 2.1.4(iii) *and Theorem* 2.2.3 *remain valid. Then,*

$$E[V(p, y(p) - w(p), \omega) \mid \mathcal{F}_{k_0}] \le r(p, \omega), \quad \textit{for } k_0 \le p \le k,$$

and

$$E[V(k, y(k) - w(k), \omega) \mid \mathcal{F}_{k_0}] \le r(k, \omega), \quad \textit{for } k \ge k_0, \tag{2.2.75}$$

provided

$$V(k_0, y_0 - w_0, \omega) \le u_0.$$

LV in Theorem 2.2.3 *is* $LV(p, y(p) - w(p), \omega)$, *and* $r(k, \omega)$ *is as defined by* (2.2.1).

(ii) *If the conditions of Corollary* 2.1.4(iv) *and Theorem* 2.2.3 *hold, then*

$$E[V(p, \Phi(k, p)(y(p) - w(p)), \omega) \mid \mathcal{F}_{k_0}] \le r(p, \omega), \quad \textit{for } k_0 \le p \le k,$$

and hence

$$E[V(k, y(k) - w(k), \omega) \mid \mathcal{F}_{k_0}] \le r(k, \omega) \quad \text{for } k \ge k_0, \tag{2.2.76}$$

provided

$$V(k_0, \Phi(k, k_0)(y_0 - w_0), \omega) \le u_0$$

and $r(k, \omega)$ *is the solution process of* (2.2.1).

(iii) *If the assumptions of Corollaries* 2.1.4(v) *and* 2.2.2(b) *are valid, then*

$$\alpha E[\|y(k,k_0,y_0) - w(k,k_0,w_0)\|^2 \,|\, \mathcal{F}_{k_0}] \\ \leq r(k,k_0,\omega), \quad \textit{for } k \geq k_0, \tag{2.2.77}$$

whenever

$$\alpha\|y_0 - w_0\|^2 \leq u_0.$$

We note that

$$\begin{cases} 2\alpha E[(y-w)^T[f(p,y,\omega) + \sigma(p,y,\omega)z(p,\omega) - h(p,w,\omega) \\ \quad - \epsilon(p,w,\omega)\xi(p,\omega) - F(p,y-w)] \leq \mu(p,\omega)\|y-w\|^2 \,|\, \mathcal{F}_p] \\ \alpha E[\|f(p,y,\omega) + \sigma(p,y,\omega)z(p,\omega) - h(p,w,\omega) \\ \quad - \epsilon(p,w,\omega)\xi(p,\omega) - F(p,y-w)k^2 \,|\, \mathcal{F}_p] \leq \nu(p,\omega)\|y-w\|^2 \\ 2\mu(p,\omega) + \nu(p,\omega) \geq -1 \end{cases} \tag{2.2.78}$$

and $r(k,\omega)$ *is the solution process of* (2.2.11).

(iv) *If the assumptions of Corollary* 2.1.4(vi) *and Theorem* 2.2.3 *are satisfied, then*

$$E[ky(k,k_0,y_0) - w(k,k_0,w_0)k^2|F_{k_0}] \leq r(k,\omega), \quad \textit{for } k \geq k_0, \tag{2.2.79}$$

when

$$\alpha(k_0)\|\Phi(k,k_0)(y_0 - w_0)\|^2 \leq u.$$

We remark that

$$\begin{cases} m^T(k,p,y-w)\Phi(k,p)E[(f(p,y,\omega) + \sigma(p,y,\omega)z(p,\omega) \\ \qquad - h(p,w,\omega) - \epsilon(p,w,\omega)\xi(p,\omega) - F(p,y-w)) \\ \quad \leq \mu(p,\omega)\|m(k,p,y-w)\|^2 \,|\, \mathcal{F}_p] \\ \|\Phi(k,p)E[(f(p,y,\omega) + \sigma(p,y,\omega)z(p,\omega) - h(p,w,\omega) \\ \qquad - \epsilon(k,w,\omega)\xi(p,\omega) - F(p,y-w))\|^2 \,|\, \mathcal{F}_p] \\ \quad \leq \nu(p,\omega)\|m(k,p,y-w)\|^2 \\ 2\mu(p,\omega) + \nu(p,\omega) \geq -1 \end{cases}$$

and $r(p,\omega)$ *is the solution process of the comparison equation* (1.2.20).

(v) *If the assumptions of Corollaries* 2.1.4(vii) *and* 2.2.3 *are fulfilled, then*

$$V(k, y(k) - w(k), \omega) \leq r(k, \omega), \quad \textit{for } k \geq k_0, \tag{2.2.80}$$

provided that $V(k_0, y_0 - w_0, \omega) \leq u_0(\omega)$.

Theorem 2.2.4. *Assume that all the hypotheses of Corollary* 2.1.4(viii) *are satisfied with* $h = F$ *and* $\epsilon(p, y, \omega) \equiv 0$. *Further, assume that* L *in* (2.1.48) *satisfies*

$$\begin{aligned} &E[V(p, m(k, p, y - w), \omega) + \Delta V(p, m(k, p, y - w), \omega) \,|\, \mathcal{F}_p] \\ &\quad \leq G(p, V(p, m(k, p, y - w, \omega))) \end{aligned} \tag{2.2.81}$$

and

$$E[V(k_0, m(k, k_0, y_0 - m_0, \omega), \omega) \,|\, \mathcal{F}_{k_0}] \leq u_0(\omega), \tag{2.2.82}$$

where $x(k, k_0, z_0) = \bar{x}(k)$ *is the solution process of either* (2.1.2) *or* (2.1.3) *depending on the choice of* z_0. *Then,*

$$E[V(k, y(k) - w(k), \omega) \,|\, \mathcal{F}_{k_0}] \leq r(k, k_0, u_0(\omega), \omega), \quad k \geq k_0. \tag{2.2.83}$$

Proof. Let $y(k, \omega)$ and $w(k, \omega)$ be the solution processes of (1.1.1) and (2.1.3), respectively. Let $m(k) = x(k, k_0, x_0)$ be a solution process of either (2.1.2) or (2.1.3) depending upon the choice of z_0. Set

$$\begin{aligned} v(p + 1, \omega) &= V(p + 1, x(k, p + 1, y(p + 1, \omega) - m(p + 1)), \omega), \\ w(k_0, \omega) &= V(k_0, m(k, k_0, y_0 - m_0), \omega). \end{aligned}$$

By following the proofs of Corollaries 1.2.2 and 2.1.1, the proof of the theorem can be completed.

To demonstrate the scope of Theorem 2.2.3, we present an example.

Example 2.2.7. We consider the stochastic difference systems (2.1.29), (2.2.46) and

$$\Delta m(k) = \hat{A}(k, e) m(k - 1), \quad m(k_0) = m_0. \tag{2.2.84}$$

It is further assumed that $h(k, w, \omega) = \hat{A}(k, e) w$ and $\epsilon(k, w, \omega) \equiv 0$ in (2.2.71). Let $V(k, x, \omega) = \|x\|^2$. By following the discussion in

Corollary 2.2.5(iv), we compute $LV(p, m(k, p, y(p) - m(p), \omega))$ as follows. Since $m(k) = \Phi(k, k_0, e)m_0(\omega)$, we assume that

$$\begin{aligned}
&2E[m^T(k, p, y - w)[\hat{A}(p, e)y + R(p, y, \omega)y \\
&\qquad + \sigma(p, y, \omega)z(p) - \hat{A}(p, e)w - F(p, y - w)] \mid \mathcal{F}_p] \\
&\quad \leq \mu(p, \omega)V(p, m(k, p, y - w)), \\
&E[\|\Phi(k, p)[\hat{A}(p, e)y + R(p, y, \omega)y + \sigma(p, y, \omega)z(p) \\
&\qquad - \hat{A}(p, e)w - F(p, y - w)]\|^2 \mid \mathcal{F}_p] \\
&\quad \leq \nu(p, \omega)V(p, m(k, p, y - w), \omega)
\end{aligned}$$

and

$$2\mu(p, \omega) + \nu(p, \omega) \geq -\gamma,$$

and hence

$$\begin{aligned}
&E[LV(p, m(k, p, y - w), \omega) \mid \mathcal{F}_p] \\
&\quad \leq [2\mu(p, \omega) + \nu(p, \omega)]V(p, m(k, p, y - w), \omega). \qquad (2.2.85)
\end{aligned}$$

Therefore, the comparison equation is

$$\Delta u(k) = [2\mu(k, \omega) + \nu(k, \omega)]u, \quad u(k_0, \omega) = u_0(\omega), \qquad (2.2.86)$$

where $u_0(\omega) = \|\Phi(k, k_0)(y_0 - w_0)\|$. By an application of Corollary 2.2.5(iv), we obtain

$$\begin{aligned}
&E[\|y(k, \omega) - m(k)\|^2 \mid \mathcal{F}_{k_0}] \\
&\quad \leq \|\Phi(k, k_0)(y_0 - v_0)\|^2 \exp\left[\sum_{p=k_0}^{k-1} 2\mu(p, \omega) + \nu(p, \omega)\right]. \qquad (2.2.87)
\end{aligned}$$

2.3 Notes and Comments

By developing a Lyapunov-like function difference operator generated along the direction of nonlinear nonstationary systems of iterative processes under Gaussian process perturbations with three different characterizations: (i) in their absence, (ii) in their presence and

(iii) with discretization round-off aggregate errors, we obtain results parallel to Theorem 1.1.1 in Section 1.1 as Remarks 2.1.3 and 2.1.4 and Corollary 2.1.1. Several basic mathematical results are derived in Section 2.1 to study qualitative and quantitative properties of iterative processes under the influence of Gaussian processes, which are based on the work of Ladde and Sambandham [31, 34]. A highly generalized variation of constants formula, Theorem 2.1.1, for nonlinear nonstationary iterative processes and its variants, Theorems 2.1.2, 2.1.3 and 2.1.4, and several corresponding special cases together with examples are outlined in Section 2.1. The results generalize and extend the results of a discrete-time deterministic system derived by Lakshmikantham and Trigante [39]. The very general system of random difference inequalities and comparison theorems and very general variational comparison theorems using the energy function with respect to systems of iterative processes with random parameters are based on the works of Ladde and Slijak [37] and Ladde and Sambandham [34], which are presented in Section 2.2. A very general variational comparison, Theorem 2.2.1, and its variants Theorems 2.2.2, 2.2.3 and 2.2.4, and corresponding special cases and examples also appeared in Section 2.2. These results are motivated by the ideas of Ladde [20]. The results in Section 2.2 are generalizations and extensions of the work of Ladde and Sambandham [33, 34], while the deterministic discrete-time comparison results are from the works of Lakshmikantham and Trigiante [39] and Lakshmikantham and Leela [38].

Chapter 3

Stochastic Systems of Difference Equations: Qualitative Analysis

3.0 Introduction

In this chapter, we systematically study the qualitative and quantitative properties of the solution process of systems of iterative processes under random parametric and Gaussian process perturbations. This is achieved by employing a Lyapunov-like function or an energy function approximation in the framework of generalized variation of parameters method and Lyapunov's Second Method. Sections 3.1, 3.2 and 3.3 cover stability, error and relative stability analysis, respectively. Each one of these sections is composed of subsections covering the study of systems of iterative processes with random parameters and Gaussian processes, respectively. This is achieved by introducing very general results, and some interesting results are discussed. Several examples are presented to illustrate the general results. Furthermore, a substantial effort has been made to demonstrate the sufficiency conditions in terms of system parameters, thereby addressing the issue of "stochastic versus deterministic."

Section 3.1 deals with the pth moment stability results via the methods of comparison and variation of constants. Moreover, the results provide estimates of the solution processes. By employing the methods of comparison and variation of constants, estimates for the absolute pth mean deviation of a solution process of a system of difference equations with Gaussian process parameters associated with a solution process of the corresponding deterministic system of difference equations are developed in Section 3.2. The pth mean stability results of solution processes of a system of difference equations with Gaussian process parameters relative to a solution process of a corresponding deterministic system of difference equations are proved in Section 3.3.

3.1 Stability Analysis

3.1.1 *Stability analysis: Iterative systems under random parameter — I*

Let $y(k,\omega) = y(k,k_0,y_0(\omega),\omega)$ be any solution process of (1.1.1), and let $x(k,\omega) = x(k,k_0,y_0(\omega))$ be the solution process of (1.1.3) through $(k_0,y_0(\omega))$. Furthermore, let $\bar{x}(k) = x(k,k_0,z_0)$ be the solution process of either (1.1.2) or (1.1.3) depending on the choice of z_0. Without loss of generality, we assume that $f(k,0,\omega) \equiv 0$ w.p. 1 and $F(k,0) \equiv 0$, for all $k \geq k_0$. $y(k,\omega) \equiv 0$, $x(k,\omega) \equiv 0$ and $\bar{x}(k) \equiv 0$ are the unique solutions of the respective initial value problems.

Definition 3.1.1. The trivial solution process of (1.1.1) is said to be

(DSM$_1$) *stable in the qth moment* if for each $\epsilon > 0$, $k_0 \in I(k_0)$ and $q \geq 1$, there exists a positive function $\delta(k_0,\epsilon)$ such that the inequality $\|y_0(\omega)\|_q \leq \delta$ implies

$$\|y(k,\omega)\|_q < \epsilon, \quad k \geq k_0,$$

where $\|y(k,\omega)\|_q = (E\|y(k,\omega)\|^q)^{1/q}$;

(DSM$_2$) *asymptotically stable in the qth moment* if it is stable in the qth moment and if for any $\epsilon > 0$, $k_0 \in I(k_0)$, there exist $\delta_0(k_0)$ and $T = T(k_0,\epsilon)$ such that the inequality $\|y_0(\omega)\|_q \leq \delta_0$ implies

$$\|y(k,\omega)\|_q < \epsilon, \quad k \geq k_0 + T.$$

Remark 3.1.1. We note that, depending on the mode of convergence in the probabilistic analysis, one can formulate other definitions of stability and boundedness. For the differential equations, we refer to Ladde and Lakshmikantham [25].

For our further use, we formulate the concept of relative stability.

Definition 3.1.2. The two difference systems (1.1.1) and (1.1.60) are said to be

(DRM_1) *relatively stable in the qth moment* if for each $\epsilon > 0$, $k_0 \in I(k_0)$ and $q \geq 1$, there exists a positive function $\delta = \delta(k_0, \epsilon)$ such that the inequality $\|y_0(\omega) - w_0\|_q \leq \delta$ implies

$$\|y(k, \omega) - w(k)\|_q < \epsilon, \quad k \geq k_0;$$

(DRM_2) *relatively asymptotically stable in the qth moment* if it is stable in the qth moment and if for any $\epsilon > 0$, $k_0 \in I(k_0)$, there exist $\delta_0 = \delta_0(k_0)$ and $T = T(k_0, \epsilon)$ such that the inequality $\|y_0 - w_0\| \leq \delta_0$ implies

$$\|y(k, \omega) - w(k)\|_q < \epsilon, \quad k \geq k_0 + T.$$

Remark 3.1.2. Based on Definition 3.1.2, a definition relative to (1.1.1) and (1.1.2) or (1.1.3) can be formulated analogously.

To study the stability analysis of (1.1.1) by an application of the comparison method, we require the stability of the comparison difference system (1.2.20). The stability concepts in Definition 3.1.1 relative to (1.2.20) are denoted by (DSM_1^*) and (DSM_2^*). In the present framework, we need a joint stability property of (1.2.20) and (1.1.2), or (1.2.20) and (1.1.3). We remark that this does not imply that each of the systems (1.2.20) and (1.1.2) (or (1.2.20) and (1.1.3)) possesses the same kind of stability property. This property offers more flexibility in applications than the existing approaches.

Let $m(k, \omega) = m(k, k_0, y_0(\omega))$ and $u(k, k_0, u_0)$ be the solutions of (1.1.3) and (1.2.20) through $(k_0, y_0(\omega))$ and (k_0, u_0), respectively. Then, we define

$$\nu(k, k_0, y_0(\omega), \omega) = u(k, k_0, m(k, k_0, y_0(\omega)), \omega) \tag{3.1.1}$$

and note that $\nu(k_0, k_0, y_0(\omega), \omega) = V(k_0, y_0(\omega), \omega)$, and V is as defined in Theorem 1.1.1. We now formulate stability concepts relative to (1.1.2) and (1.1.20).

Definition 3.1.3. The trivial solution processes $m = 0$ and $u = 0$ of (1.1.2) and (1.1.20) are said to be

(DJM$_1$) *jointly stable in the mean* if for $\epsilon > 0$, $k_0 \in I(k_0)$, there exists a $\delta = \delta(k_0, \epsilon) > 0$ such that

$$E\|y_0(\omega)\|_p \leq \delta \text{ implies } \sum_{i=1}^{m} E(\nu_i(k, k_0, y_0(\omega), \omega) < \epsilon,$$

$$k \geq k_0;$$

(DJM$_2$) *jointly asymptotically stable in the mean* if it is jointly stable in the mean and if for any $\epsilon > 0$, $k_0 \in I(k_0)$, there exist $\delta_0 = \delta_0(k_0) > 0$ and $T = T(k_0, \epsilon) > 0$ such that the inequality $E\|y_0(\omega)\|_q \leq \delta_0$ implies

$$\sum_{i=1}^{m} E[\nu_i(k, k_0, y_0(\omega), \omega)] < \epsilon, \quad t \geq t_0 + T.$$

The joint relative stability of (1.1.20) and (1.1.3) ((1.1.20) and (1.1.3)) is defined as follows.

Definition 3.1.4. The systems (1.1.1), (1.1.60), (1.1.2) or (1.1.3) are said to be

(DJR$_1$) *jointly relatively stable in the mean* if for each $\epsilon > 0$, $k_0 \in I(k_0)$, there exists $\delta = \delta(k_0, \epsilon) > 0$ such that the inequality $E\|y_0(\omega) - w_0\| \leq \delta$ implies

$$\sum_{i=1}^{m} E[\nu_i(k, k_0, y_0(\omega), \omega) - w_0)] < \epsilon, \quad k \geq k_0.$$

We present some stability criteria that assure the stability in the qth moment of the trivial solution processes of (1.1.1). Furthermore, some illustrations are given to show that the stability conditions are connected with the statistical properties of random rate functions of systems of difference equations. Examples are worked out to exhibit the advantage of the joint stability concepts.

Theorem 3.1.1. *Let the hypotheses of Theorem* 1.2.3 *be satisfied. Further, assume that* $F(k, 0) \equiv 0$, $f(k, 0, \omega) \equiv 0$ *and* $G(k, 0, \omega) \equiv 0$ *w.p.* 1, *and for* $(k, x) \in I(k_0 + 1) \times R^n$,

$$b(\|y\|)^q \leq \sum_{i=1}^{m} V_i(k, y, \omega) \leq a(k, \|y\|^q) \tag{3.1.2}$$

whenever $b \in \mathcal{VK}$, $a \in \mathcal{CK}$ *and* $q \geq 1$. *Then,*

(DJM_1) of (1.2.20) and (1.1.3) implies (DSM_1) of (1.1.1), and
(DJM_2) of (1.2.20) and (1.1.3) implies (DSM_2) of (1.1.1).

Proof. Let $\epsilon > 0$, $k_0 \in I(k_0)$ be given. Assume that (DJM_1) holds. Then, for $b(\epsilon) > 0$ and $k_0 \in I(k_0)$, there exists a $\delta = \delta(\epsilon, k_0)$ such that $\|y_0(\omega)\|_p \leq \delta$ implies

$$\sum_{i=1}^{m} E(\nu_i(k, k_0, y_0(\omega), \omega) < b(\epsilon^q), \quad k \geq k_0, \tag{3.1.3}$$

where

$$\nu(k, k_0, y_0(\omega), \omega) = r(k, k_0, V(k, k_0, y_0(\omega), \omega), \omega). \tag{3.1.4}$$

$r(k, k_0, u_0, \omega)$ is the solution process of (1.2.20), and $m(k, k_0, y_0(\omega))$ is the solution process of (1.1.2) through $(k_0, y_0(\omega))$. Now, we claim that if $\|y_0(\omega)\|_q \leq \delta$, then $\|y(k, \omega)\|_q < \epsilon$, $k \geq k_0$. Suppose that this is false. Then, there would exist a solution process $y(k, k_0, y_0(\omega), \omega)$ with $\|y_0(\omega)\|_q \leq \delta$ and a $k_1 > k_0$ such that

$$\|y(k_1, \omega)\|_q = \epsilon \quad \text{and} \quad \|y(k, \omega)\|_q \leq \epsilon, \quad k_0 \leq k \leq k_1. \tag{3.1.5}$$

On the other hand, by Theorem 1.2.3 in the context of Remark 1.2.2, we have

$$V(k, y(k, \omega), \omega) \leq r(k, k_0, V(k_0, m(k, \omega), \omega), \omega), \quad k \geq k_0. \tag{3.1.6}$$

From (3.1.2), (3.1.4), (3.1.6) and using the convexity of b, we obtain

$$\begin{aligned} b(E\|y(k, \omega)\|^p) &\leq \sum_{i=1}^{m} EV_i(k, y(k, \omega), \omega) \\ &\leq \sum_{i=1}^{m} \nu_i(k, k_0, y_0(\omega), \omega), \end{aligned} \tag{3.1.7}$$

for $k \geq k_0$. Equations (3.1.3), (3.1.5) and (3.1.7) lead to the contradiction

$$\begin{aligned} b(\epsilon^p) &\leq \sum_{i=1}^{m} E(V_i(k_1, y(k_1, \omega), \omega)) \\ &\leq \sum_{i=1}^{m} \nu_i(k_1, k_0, y_0(\omega), \omega) < b(\epsilon^p). \end{aligned}$$

This proves (DSM_1). The proof of (DSM_2) can be constructed analogously. This completes the proof of the theorem.

The following example illustrates the scope and usefulness of the joint stability concept and Theorem 3.1.1.

Example 3.1.1. Consider Example 1.2.3. We further assume that $H(k, 0, \omega) \equiv 0$ w.p. 1. Furthermore, let λ in (1.2.53) satisfy

$$E\left[\exp\left[\sum_{p=k_0}^{k-1}(\lambda(p,\omega))\right]\right] \leq M \quad \text{and} \quad \left(\frac{\sum_{p=k_0}^{k-1}\lambda(p,\omega)}{k-k_0}\right) \to \alpha$$

as $(k - k_0) \to \infty$, for some negative numbers α; M is a positive number; and $\|y_0(\omega)\|_2 < \infty$. It is clear that

$$\nu(k, k_0, y_0(\omega), \omega) = |m(k,\omega)|^2 \exp \sum_{s=k_0}^{k} (\lambda(s,\omega)).$$

With this, together with the assumptions about H, $y_0(\omega)$ and $\lambda(k, \omega)$, it follows that the trivial solution processes $m \equiv 0$ and $u \equiv 0$ of (1.2.51) and (1.2.53) are jointly stable in the mean.

Moreover, from the conditions on λ and $y_0(\omega)$, one can conclude that $m \equiv 0$ and $u \equiv 0$ of (1.2.51) and (1.2.53) are jointly asymptotically stable in the mean. From this and with an application of Theorem 1.2.3, one can conclude that the trivial solution process of (1.2.49) is asymptotically stable in the second moment. We remark that the trivial solution of (1.2.51) is unbounded. However, the asymptotic stability is guaranteed by random environmental perturbations characterized by the above conditions.

Example 3.1.2. Consider Example 1.2.4, where we replace (1.2.59) by

$$LV(p, x(k,p,y), \omega) \leq (\bar{\alpha}(p) + \eta(p,\omega))V(p, x(k,p,y), \omega), \qquad (3.1.8)$$

where $\lambda(p,\omega) = \bar{\alpha}(p) + \eta(p,\omega)$ and $\eta(p,\omega)$ is a stationary Gaussian process with mean $E(\eta(p,\omega)) = 0$ and covariance function $C(k - p) = E(\eta(k,\omega)\eta(p,\omega))$. Let $H(k, 0, \omega) \equiv 0$ w.p. 1. By following the discussion in Example 1.2.2, we have

$$\Delta u(p,\omega) = (\bar{\alpha}(p) + \eta(p,\omega))u(p,\omega), u(k_0,\omega) = u_0(\omega) \tag{3.1.9}$$

and

$$V(k, y(k,\omega), \omega) \leq V(k_0, m(k,\omega), \omega) \exp\left[\sum_{s=k_0}^{k} (\bar{\alpha}(s) + \eta(s,\omega))\right]. \tag{3.1.10}$$

Let $y_0(\omega)$ and $\eta(s,\omega)$ be independent. By taking the expectation on both sides of (3.1.10), we obtain

$$E|y(k,\omega)|^2 \leq E|m(k,\omega)|^2 E\left(\exp\sum_{s=k_0}^{k} (\eta(s,\omega) + \bar{\alpha}(s))\right). \tag{3.1.11}$$

From the properties of the Gaussian process, we have

$$E\left[\exp\sum_{s=k_0}^{k} (\eta(s,\omega) + \bar{\alpha}(s))\right]$$

$$= \exp\sum_{s=k_0}^{k}\left[(\bar{\alpha}(s)) + \frac{1}{2}\sum_{u=k_0}^{k}\sum_{s=k_0}^{k} C(u-s)\right].$$

The trivial solutions of (1.2.58) and (3.1.9) are jointly stable in mean if $\|y_0\|_2 < \infty$, and

$$\frac{1}{(k-k_0)}\left[\sum_{s=k_0}^{k}\bar{\alpha}(s) + \frac{1}{2}\sum_{u=k_0}^{k}\sum_{s=k_0}^{k} C(u-s)\right] < -1, \quad \text{for } k \geq k_0. \tag{3.1.12}$$

From this and (3.1.10), we conclude that the trivial solution of (1.2.56) is stable in mean square. Moreover, it is asymptotically stable in the second moment.

Remark 3.1.3. Example 3.1.2 shows that the trivial solution process $m \equiv 0$ of $\Delta m(k) = m(k), m(k_0) = m_0$ is unstable (exponentially with base 2) in the mean-square sense. However, the joint stability guaranteed by Theorem 3.1.1 provides greater advantage to study the stability problems.

In the following, we illustrate the use of the variation of constants method to study the qth moment stability of the trivial solution processes of (1.1.1) and (1.1.55).

Theorem 3.1.2. *Let the hypothesis of* Theorem 1.1.3 *be satisfied. Further, assume that* V, $x(k,\omega)$, $\Phi(k,s,y(k,\omega))$, $R(k,y(k,\omega),\omega)$ *and* $F(k,y)$ *satisfy:*

(D1) $b(\|x\|^q) \le \|V(k,x,\omega)\| \le a(\|x\|^q)$, *for* $(k,x) \in I(k_0+1) \times R^n$, *where* $b \in \mathcal{VK}$ *and* $a \in \mathcal{CK}$;

(D2) $F(k,0) \equiv 0, f(k,0,\omega)$ *and* $R(k,0,\omega) \equiv 0$ *w.p.* 1, *for* $k \in I(k_0+1)$;

(D3) $\|D^*W(k,s,y,\omega)\| \le \lambda(s,\omega)\|V(s-1,y,\omega)\|$, *for* $k_0 \le s \le k$, $\|y\|^q < \rho$, *where* ρ *is some positive real,* $\mathcal{D}^*W(k,s,y,\omega)$ *is defined by*

$$\mathcal{D}^*W(k,s,y,\omega) = W(k,k_0,x(k,k_0,y(k,\omega)),\omega) \times \psi^{-1}(s,k_0,p(s,\omega),p(s-1,\omega))R(s,y,\omega), \tag{3.1.13}$$

where $p(s,\omega)$ is as defined in (1.1.43), $\lambda \in M[I(k_0), R(\Omega, R_+)]$ *and satisfies the relation*

$$E\left[\exp \sum_{s=k_0+1}^{k} \lambda(s,\omega)\right] \le \exp\left[\sum_{s=k_0+1}^{k} \hat{\lambda}(s)\right], \tag{3.1.14}$$

where λ *and* y_o *are independent random variables and* $\sum_{k=1}^{\infty} \hat{\lambda}(k) < \infty$;

(D4) $\|V(k_0,x(k,\omega),\omega)\| \le \alpha[\|y_0\|^q]$ *whenever* $E[\|y_0\|^q] < \rho$, *for some* $\rho > 0$ *and where* $\alpha \in \mathcal{CK}$ *Then, the trivial solution process of* (1.1.1) *is stable in the qth mean.*

Proof. Let $y(k,\omega)$ be a sample solution process of (1.1.1), and let $x(k,s,y(s,\omega))$ and $x(k,\omega) = x(k,k_0,y_0(\omega))$ be the sample solution processes of (1.1.3) through $(s,y(s))$ and $(k_0,y_0(\omega))$, respectively, for $k_0 \le s \le k$ and $k_0 \in I(k_0)$. From hypothesis (D2), we have $x(k,k_0,0) \equiv 0$ and $y(k,k_0,0,\omega) \equiv 0$ w.p. 1. From Theorem 1.1.3 and conditions (D3) and (D4), we have

$$\|V(k, y(k,\omega),\omega)\| \le \alpha[\|y_0\|^q] + \sum_{s=k_0+1}^{k} \lambda(s-1)\|V(s, y(s,\omega),\omega)\| \tag{3.1.15}$$

if $E[\|y(s,\omega)\|^q] < \rho$. By setting $m(k,\omega) = \|V(k, y(k,\omega),\omega)\|$ and using Lemma A.1.2, (3.1.15) reduces to

$$\Delta R(k,\omega) \le \lambda(k)R(k,\omega), \tag{3.1.16}$$

where

$$R(k,\omega) = \sum_{s=k_0+1}^{k} \lambda(s-1)m(s,\omega) + \alpha(\|y_0\|^p)$$

as long as $E(\|y(s,\omega)\|^q) < \rho$, for $k_0 \le s \le k$. This, together with an application of Corollary 1.2.2, yields

$$R(k,\omega) \le \alpha(\|y_0(\omega)\|^q) \exp \sum_{s=k_0+1}^{k} \lambda(s-1,\omega),$$

and hence

$$\|V(k, y(k,\omega),\omega)\| \le \alpha(\|y_0(\omega)\|^q) \exp \sum_{s=k_0+1}^{k} \lambda(s-1,\omega) \tag{3.1.17}$$

as long as $E[\|y(k,\omega)\|^q] < \rho$, for $k_0 \le s \le k$. From the nature of α, $y_0(\omega)$ and (3.1.14), (3.1.17) reduces to

$$E[\|V(k, y(k,\omega),\omega)\|] \le \alpha(E[\|y_0(\omega)\|^q]) \exp \sum_{s=k_0+1}^{k} \hat{\lambda}(s)$$

as long as $E(\|y(s,\omega)\|^q) < \rho$, for $k_0 \le s \le k$. From this, along with condition (D1) and the property of $\hat{\lambda}$, we have

$$b(E[\|y(k,\omega)\|^q]) \le K\alpha(E[\|y_0\|^q]) \tag{3.1.18}$$

as long as $E(\|y(k,\omega)\|^q) \leq \rho$, for some positive number K, $k \geq k_0$. To conclude $E[\|y(k,\omega)\|^q] \leq \rho$, for all $k \geq k_0$, and stability property (DSM_1) of (1.1.1), first we pick $y_0(\omega)$ such that

$$K\alpha(E[\|y_0(\omega)\|^q]) < b(\rho).$$

For this choice of y_0, we claim that $E[\|y(k,\omega)\|^q] < \rho$, for all $k \geq k_0$.

If this claim is false, then there exists $\bar{k} \in I(k_0)$ such that $\bar{k} > k_0$, $E[\|y(s,\omega)\|^q] < \rho$, for $k_0 \leq s \leq \bar{k}$ and $E[\|y(\bar{k},\omega)\|^q] = \rho$. For $k \in [k_0, \bar{k}]$, (3.1.18) is valid, and hence using condition (D1) and (3.1.18), we get

$$b(\rho) \leq b(E\|y(\bar{k},\omega)\|^q) < b(\rho).$$

This contradiction establishes the claim that $E[\|y(k,\omega)\|^q] < \rho$. Hence, (3.1.18) is valid for all $k \geq k_0$. To prove (DSM_1) of (1.1.1), we pick any $\epsilon > 0$ and choose $y_0(\omega)$ such that

$$K\alpha(E[\|y_0(\omega)\|^q]) < b(\epsilon^q). \tag{3.1.19}$$

This implies that

$$E[\|y_0(\omega)\|^q] < \alpha^{-1}\left(K^{-1}b(\epsilon^q)\right).$$

From (3.1.18) and (3.1.19), we have

$$E(\|y(k,\omega)\|^q) < \epsilon^q, \quad k \geq k_0,$$

whenever

$$(E[\|y_0(\omega)\|^q])^{1/q} < \delta(\epsilon),$$

where

$$\delta(\epsilon) = \left(\alpha^{-1}\left(K^{-1}b(\epsilon^q)\right)\right)^{1/q} > 0.$$

This is based on the property of the function b, α and α^{-1}. This completes the proof of the theorem.

In the following theorem, we present sufficient conditions for the qth moment asymptotic stability of the trivial solution of (1.1.1).

Theorem 3.1.3. *Assume that the hypotheses of Theorem* 3.1.2 *hold, except that* (D3) *and* (D4) *are replaced by*

(D5)

$$\mathcal{D}\|^*W(k,s,y,\omega)\| \le \mu(k-s)[\|V(k_0,x(k,\omega),\omega)\| + \sum_{r=k_0+1}^{k} \lambda_2(r,\omega)\eta(k-r)\|V(r-1,y(r-1,\omega),\omega)\|] + \lambda_1(s,\omega)\eta(k-s)\|V(s-1,y,\omega)\|, \tag{3.1.20}$$

for $k_0 \le s \le k$, $|y|^q \le \rho$, *for some* $\rho > 0$, *and*

(D6)

$$\|V(k_0,x(k,\omega),\omega)\| \le \alpha(\|y_0(\omega)\|^q)\beta(k-k_0), k \ge k_0,$$

provided $E\|y_0(\omega)\|^q \le \rho$, *where* μ, η, β *are nonnegative sequences defined on* $I(k_0)$ *and satisfy the following relation:*

$$\lim_{k\to\infty} \sum_{s=k_0+1}^{k} \eta(k-s)\beta(s-k_0) = 0, \tag{3.1.21}$$

$\lim_{r\to\infty}\beta(r) = 0$, μ *is summable on* $I(k_0)$ *with a sum* $\bar{\mu}$, η *is bounded on* $I(k_0)$ *and* λ *defined by* $\lambda = \lambda_1 + \lambda_2\bar{\mu}$ *satisfies all the conditions in* (1.1.14) *with* $q = 2$. *Then, the trivial solution process of* (1.1.1) *is asymptotically stable in the qth moment.*

Proof. By employing conditions (D5) and (D6) and following the proof of Theorem 3.1.2, we arrive at

$$\|V(k,y(k,\omega),\omega)\| \le \alpha[\|y_0(\omega)\|^q]\beta(k-k_0) + \sum_{s=k_0+1}^{k} (\lambda_1(s,\omega)\eta(k-s)\|V(s-1,y(s-1,\omega),\omega)\| + \mu(k-s)\|V(k_0,x(k,\omega),\omega)\|) + \sum_{s=k_0+1}^{k} \mu(k-s) \times \left(\sum_{r=k_0+1}^{k} \lambda_2(r,\omega)\eta(k-r)\|V(r-1,y(r-1,\omega),\omega)\|\right) \tag{3.1.22}$$

as long as $E\|y(k,\omega)\|^q \le \rho$. By setting

$$m(k,\omega) = \|V(k, y(k,\omega), \omega)\|,$$
$$n(k,\omega) = \alpha(\|y_0(\omega)\|^q)\beta(k - k_0)(1 + \bar{\mu})$$

and

$$\nu(k,s,\omega) = (\lambda_1(s,\omega) + \bar{\mu}\lambda_2(s,\omega))\eta(k-s),$$

(3.1.22) can be rewritten as

$$m(k,\omega) \le n(k,\omega) + \sum_{s=k_0+1}^{k} \nu(k,s,\omega) m(s-1,\omega),$$

as long as $E\|y(k,\omega)\|^q \le \rho$, where $\bar{\mu} = \sum_{s=k_0+1}^{k} \mu(k-s)$. Using the estimate in Lemma A.1.3, we get

$$m(k,\omega) \le n(k,\omega) + \sum_{s=k_0+1}^{k} \nu(k,s,\omega) n(s,\omega) \times \left[\exp\left(\sum_{\tau=s+1}^{k} \nu(k,\tau,\omega)\right)\right]. \tag{3.1.23}$$

From the definitions of $\nu(k,s,\omega)$, $m(k,\omega)$ and $n(k,\omega)$, (3.1.23) becomes

$$\|V(k, y(k,\omega), \omega)\| \le \alpha(\|y_0(\omega)\|^q)(1 + \bar{\mu}) \times \left[\beta(k-k_0) + \exp\left(\sum_{s=k_0+1}^{k} \nu(k,s,\omega)\right) \times \sum_{s=k_0+1}^{k} \eta(k-s)\beta(s-k_0)\lambda(s,\omega)\right] \tag{3.1.24}$$

as long as $E\|y(k,\omega)\|^q < \rho$, where $\lambda(k,\omega) = [\lambda_1(k,\omega) + \bar{\mu}\lambda_2(k,\omega)]$. From (3.1.24), condition (D1) and the properties of functions α, λ

and the Schwartz inequality, we obtain

$$b(E\|y(k,\omega)\|^q)$$
$$\leq \alpha(E\|y_0(\omega)\|^q)(1+\bar{\mu})\Bigg(\beta(k-k_0)+K\exp$$
$$\times\left[\sum_{s=k_0+1}^{k}\hat{\lambda}(s)\right]\sum_{s=k_0+1}^{k}\eta(k-s)\beta(s-k_0)\Bigg) \qquad (3.1.25)$$

as long as $E\|y(k,\omega)\|^q \leq \rho$ and $K = \sup_{\tau_1 \geq k_0}[\mathrm{Cor}(\lambda(s,\omega)\lambda(\tau,\omega))^{1/2}]$. This is guaranteed by (3.1.14). From (3.1.25) and the properties of the underlying function, one can easily conclude that $E\|y(k,\omega)\|^q \leq \rho$, for all $k \geq k_0$. Hence, property (DSM_1) of the trivial solution of (1.1.1) can be concluded by following the argument of Theorem 3.1.2. To conclude (DSM_2), it is obvious from (3.1.25) and the nature of β and (3.1.21) that $b(E\|y(k,\omega)\|^q)$ tends to zero as $k \to \infty$. Hence, one can manipulate to verify that technical definitions of (DSM_2). The details are left to the reader. This completes the proof of the theorem.

Remark 3.1.4. We note that Theorems 3.1.2 and 3.1.3 also provide sufficient conditions for the stability w.p. 1 of the trivial solution processes of (1.1.1). This fact follows from relations (3.1.17) and (3.1.24).

Remark 3.1.5. The conditions (D4) and (D6) in Theorems 3.1.2 and 3.1.3 imply uniform stability and asymptotic stability, respectively, in the qth moment of the trivial solution process of (1.1.3).

To appreciate the assumptions of Theorem 3.1.3, we present the following result, which is applicable to other problems.

Corollary 3.1.1. *Let the hypotheses of Theorem* 3.1.3 *be satisfied, except that the relations* (3.1.14) *and* (3.1.21) *are replaced by*

$$\eta(k-s)\beta(s-1-k_0) \leq K\beta(k-k_0), \quad k \geq k_0, \qquad (3.1.26)$$

and

$$\lim_{k\to\infty}\beta(k-k_0)E\left[\exp\left[K\sum_{s=k_0+1}^{k}\lambda(s,\omega)\right]\right] = 0, \qquad (3.1.27)$$

where K is some positive constant. Then, the trivial solution (1.1.1) *is asymptotically stable in the qth moment.*

Proof. By following the proof of Theorem 3.1.3, we arrive at (3.1.22). Now, by using (3.1.26), (3.1.22) can be rewritten as

$$m(k,\omega) \leq \alpha(\|y_0(\omega)\|^q)(1+\bar{\mu}) + \sum_{s=k_0+1}^{k} \nu(k,s,\omega)m(s-1,\omega) \tag{3.1.28}$$

as long as $E\|y(k,\omega)\|^q < \rho$, where

$$m(k,\omega) = \|V(k,y(k,\omega),\omega)\|/\beta(k-k_0)$$

and

$$\nu(k,s,\omega) = (\lambda_1(s,\omega) + \bar{\mu}\lambda_2(s))K\beta(k-k_0).$$

By applying Lemma A.1.2 to (3.1.28), we get

$$m(k,\omega) \leq (1+\bar{\mu})\alpha(\|y_0(\omega)\|^p)\exp\left[\sum_{s=k_0+1}^{k} \nu(k,s,\omega)\right].$$

which implies

$$b(\|y(k,\omega)\|^q) \leq (1+\bar{\mu})\alpha(\|y_0(\omega)\|^q)\beta(k-k_0)\exp\left(\sum_{s=k_0+1}^{k} \nu(k,s,\omega)\right).$$

By taking the expectation on both sides and using the properties of b, α and $y_0(\omega)$, we have

$$\begin{aligned} b(E\|y(k,\omega)\|^q) \leq (1+\bar{\mu})&\alpha(E[\|y_0(\omega)\|^q])\beta(k-k_0)E \\ &\times \left[\exp\sum_{s=k_0+1}^{k} \nu(k,s,\omega)\right]. \end{aligned} \tag{3.1.29}$$

The rest of the proof can be completed by using the argument in the proof of Theorem 3.1.3. Thus, the proof of the corollary is complete.

Remark 3.1.6. If $\eta(r) = \beta(r) = \exp(-\gamma r)$, for some $\gamma > 0$, then these functions are admissible in Corollary 3.1.1.

Remark 3.1.7. For the qth moment asymptotic stability, condition (D5) can be further modified. This is the content of the following corollary.

Corollary 3.1.2. *Suppose that all the conditions of Theorem* 3.1.3 *hold, except that condition (D5) is replaced by*

(D5*)

$$\begin{aligned}
&\|\mathcal{D}^* W(k, s, y, \omega)\| \\
&\quad \leq \mu(k-s)\Big[\|V(k_0, x(k, \omega), \omega)\| \\
&\qquad + \sum_{r=k_0+1}^{k} \eta(k-r)\|V(r-1, y(r-1, \omega), \omega)\| \\
&\qquad + \eta(k-s)\|V(s-1, y, \omega)\|\Big],
\end{aligned}$$

for $k_0 \leq s \leq k$ *and* $\|y\|^q < \rho$. *Further, assume that*

$$\sum_{u=s+1}^{k} \eta(k-u) \leq K, \quad \text{for } k_0 \leq s \leq k,$$

for some $K > 0$. *Then, the conclusion of Theorem* 3.1.3 *remains valid.*

We discuss a few examples to illustrate the fruitfulness of our result.

Example 3.1.3. Consider Example 1.1.1. We assume that the solution process $x(k, k_0, y_0(\omega)) = x(k, \omega) = \Phi(k, k_0)y_0(\omega)$ fulfills the relation

$$\|\Phi(k, s)\| \leq \eta(k-s)$$

and A and $\hat{A}$ satisfy

$$\|A(s, \omega) - \hat{A}(s)\| \leq \bar{\lambda}(s, \omega), \tag{3.1.30}$$

for some $\eta, \bar{\lambda} > 0$. Let $V(k, x) = \|x\|^2$. Then,

$$\mathcal{D}^* W(k, s, y, \omega) = \left[2x_0^T \Phi(k, k_0) + \sum_{r=k_0+1}^{k} R^T(r, y, \omega)\Phi^T(k, r) \right] \Phi(k, s) R(s, y, \omega).$$

Therefore,

$$\begin{aligned}
&|\mathcal{D}^* W(k, s, y, \omega)| \\
&\quad \leq \left| \left[2(\Phi(k, k_0)x_0)^T + \sum_{r=k_0+1}^{k} R^T(r, y, \omega)\Phi^T(k, r) \right] \right. \\
&\qquad \left. \times \ \Phi(k, s) R(s, y, \omega) \right| \\
&\quad \leq 2 \left| (\Phi(k, k_0)x_0)^T \Phi(k, s) R(s, y, \omega) \right| \\
&\qquad + \left| \left(\sum_{r=k_0+1}^{k} R^T(r, y, \omega)\Phi^T(k, r) \right) \Phi(k, s) R(s, y, \omega) \right| \\
&\quad \leq 2\|\Phi(k, k_0)x_0\| \|\Phi(k, s)\| \|R(s, y, \omega)\| \\
&\qquad + \sum_{r=k_0+1}^{k} \|R^T(r, y, \omega)\| \ \|\Phi^T(k, r)\| \|\Phi(k, s)\| \|R(s, y, \omega)\| \\
&\quad \leq \|\Phi(k, s)\| \left(2\|\Phi(k, k_0)x_0\| \|R(s, y, \omega)\| \right. \\
&\qquad \left. + \sum_{r=k_0+1}^{k} \|R(r, y, \omega)\| \|\Phi^T(k, r)\| \|\Phi(k, s)\| \|R(s, y, \omega)\| \right) \\
&\quad \leq \|\Phi(k, s)\| \left(\|\Phi(k, k_0)x_0\|^2 + \|R(s, y, \omega)\|^2 \right.
\end{aligned}$$

$$
\begin{aligned}
&+\frac{1}{2}\sum_{r=k_0+1}^{k}\|\Phi(k,r)\|(\|R(r,y,\omega)\|^2+\|R(s,y,\omega)\|^2)\Bigg)\\
&\le \eta(k-s)\Bigg(\|x(k,k_0,x_0)\|^2+\bar{\lambda}^2(s)\|y(s,\omega)\|^2\\
&+\frac{1}{2}\sum_{r=k_0+1}^{k}\|\Phi(k,r)\|(\bar{\lambda}^2(r)\|y(r,\omega)\|^2+\bar{\lambda}^2(s)\|y(s,\omega)\|^2)\Bigg)\\
&\le \eta(k-s)(V(k_0,x(k,\omega),\omega)+\bar{\lambda}^2(s)V(s,y(s,\omega),\omega)\\
&+\frac{1}{2}\sum_{r=k_0+1}^{k}\eta(k-r)(\bar{\lambda}^2(r)V(r,y(r,\omega),\omega)\\
&+\bar{\lambda}^2(s)V(s,y(s,\omega),\omega))\\
&\le \eta(k-s)(V(k_0,x(k,\omega),\omega)\\
&+\bar{\lambda}^2(s)\left[1+\frac{1}{2}\sum_{r=k_0+1}^{k}\eta(k-r)\right]V(s,y(s,\omega),\omega)\\
&+\frac{1}{2}\sum_{r=k_0+1}^{k}\eta(k-r)\bar{\lambda}^2(r)V(r,y(r,\omega),\omega))\\
&\le \eta(k-s)\Bigg[V(k_0,x(k,\omega),\omega)+\lambda_1(s,\omega)V(s,y(s,\omega),\omega)\\
&\qquad+\sum_{r=k_0+1}^{k}\lambda_2(r,\omega)\eta(k-r)V(r,y(r,\omega),\omega)\Bigg],
\end{aligned}
\tag{3.1.31}
$$

where $\lambda_1(s,\omega)=\bar{\lambda}^2(s)(1+(1/2)\bar{\eta})$ and $\lambda_2(s,\omega)=(1/2)\bar{\lambda}^2(s)$, and it is assumed that $\sum_{s=\omega}^{\infty}\eta(s)=\bar{\eta}$. Further,

$$
\begin{aligned}
V(k_0,x(k,\omega),\omega)&=\|x(k,\omega)\|^2=\|\Phi(k,k_0)\|^2\|y_0(\omega)\|^2\\
&\le[\eta(k-k_0)]^2\|y_0(\omega)\|^2.
\end{aligned}
\tag{3.1.32}
$$

We assume that

$$\lim_{k\to\infty}\left[\sum_{s=k_0+1}^{k}[\eta(k-s)\eta(s-k_0)]^2\right]=0. \tag{3.1.33}$$

Furthermore, we assume that

$$E\left[\exp\left[\sum_{s=k_0+1}^{k}[\bar{\lambda}(s)]^2\right]\right]$$

is bounded for $k \geq k_0$. From (3.1.31) and (3.1.32), we see that all the hypotheses of Theorem 3.1.3 are satisfied with $\beta(k-k_0) = [\eta(k-k_0)]^2$, $\mu = \eta$. Therefore, the trivial solution of (1.1.55) is asymptotically stable in the qth moment with $q = 2$.

3.1.2 *Stability analysis: Ito–Doob-type stochastic iterative systems — II*

In this section, we present some stability criteria that assure the stability in the qth moment of the trivial solution processes of (2.1.1). Furthermore, some illustrations are given to show that the stability conditions are connected with the influence of random environmental perturbations on the system. Examples are worked out to exhibit the advantages of the joint stability concepts.

Theorem 3.1.4. *Let the hypotheses of Theorem* 2.2.1 *be satisfied. Further, assume that* $F(k,0) \equiv 0$, $f(k,0,\omega) \equiv 0$, $\sigma(k,0,\omega) \equiv 0$ *and* $G(k,0,\omega) \equiv 0$ *w.p.* 1, *and for* $(k,y) \in I(k_0+1) \times R^n$,

$$b(\|y_k\|^q) \leq \sum_{i=1}^{m} V_i(k,y,\omega) \leq a(\|y_k\|^q) \tag{3.1.34}$$

whenever $b \in \mathcal{VK}$, $a \in \mathcal{CK}$ *and* $q \geq 1$. *Then,*

(a) (DJM_1) *of* (2.2.1) *and* (2.1.2) *implies* $(DSM1)$ *of* (2.1.1), *and*
(b) $(DJM2)$ *of* (2.2.1) *and* (2.1.2) *implies* $(DSM2)$ *of* (2.1.1).

Proof. Let $\epsilon > 0$, $k_0 \in I(k_0)$ be given. Assume that (DJM_1) holds. Then, for $b(\epsilon) > 0$ and $k_0 \in I(k_0)$, there exists a $\delta = \delta(\epsilon, k_0)$ such that $\|y_0\|_p \leq \delta$ implies

$$\sum_{i=1}^{m} E[\nu_i(k, k_0, y_0(\omega), \omega)] < b(\epsilon^q), \quad k \geq k_0, \tag{3.1.35}$$

where

$$\nu(k, k_0, y_0(\omega), \omega) = r(k, k_0, V(k, k_0, y_0(\omega), \omega), \omega), \tag{3.1.36}$$

$r(k, k_0, u_0, \omega)$ is the maximal solution process of (2.2.1) and $m(k, k_0, y_0(\omega))$ is the solution process of (2.1.2) through $(k_0, y_0(\omega))$. Now, we claim that if $\|y_0(\omega)\|_q \leq \delta$, then $\|y(k, \omega)\|_q < \epsilon$, $k \geq k_0$. Suppose that this is false. Then, there would exist a solution process $y(k, k_0, y_0(\omega), \omega)$ with $\|y_0(\omega)\|_q \leq \delta$ and a $k_1 > k_0$ such that

$$\|y(k_1, \omega)\|_q = \epsilon \quad \text{and} \quad \|y(k, \omega)\|_q \leq \epsilon, \quad k_0 \leq k \leq k_1. \tag{3.1.37}$$

On the other hand, by Theorem 2.2.1 in the context of Remark 2.2.1, we have

$$V(k, y(k, \omega), \omega) \leq r(k, k_0, V(k_0, m(k, \omega), \omega), \omega), \quad k \geq k_0. \tag{3.1.38}$$

From (3.1.34), (3.1.36), (3.1.38) and using the convexity of b, we obtain

$$\begin{aligned} b(E\|y(k, \omega)\|^p) &\leq \sum_{i=1}^{m} EV_i(k, y(k, \omega), \omega) \\ &\leq \sum_{i=1}^{m} \nu_i(k, k_0, y_0(\omega), \omega), \end{aligned} \tag{3.1.39}$$

for $k \geq k_0$. Equations (3.1.35), (3.1.37) and (3.1.39) lead to the contradiction

$$\begin{aligned} b(\epsilon^p) &\leq \sum_{i=1}^{m} E[V_i(k_1, y(k_1, \omega), \omega)] \\ &\leq \sum_{i=1}^{m} \nu_i(k_1, k_0, y_0(\omega), \omega) < b(\epsilon^p). \end{aligned}$$

This proves (DSM_1). The proof of (DSM_2) can be constructed analogously. This completes the proof of the theorem.

The following example illustrates the scope and usefulness of the joint stability concept and Theorem 3.1.4.

Example 3.1.4. Let's consider

$$\Delta y(k) = -\frac{k}{k+1} y(k) + \sigma(k, y(k), \omega) z(k, \omega), \quad y(k_0) = y_0, \tag{3.1.40}$$

and

$$\Delta x(k) = -\frac{k}{k+1} x(k), \quad x_0 = y_0, \tag{3.1.41}$$

where σ satisfies all the assumptions in Example 2.2.1. We note that

$$m(k, y, p) = \frac{(p-1)!}{(k+1)!} y, \quad k \geq p \geq k_0. \tag{3.1.42}$$

We further assume that $\sigma(k, 0, \omega) \equiv 0$ w.p. 1. Furthermore, let λ in (2.2.35) satisfy

$$E\left[\prod_{p=k_0}^{k-1} (1 + \lambda(p, \omega))\right] \leq M, \tag{3.1.43}$$

where M is a positive number and $\|y_0(\omega)\|^2 < \infty$. It is clear that

$$\nu(k, k_0, y_0(\omega), \omega) = |m(k, \omega)|^2 E\left[\prod_{p=k_0}^{k-1} (1 + \lambda(p, \omega))\right].$$

With this, together with the assumptions about σ, $y_0(\omega)$ and $\lambda(k, \omega)$, it follows that the trivial solution processes $m \equiv 0$ and $u \equiv 0$ of (3.1.42) and (2.2.35) are jointly stable in the mean.

Moreover, from the conditions on λ and $y_0(\omega)$, one can conclude that $m \equiv 0$ and $u \equiv 0$ of (3.1.42) and (2.2.35) are jointly asymptotically stable in the mean. From this and with an application of Theorem 2.2.1, one can conclude that the trivial solution process of (3.1.40) is asymptotically stable in the second moment. We further remark that the trivial solution of (2.2.32) is unbounded. The asymptotic stability in the second moment is generated by the variational comparison method.

Example 3.1.5. Consider Example 2.2.6 with $\sigma(k, y, \omega) \equiv 0$. For $\alpha(p, \omega) > 0$ and $\lambda(p, \omega) > 0$, it is clear that inequality (2.2.70) can be written as

$$E[y^2(k,\omega) \mid \mathcal{F}_{k_0}] \leq \prod_{p=k_0}^{k-1} (1 - \alpha(p,\omega)) y_0^2 \prod_{p=k_0}^{k-1} \times (1 - \alpha(p,\omega))(1 + \lambda(p,\omega)). \quad (3.1.44)$$

Moreover, one can find the following estimates:

(i) For given $\alpha(p, \omega) \in (\frac{1}{2}, 1)$ and any $\lambda(p, \omega)$ satisfying $\lambda(p, \omega) \leq \frac{2\alpha(p,\omega)-1}{1-\alpha(p,\omega)}$, (3.1.44) reduces to

$$E[|y(k,\omega)|^2 \mid \mathcal{F}_{k_0}] \leq E[y_0^2] \exp\left[-\sum_{p=k_0}^{k-1} \left(3 + \lambda(p,\omega) - \frac{1 + \lambda(p,\omega)}{\alpha(p,\omega)} \right) \right]. \quad (3.1.45)$$

Moreover, the trivial solution of (2.2.63) is asymptotically mean-square stable.

(ii) For given $\lambda(p, \omega) > 0$ and any $\alpha(p, \omega)$ satisfying $\alpha(p, \omega) > \frac{1+\lambda(p,\omega)}{2+\lambda(p,\omega)}$, (3.1.44) becomes

$$E[|y(k,\omega)|^2 \mid \mathcal{F}_{k_0}] \leq E[y_0^2] \exp\left[-\sum_{p=k_0}^{k-1} \left(3 - \frac{1 + \alpha(p,\omega)}{\alpha(p,\omega)} + \lambda(p,\omega) \right) \right].$$

Moreover, $y(k, \omega) \equiv 0$ is mean-square asymptotically stable.

(iii) For given $\alpha(p, \omega) \in (0, 1)$ and $1 + \alpha(p, \omega) > \lambda(p, \omega) > \frac{1-\alpha(p,\omega)}{\alpha(p,\omega)}$, (3.1.44) can be reduced to

$$E\left[|y(k,\omega)|^2 \mid \mathcal{F}_{k_0}\right] \leq E[y_0^2] \exp\left[-\sum_{p=k_0}^{k-1} \left(1 - \lambda(p,\omega) + 2\alpha(p,\omega) - \frac{1 + \alpha(p,\omega)}{\lambda(p,\omega)} \right) \right].$$

From this, we conclude that the trivial solution of (2.2.63) is mean-square asymptotically stable.

(iv) For given $\lambda(p,\omega) > 0$ and any $\alpha(p,\omega)$ satisfying $\alpha(p,\omega) > \beta(p,\omega)$, with $\beta(p,\omega) = \max n\{\frac{1}{1+\lambda(p,\omega)}, \lambda(p,\omega) - 1\}$, (3.1.44) reduces to

$$
\begin{aligned}
&E\left[|y(k,\omega)|^2 \mid \mathcal{F}_{k_0}\right] \\
&\quad \leq E[y_0^2] \exp\left[-\sum_{p=k_0}^{k-1}\left(1 - \lambda(p,\omega) + 2\alpha(p,\omega)\right.\right. \\
&\qquad\qquad \left.\left. - \frac{\alpha(p,\omega)+1}{\lambda(p,\omega)}\right)\right].
\end{aligned}
$$

Furthermore, the trivial solution of (2.2.63) is mean-square asymptotically stable.

Remark 3.1.8. Example 2.3.2 shows that for $\alpha(p,\omega) \in (0,1)$, the trivial solution process $m \equiv 0$ of (2.2.76) is stable (exponentially) in the mean-square sense. However, the joint stability is guaranteed by Theorem 3.1.4 under the parametric conditions in (i)–(iv). Moreover, under the parametric conditions in (i) and (ii), the exponential stability in the mean-square sense is tolerated by the magnitude of environmental perturbations characterized by the parameter $\lambda(p,\omega)$, $p \geq k_0$. Under the parametric conditions (iii) and (iv), the asymptotic stability in the mean-square sense is guaranteed under the known magnitude of random perturbations.

In the following, we illustrate the use of the variation of constants method to study the qth moment stability of the trivial solution processes of (2.1.1) and (2.1.29).

Theorem 3.1.5. *Let the hypothesis of Theorem 2.1.2 be satisfied. Further, assume that V, $x(k,\omega)$, $\Phi(k,s,y(k,\omega))$, $R(k,y(k,\omega),\omega)$ and $F(k,y)$ satisfy*

(D1) $b(\|x\|^q) \leq \|V(k,x,\omega)\| \leq a(\|x\|^q)$, *for* $(k,x) \in I(k_0+1) \times R^n$, *where* $b \in \mathcal{VK}$ *and* $a \in \mathcal{CK}$.

(D2) $F(k,0) \equiv 0$, $f(k,0,\omega)$, *and* $R(k,0,\omega) \equiv 0$ *w.p.* 1, *for* $k \in I(k_0+1)$.

(D3) $\|E[D^*W(k,s,y,\omega) \mid \mathcal{F}_{k-1}]\| \le \lambda(s,\omega)\|V(s-1,y,\omega)\|$, *for* $k_0 \le s \le k$, $\|y\|^q < \rho$, *where* ρ *is some positive real,* $D^*W(k,s,y,\omega)$ *is defined by*

$$D^*W(k,s,y,\omega) = W(k,k_0,x(k,k_0,y(k,\omega)),\omega) \times \psi^{-1}(s,k_0,p(s,\omega),p(s-1,\omega))R(s,y,\omega), \tag{3.1.46}$$

where $p(s,\omega)$ *is as defined in* (2.1.20), $\lambda \in M[I(k_0), R(\Omega, R_+)]$ *and satisfies the relation*

$$E\left[\exp \sum_{s=k_0+1}^{k} \lambda(s,\omega)\right] \le \exp\left[\sum_{k=k_0+1}^{k} \hat{\lambda}(s)\right], \tag{3.1.47}$$

where λ *and* y_0 *are independent random variables and* $\sum_{k=1}^{\infty} \hat{\lambda}(k) < \infty$.

(D4) $\|V(k_0,x(k,\omega),\omega)\| \le \alpha[\|y_0\|^q]$ *whenever* $E[\|y_0\|^q] < \rho$, *for some* $\rho > 0$ *and where* $\alpha \in \mathcal{CK}$.

Then, the trivial solution process of (2.1.1) *is stable in the qth mean.*

Proof. Let $y(k,\omega)$ be a sample solution process of (2.1.1), and let $x(k,s,y(s,\omega))$ and $x(k,\omega) = x(k,k_0,y_0(\omega))$ be the sample solution processes of (2.1.3) through $(s,y(s))$ and $(k_0,y_0(\omega))$, respectively, for $k_0 \le s \le k$ and $k_0 \in I(k_0)$. From hypothesis (D2), we have $x(k,k_0,0) \equiv 0$ and $y(k,k_0,0,\omega) \equiv 0$ w.p. 1. From Theorem 2.1.2 and conditions (D3) and (D4), we have

$$E[\|V(k,y(k,\omega),\omega)\|] \le a(E[\|y_0\|^q]) + \sum_{s=k_0+1}^{k} \lambda(s-1)E[\|V(s,y(s,\omega),\omega)\|] \tag{3.1.48}$$

if $E[\|y(s,\omega)\|^q] < \rho$. By setting $m(k) = E[\|V(k,y(k,\omega),\omega)\|]$ and using Lemma A.1.2, (3.1.48) reduces to

$$\Delta R(k) \le \lambda(k-1)R(k), \tag{3.1.49}$$

where

$$R(k) = \sum_{s=k_0+1}^{k} \lambda(s-1)m(s,\omega) + a(E[\|y_0\|^p])$$

as long as $E[\|y(s,\omega)\|^q] < \rho$, for $k_0 \leq s \leq k$. This, together with an application of Corollary 1.2.2, yields

$$R(k) \leq a(E[\|y_0(\omega)\|^q]) \exp \sum_{s=k_0+1}^{k} \lambda(s-1),$$

and hence

$$E[\|V(k, y(k,\omega), \omega)\|] \leq a(E[|y_0(\omega)\|^q]) \exp \sum_{s=k_0+1}^{k} \lambda(s-1) \quad (3.1.50)$$

as long as $E[\|y(k,\omega)\|^q] < \rho$, for $k_0 \leq s \leq k$. From the nature of α, $y_0(\omega)$ and (3.1.47), (3.1.50) reduces to

$$E\left[\|V(k, y(k,\omega), \omega)\|\right] \leq a\left(E[\|y_0(\omega)\|^q]\right) \exp \sum_{s=k_0+1}^{k} \hat{\lambda}(s)$$

as long as $E[\|y(s,\omega)\|^q] < \rho$, for $k_0 \leq s \leq k$. From this, condition (D1) and the property of $\hat{\lambda}$, we have

$$b(E[\|y(k,\omega)\|^q]) \leq Ka(E[\|y_0\|^q]) \quad (3.1.51)$$

as long as $E[\|y(k,\omega)\|^q] \leq \rho$, for some positive number $K, k \geq k_0$. To conclude $E[\|y(k,\omega)\|^q] \leq \rho$, for all $k \geq k_0$ and stability property (DSM1) of (1.1.1), first we pick $y_0(\omega)$ such that

$$K\alpha(E[\|y_0(\omega)\|^q]) < b(\rho).$$

For this choice of y_0, we claim that $E[\|y(k,\omega)\|^q] < \rho$, for all $k \geq k_0$.

If this claim is false, then there exists $\bar{k} \in I(k_0)$ such that $\bar{k} > k_0$, $E[\|y(s,\omega)\|^q] < \rho$, for $k_0 \leq s \leq \bar{k}$, and $E[\|y(\bar{k},\omega)\|^q] = \rho$. For $k \in$

$[k_0, \bar{k}]$, (3.1.51) is valid, and hence using condition (D1) and (3.1.51), we get

$$b(\rho) \leq b(E[\|y(\bar{k}, \omega)\|^q]) < b(\rho).$$

This contradiction establishes the claim that $E[\|y(k,\omega)\|^q] < \rho$. Hence, (3.1.51) is valid for all $k \geq k_0$. To prove (DSM1) of (1.1.1), we pick any $\epsilon > 0$ and choose $y_0(\omega)$ such that

$$K_\alpha(E[\|y_0(\omega)\|^q]) < b(\epsilon^q). \tag{3.1.52}$$

This implies that

$$E[\|y_0(\omega)\|^q] < \alpha^{-1}\left(\frac{1}{K}b(\epsilon^q)\right).$$

From (3.1.51) and (3.1.52), we have

$$E(\|y(k,\omega)\|^q) < \epsilon^q, \quad k \geq k_0,$$

whenever

$$(E[\|y_0(\omega)\|^q])^{1/q} < \delta(\epsilon),$$

where

$$\delta(\epsilon) = \left(\alpha^{-1}\left(\frac{1}{K}b(\epsilon^q)\right)\right)^{1/q} > 0.$$

This is based on the property of the function b, α, and α^{-1}. This completes the proof of the theorem.

In the following theorem, we present sufficient conditions for the qth moment asymptotic stability of the trivial solution of (1.1.1).

Theorem 3.1.6. *Assume that the hypotheses of Theorem* 3.1.5 *hold, except that (D3) and (D4) are replaced by*

(D5)

$$\begin{aligned}\|E[D^*W(k,s,y,\omega)]\| &\leq \mu(k-s)E[\|V(k_0,x(k,\omega),\omega)\|]\\ &+ \sum_{r=k_0+1}^{k} \lambda_2(r)\eta(k-r)E[\|V(r-1,y(r-1,\omega),\omega)\|]\\ &+ \lambda_1(s)\eta(k-s)E[\|V(s-1,y,\omega)\|],\end{aligned} \tag{3.1.53}$$

for $k_0 \leq s \leq k$, $ky^q \leq \rho$, for some $\rho > 0$, and

(D6)

$$E[\|V(k_0,x(k,\omega),\omega)\|] \leq \alpha(E[\|y_0(\omega)\|^q])\beta(k-k_0), \quad k \geq k_0,$$

provided $E[\|y_0(\omega)\|^q] \le \rho$, where μ, η and β are nonnegative sequences defined on $I(k_0)$ and satisfy the following relation:

$$\lim_{k\to\infty} \sum_{s=k_0+1}^{k} \eta(k-s)\beta(s-k_0) = 0, \tag{3.1.54}$$

$\lim_{r\to\infty}\beta(r) = 0$, μ is summable on $I(k_0)$ with a sum $\bar{\mu}$, η is bounded on $I(k_0)$, and λ defined by $\lambda = \lambda_1 + \lambda_2\bar{\mu}$ satisfies all the conditions in (3.1.47) with $q = 2$. Then, the trivial solution process of (2.1.1) is asymptotically stable in the qth moment.

Proof. By employing conditions (D5) and (D6) and following the proof of Theorem 3.1.5, we arrive at

$$\begin{aligned}
E[\|V(k,y(k,\omega),\omega)\|] &\le \alpha(E[\|y_0(\omega)\|^q])\beta(k-k_0) \\
&\quad + \sum_{s=k_0+1}^{k} (\lambda_1(s)\eta(k-s)E[\|V(s-1,y(s-1,\omega),\omega)\|] \\
&\quad + \mu(k-s)E[\|V(k_0,x(k,\omega),\omega)\|] + \sum_{s=k_0+1}^{k} \mu(k-s) \\
&\quad \times \left(\sum_{r=k_0+1}^{k} \lambda_2(r)\eta(k-r)E[\|V(r-1,y(r-1,\omega),\omega)\|] \right)
\end{aligned} \tag{3.1.55}$$

as long as $E\|y(k,\omega)\|^q \le \rho$. By setting

$$\begin{aligned}
m(k) &= E[\|V(k,y(k,\omega),\omega)\|], \\
n(k) &= \alpha(E\|y_0(\omega)\|^q)\beta(k-k_0)(1+\bar{\mu})
\end{aligned}$$

and

$$\nu(k,s) = (\lambda_1(s) + \bar{\mu}\lambda_2(s))\eta(k-s),$$

(3.1.55) can be rewritten as

$$m(k) \le n(k) + \sum_{s=k_0+1}^{k} \nu(k,s)n(s-1)$$

as long as $E\|y(k,\omega)\|^q \leq \rho$, where $\bar{u} = \sum_{s=k_0+1}^{k} \mu(k-s)$. Using the estimate in Lemma A.1.3, we get

$$m(k) \leq n(k) + \sum_{s=k_0+1}^{k} \nu(k,s)n(s)\left[\exp\left(\sum_{\tau=s+1}^{k} \nu(k,\tau)\right)\right]. \quad (3.1.56)$$

From the definitions of $\nu(k,s), m(k)$ and $n(k)$, (3.1.56) becomes

$$\begin{aligned} E[\|V(k,y(k,\omega),\omega)\|] &\leq \alpha(E[\|y_0(\omega)\|^q])(1+\bar{\mu}) \\ &\times \left[\beta(k-k_0) + \exp\left(\sum_{s=k_0+1}^{k} \nu(k,s)\right)\right. \\ &\left.\times \sum_{s=k_0+1}^{k} \eta(k-s)\beta(s-k_0)\lambda(s)\right] \end{aligned} \quad (3.1.57)$$

as long as $E\|y\|^q < \rho$, where $\lambda(k) = [\lambda_1(k) + \bar{\mu}\lambda_2(k)]$. From (3.1.57), condition (D1) and the properties of functions α, λ and the Schwartz inequality, we obtain

$$b(E\|y(k,\omega)\|^q) \leq \alpha(E\|y_0(\omega)\|^q)(1+\bar{\mu})\beta(k-k_0) \quad (3.1.58)$$

as long as $E\|y(k,\omega)\|^q \leq \rho$. This is guaranteed by (3.1.47). From (3.1.58) and the properties of the underlying function, one can easily conclude that $E\|y(k,\omega)\|^q \leq \rho$, for all $k \geq k_0$. Hence, property (DSM_1) of the trivial solution of (1.1.1) can be concluded by following the argument of Theorem 3.1.5. To conclude (DSM_2), it is obvious from (3.1.58) and the nature of β and (3.1.54), that $b(E\|y(k,\omega)\|^q)$ tends to zero as $k \to \infty$. Hence, one can manipulate to verify the technical definitions of (DSM_2). The details are left to the reader. This completes the proof of the theorem.

Remark 3.1.9. We note that Theorems 3.1.5 and 3.1.6 also provide sufficient conditions for the qth mean stability of the trivial solution processes of (2.1.1). This fact follows from relations (3.1.50) and (3.1.57).

Remark 3.1.10. The conditions (D4) and (D6) in Theorems 3.1.5 and 3.1.6 imply uniform stability and asymptotic stability, respectively, in the qth moment of the trivial solution process of (2.1.3).

To appreciate the assumptions of Theorem 3.1.6, we present the following result, which is applicable to other problems.

Corollary 3.1.3. *Let the hypothesis of Theorem* 3.1.6 *be satisfied, except that the relations* (3.1.47) *and* (3.1.47) *are replaced by*

$$\eta(k-s)\beta(s-1-k_0) \leq K\beta(k-k_0), \quad k \geq k_0, \tag{3.1.59}$$

and

$$\lim_{k\to\infty} \beta(k-k_0)E\left[\exp\left(K\sum_{s=k_0+1}^{k}\lambda(s,\omega)\right)\right] = 0, \tag{3.1.60}$$

respectively, where K *is some positive constant. Then, the trivial solution* (2.1.1) *is asymptotically stable in the qth moment.*

Proof. By following the proof of Theorem 3.1.6, we arrive at (3.1.55). Now, by using (3.1.59), (3.1.55) can be rewritten as

$$m(k) \leq \alpha(E[\|y_0(\omega)\|^q])(1+\bar{\mu}) + \sum_{s=k_0+1}^{k}\nu(k,s)m(s-1) \tag{3.1.61}$$

as long as $E\|y(k,\omega)\|^q < \rho$, where

$$m(k) = E\|V(k,y(k,\omega),\omega)\|]/\beta(k-k0)$$

and

$$\nu(k,s) = (\lambda_1(s)+\bar{\mu}\lambda_2(s))K\beta(k-k_0).$$

By applying Lemma A.1.2 to (3.1.61), we get

$$m(k) \leq (1+\bar{\mu})\alpha(E[\|y_0(\omega)\|^p])\exp\left(\sum_{s=k_0+1}^{k}\nu(k,s)\right),$$

which implies

$$b(E\|y(k,\omega)\|^q) \leq (1+\bar{\mu})\alpha(E[\|y_0(\omega)\|^q])\beta(k-k_0)E$$
$$\times\left[\exp\sum_{s=k_0+1}^{k}\nu(k,s)\right]. \tag{3.1.62}$$

The rest of the proof can be completed by using the argument in the proof of Theorem 3.1.6. Thus, the proof of the corollary is complete.

Remark 3.1.11. If $\eta(r) = \beta(r) = \exp(-\gamma r)$, for some $\gamma > 0$, then these functions are admissible in Corollary 3.1.3.

Remark 3.1.12. For the qth moment asymptotic stability, condition (D5) can be further modified. This is the content of the following corollary.

Corollary 3.1.4. *Suppose that all the conditions of Theorem 3.1.6 hold, except that condition (D5) is replaced by*

(D5*)

$$\begin{aligned}\|E[\mathcal{D}^*W(k,s,y,\omega)\|] &\le \mu(k-s)E[\|V(k_0,x(k,\omega),\omega)\|] \\ &\quad + \sum_{r=k_0+1}^{k} \eta(k-r)E[\|V(r-1,y(r-1,\omega),\omega)\|] \\ &\quad + \eta(k-s)E[\|V(s-1,y,\omega)\|],\end{aligned}$$

for $k_0 \le s \le k$ *and* $\|y\|^q < \rho$. *Further, assume that*

$$\sum_{u=s+1}^{k} \eta(k-u) \le K, \quad \text{for } k_0 \le s \le k,$$

for some $K > 0$. *Then, the conclusion of Theorem 3.1.6 remains valid.*

We discuss an example to illustrate the fruitfulness of our result.

Example 3.1.6. Consider Example 2.1.1. We assume that the solution process $x(k,k_0,y_0(\omega)) = x(k,\omega) = \Phi(k,k_0)y_0(\omega)$ fulfills the following relation:

$$\|\Phi(k,s)\| \le \exp[-\alpha(t-s)], \text{ for } t \ge s \ge t_0. \tag{3.1.63}$$

Let $V(t,x) = \|x\|^2$. From Corollary 2.1.2(vi) and imitating the argument used in Example 2.2.3, we have

$$\begin{aligned}LV(p,m(k,p,y),\omega) &= 2y^T\Phi^T(k,p)\Phi(k,p)(A(p,\omega)-\hat{A}(p))y \\ &\quad + y^T\Phi(k,p)(A(p,\omega)-\hat{A}(p))^T \\ &\quad \times (A(p,\omega)-\hat{A}(p))\Phi(k,p)y\end{aligned}$$

$$+\operatorname{tr}(\Phi^T(k,p)\sigma(p,y,\omega)\sigma^T(p,y,\omega)\Phi(k,p))$$
$$\leq \nu(p,\omega)V(p,m(k,p,y),\omega), \tag{3.1.64}$$

where $\nu(p,\omega) = -2\lambda(p,\omega) + M(p,\omega) + \gamma(p,\omega)$ and $\gamma(p,\omega)$, $M(p,\omega)$ and $\lambda(p,\omega)$ are determined by the following:

$$\begin{cases} \operatorname{tr}(\Phi(k,p)\sigma(p,y,\omega)\sigma^T(p,y,\omega)\Phi(k,p)) \leq \gamma(p,\omega)V(p,m(k,p,y)), \\ y^T(A(p,\omega)-\hat{A}(p))^T(A(p,\omega)-\hat{A}(p))y \leq M(p,\omega)V(p,m(k,p,y)), \\ y^T\Phi^T(k,p)\Phi(k,p)(A(p,\omega)-\hat{A}(p))y \\ \qquad \leq -\lambda(p,\omega)V(p,m(k,m(k,p,y))). \end{cases}$$

In this case, the comparison equation is (2.2.52). Thus, we obtain

$$E[\|y(k,k_0,y_0)\|^2] \leq \|\Phi(k,k_0)y_0\|^2 \exp\left[\sum_{s=k_0+1}^{k} V(s,\omega)\right].$$

The trivial solution process of (2.1.32) is mean-square stable if $\lim_{k\to\infty}\sum_{s=k_0+1}^{k}\gamma(s,\omega) \to -\infty$ a.s.

3.2 Error Estimates

3.2.1 *Error estimates: Iterative systems under random parameters — I*

In this section, we develop error estimates on the pth moment derivation of a solution process of (1.1.1) with the solution processes of (1.1.60), in particular (1.1.2) and (1.1.3). We employ the comparison method and the variation of constants method to derive the error estimates.

Now, we present a few error estimate results by employing the comparison theorems.

Theorem 3.2.1. *Let the hypothesis of Theorem* 1.2.5 *be satisfied. Further, assume that*

$$b(\|x\|^p) \leq \sum_{i=1}^{m} V_i(k,x,\omega), \tag{3.2.1}$$

where $b \in \mathcal{CK}$, and it is a convex function with $p \geq 1$. Then,

$$b(E(\|y(k,\omega) - w(k)\|^p) \leq \sum_{i=1}^{m} E(r_i(k, k_0, V(k_0, m(k, k_0, y_0 - w_0), \omega), \omega)), \quad k \geq k_0, \tag{3.2.2}$$

where $w(k) \equiv w(k, k_0, w_0)$ is the solution process of (1.1.60).

Proof. By choosing $u_0 = V(k_0, m(k) - w(k), \omega)$, (1.2.83) reduces to

$$\sum_{i=1}^{m} V_i(k, y(k,\omega) - w(k), \omega) \leq \sum_{i=1}^{m} r_i(k, k_0, V(k_0, m(k) - w(k), \omega), \omega).$$

This, together with (3.2.1) and the convexity of b, yields

$$b(E[\|y(k,\omega) - w(k)\|]^p) \leq \sum_{i=1}^{m} E[r_i(k, k_0, V(k_0, m(k, k_0, y_0 - w_0), \omega)].$$

This completes the proof of the theorem.

Example 3.2.1. From Example 1.2.6 and Theorem 3.2.1 with $h(k, x, \omega) = F(k, x, \omega)$ in (1.1.3), we have

$$E[\|y(k,\omega) - \bar{x}(k)\|^2] \leq E\left[\|m(k, k_0, m_0 - z_0)\|^2 \times \exp\left[\sum_{s=k_0}^{k-1} (\mu(s,\omega) + \nu(s,\omega))\right]\right]. \tag{3.2.3}$$

We further assume that

$$\|\Phi(k, k_0, e)\|^2 \leq \exp\left[\sum_{s=k_0}^{k-1} (\alpha_p(s) - 1)\right]. \tag{3.2.4}$$

This implies

$$\|m(k, k_0, y_0 - z_0)\|^2 \leq \|y_0(\omega) - z_0\|^2 \exp\left[\sum_{s=k_0}^{k-1} (\alpha_p(s) - 1)\right]. \tag{3.2.5}$$

From (3.2.5), (3.2.3) reduces to

$$E(\|y(k,\omega)-\bar{x}(k)\|^2) \le E\left[\|x_0(\omega)-z_0\|^2 \exp\sum_{s=k_0}^{k-1}(\alpha_p(s)+\mu(s,\omega)+\nu(s,\omega)-1)-1\right]. \tag{3.2.6}$$

This, together with the independence of the initial data for the rate coefficient processes, we get

$$E[\|y(k,\omega)-\bar{x}(k)\|^2] \le (E(\|y_0(\omega)-z_0\|^2))E \times \left[\exp\sum_{s=k_0}^{k}(\alpha_p(s)+\mu(s,\omega)+\nu(s,\omega)-1)\right]^2. \tag{3.2.7}$$

We note that, depending on the nature of stochastic processes μ, ν, and random variables $y_0(\omega)$ and z_0, we can obtain specific estimates (3.2.7). The estimate for the absolute mean deviation can be given in terms of the statistical properties of stochastic processes in (3.2.7). Furthermore, from (3.2.7), one can obtain the estimates for the absolute pth mean deviation with $p \ge 1$.

To further illustrate this result, we present another example with regard to the linear systems of difference equations (1.1.55), (1.1.56) and (1.1.57).

Example 3.2.2. From Example 1.2.6 and Theorem 3.2.1, we have

$$E[\|y(k,\omega)-\bar{x}(k)\|^2] \le E\left[\|x(k,\omega)-\bar{x}(k)\|^2 \exp\left[\sum_{s=k_0}^{k}(\mu(s,\omega)+\nu(s,\omega))\right]\right]. \tag{3.2.8}$$

The linearity of the system gives

$$\|\Phi(k,s)\| \le \exp\left[\sum_{\tau=s}^{k}(\alpha_p(\tau)-1)\right]. \tag{3.2.9}$$

(3.2.8) together with (3.2.9) gives

$$E[\|y(k,\omega) - \bar{x}(k)\|^2] \\ \leq E\left[\|y_0(\omega) - z_0\|^2 \exp\left[\sum_{s=k_0}^{k-1} (\alpha_p(s) + (\mu(s,\omega) + \nu(s,\omega) - 1)\right]\right]. \tag{3.2.10}$$

From this, again depending on the distribution of the processes α, β, $y_0(\omega)$ and z_0, one can obtain the estimates for the absolute mean deviation of the solution process $y(k,\omega)$ of (1.1.55) with either (1.1.56) or (1.1.57). For example, if z_0, $y_0(\omega)$ and $\alpha(s,\omega)$ are mutually independent, then (3.2.10) reduces to

$$E[\|y(k,\omega) - \bar{x}(k)\|^2] \\ \leq E[\|y_0(\omega) - z_0\|^2]E\left[\exp\sum_{s=k_0}^{k-1} (\alpha_p(s) + \mu(s,\omega) + \nu(s,\omega) - 1)\right]. \tag{3.2.11}$$

If we further assume that $\alpha(s,\omega)$ is a Gaussian process with mean zero and covariance $C(k,s)$ and

$$K(k,s) = E\left(\sum_{s=k_0}^{k-1} \beta(s,\omega) \exp\left[\sum_{\tau=s}^{k-1} \mu(s,\omega) + \nu(s,\omega) - 1\right]\right), \tag{3.2.12}$$

then (3.2.11) reduces to

$$E[\|y(k,\omega) - \bar{x}(k)\|^2] \\ \leq E[\|y_0(\omega) - z_0\|^2] \exp\left[\sum_{s=k_0}^{k-1} (\alpha_p(s) + C(k,s) - 2)\right]. \tag{3.2.13}$$

Next, we present a few error estimate results by employing the method of variation of parameters.

Theorem 3.2.2. *Suppose that all the hypotheses of Theorem* 1.1.4 *hold. Further, assume the following:*

(i)

$$b(\|x\|^p) \leq \sum_{i=1}^{m} V_i(k, x, \omega) \leq a(\|x\|^p), \tag{3.2.14}$$

and

(ii)

$$\sum_{i=1}^{m} \|\mathcal{D}^* W_i(k, s, y, z, \omega)\| \leq g_1(s, \omega) C(\|y - z\|) + g_2(s, \|z\|, \omega), \tag{3.2.15}$$

where $\mathcal{D}^* W_i(k, s, y, z, \omega)$ *is the ith component of*

$$\begin{aligned} &\mathcal{D}^* W(k, s, y, z, \omega) \\ &\quad = W(k, k_0, x(k, k_0, p(k, w)) - x(k, k_0, z_0), \omega)\psi^{-1} \\ &\qquad \times (s, k_0, p(s, \omega), p(s-1, \omega))R(s, y, \omega), \end{aligned}$$

where ψ *and* W *are as defined in* (1.1.40) *and* (1.1.42), *respectively;* $a \in \mathcal{CK}$, $b \in \mathcal{VK}$, $c \in \mathcal{K}$, b *is invertible;* $g_1 \in M[I(k_0), R[\Omega, R_+]]$, $g_2 \in M[I(k_0) \times R_+, R[\Omega, R_+]]$ *and* $a(0) = 0 = b(0)$. *Then,*

$$b(\|y(k, \omega) - \bar{x}(k\|^p) \leq a(\|y(k, \omega) - \bar{x}(k)\|^p) + r(k, \omega), \quad k \geq k_0, \tag{3.2.16}$$

where $r(k, \omega)$ *is the solution process of*

$$\Delta u(k, \omega) = G(k, u(k-1, \omega), \omega), \quad u(k_0, \omega) = u_0(\omega) \geq 0 \tag{3.2.17}$$

and

$$\begin{aligned} G(k, u, \omega) &= C((b^{-1}(\|x(k-1, \omega) - \bar{x}(k-1)\|^p + u))^{1/p}) g_1(k, \omega) \\ &\quad + g_2(k, \|\bar{x}(k)\|, \omega); \end{aligned}$$

$y(k, \omega)$ *is the solution process of* (1.1.1); *and* $\bar{x}(k) = x(k, k_0, z_0)$ *is the solution process of* (1.1.2) *or* (1.1.3), *respectively, when* $z_0 = m_0$ *or* $z_0 = y_0(\omega)$. *Moreover,*

$$\begin{aligned} &b(E[\|y(k, \omega) - \bar{x}(k)\|^p] \\ &\quad \leq a(E[\|y(k, \omega) - \bar{x}\|^p]) + E[r(k-1, \omega)], \quad k \geq k_0. \end{aligned} \tag{3.2.18}$$

Proof. Let $y(k,\omega) = y(k,k_0,y_0(\omega))$ and $x(k,\omega) = x(k,k_0,y_0(\omega))$ be solution processes of (1.1.1) through $(k_0,y_0(\omega))$ and $(k_0,x_0(\omega))$, respectively. Let $\bar{x}(k) = x(k,k_0,z_0)$ be the solution as defined in the theorem, where z_0 is either m_0 or $x_0(\omega)$. From Theorem 1.1.4, (3.2.14) and (3.2.15), we obtain

$$\begin{aligned} b(\|y(k,\omega)-\bar{x}(k)\|^p) &\le a(\|x(k,\omega)-\bar{x}(k)\|^p) \\ &\quad + \sum_{s=k_0+1}^{k} [C(\|y(s-1,\omega)-\bar{x}(s-1)\|)g_1(s,\omega) \\ &\qquad + g_2(s,\|\bar{x}(s-1)\|,\omega)]. \end{aligned} \tag{3.2.19}$$

Set

$$m(k,\omega) = \begin{cases} \sum_{s=k_0+1}^{k}[C(\|y(s-1,\omega)-\bar{x}(s-1))\|)g_1(s,\omega) \\ \quad +g_2(s,\|\bar{x}(s-1)\|,\omega)], & k \ge k_0+1, \\ 0, & \text{for } k = k_0. \end{cases} \tag{3.2.20}$$

Therefore, (3.2.19) reduces to

$$b(\|y(k,\omega)-\bar{x}(k)\|^p) \le a(\|x(k,\omega)-\bar{x}(k)\|^p) + m(k,\omega). \tag{3.2.21}$$

From (3.2.20) and (3.2.21), followed by algebraic simplification, we have

$$\begin{aligned} \Delta m(k,\omega) &= m(k,\omega) - m(k-1,\omega) \\ &= C(\|y(k-1,\omega)-\bar{x}(k-1)\|)g_1(k,\omega) \\ &\quad + g_2(k,\|\bar{x}(k-1)\|,\omega) \\ &\le h(a(\|x(k-1,\omega)-\bar{x}(k-1)\|^p) + m(k-1,\omega))g_1(k,\omega) \\ &\quad + g_2(k,\|\bar{x}(k-1)\|,\omega), \end{aligned} \tag{3.2.22}$$

where $h(s) = C((b^{-1}(s))^{1/p})$. In this case, the comparison equation is

$$\Delta u(k,\omega) = G(k,u(k-1,\omega),\omega), \quad u(k_0,\omega) = u_0(\omega), \tag{3.2.23}$$

where G is as defined in (3.2.17) and is rewritten as $G(k,u,\omega) = h(a(\|x(k-1,\omega) - \bar{x}(k-1)\|^p) + u)\lambda_1(k,\omega) + g_2(k, \|\bar{x}(k-1)\|, \omega)$. From the monotone property of C and b^{-1}, we conclude that the random process G in (3.2.23) satisfies the following relation:

$$G(k,r,\omega) - G(k,u,\omega) \geq -(r-u), \qquad \text{for } r \geq u. \tag{3.2.24}$$

Hence, from (3.2.22), (3.2.23), (3.2.24) and an application of Corollary 1.2.2, we conclude that that

$$m(k,\omega) \leq r(k,\omega), \qquad \text{for } k \geq k_0, \tag{3.2.25}$$

whenever

$$m(k_0,\omega) \leq u_0(\omega).$$

From (3.2.25), (3.2.21) reduces to (3.2.16). The proof of (3.2.18) follows by taking the expectation of both sides of (3.2.16) properties of a and b.

Example 3.2.3. To illustrate the scope of Theorem 3.2.2, we consider systems (1.1.52), (1.1.53) and (1.1.54). We assume that the fundamental solution $\Phi(k,k_0,y_0)$ associated with the variational system of (1.1.52) satisfies the following relation:

$$\|\psi(k,k_0,x,y)\psi^{-1}(s,k_0,x,\bar{y})\| \leq L, \tag{3.2.26}$$

for $k \geq s \geq k_0$ and $x, \bar{x}, y, \bar{y} \in \mathbb{R}^n$, where L is some positive constant and ψ is as defined in (1.1.40). Let us assume that $V(k,x,\omega) = x$ and $R(s,y,\omega)$ satisfies the Lipschitz condition

$$\|R(s,y,\omega) - R(s,z,\omega)\| \leq \lambda_1(s,\omega)\|y-z\| \tag{3.2.27}$$

and the growth condition

$$\|R(s,z,\omega)\| \leq \lambda_2(s,\|z\|,\omega), \tag{3.2.28}$$

where $\lambda_1 \in M[I(k_0), R[\Omega, R_+]]$ and $\lambda_2 \in M[I(k_0) \times R_+, R[\Omega, R_+]]$. Under these conditions,

$$\begin{aligned}
&\mathcal{D}^*W(k,s,y,z,\omega)\\
&\quad = \psi(k,k_0,p(k,\omega),x_0(\omega))\psi^{-1}(s,k_0,p(s,\omega),p(s-1,\omega))R(s,y,\omega)\\
&\quad = \psi(k,k_0,p(k,\omega),x_0(\omega))\psi^{-1}(s,k_0,p(s,\omega),p(s-1,\omega))
\end{aligned}$$

$$\times (R(s,y,\omega) - R(s,z,\omega)) + \psi(k,k_0,p(k,\omega),x_0(\omega))$$
$$\times \psi^{-1}(s,k_0,p(s,\omega),p(s-1,\omega))R(s,z,\omega)$$

satisfies condition (3.2.15) with g_1, g_2 and C defined on L, λ_1, λ_2 and the norms in (3.2.26), (3.2.27) and (3.2.28). Obviously, V satisfies (3.2.14) with a, b and C being linear functions, and $p = 1$ or 2 depending on the norm of x. In this case, the function G in the comparison equation (3.2.17) takes the form

$$G(k,u,\omega) = \begin{cases} (L_1\|x(k-1,\omega) - \overline{x}(k-1)\| + L_2 u)g_1(k,\omega) \\ \quad +g_2(k,k\overline{x}(k-1)k,\omega), & \text{for } p = 1, \\ (L_1^*\|x(k-1,\omega) - \overline{x}(k-1)\|^2 + L_2^* u)^{1/2} g_1(k,\omega) \\ \quad +g_2(k,\|\overline{x}(k-1)\|,\omega), & \text{for } p = 2, \end{cases}$$

where L_1, L_2, L_1^* and L_2^* are some positive constants. For $p = 1$, the solution process of (3.2.17) is

$$r(k,\omega) = \begin{cases} u_0(\omega), & \text{for } k = k_0, \\ u_0(\omega)\prod_{s=k_0+1}^{k}(1 + L_2 g_1(s,\omega)) + \sum_{s=k_0+1}^{k}(L_1 g_1(s,\omega) \\ \quad |x(s-1,\omega) - \overline{x}(s-1)\| + g2(s,\|\overline{x}(s-1),\omega)\|) \\ \quad \prod_{\tau=s+1}^{k}(1 + L_2 g_1(s,\omega)), & \text{for } k \geq k_0 + 1. \end{cases} \tag{3.2.29}$$

Thus, by the application of Theorem 3.2.2, we have $\|y(k,\omega) - \bar{x}(k)\| \leq L_3\|x(k,\omega) - \bar{x}(k)\| + L_4 r(k,\omega), \quad k \geq k_0$, where L_3 and L_4 are some positive constants. Hence,

$$E[\|y(k,\omega) - \bar{x}(k)\|] \leq L_3 E[\|x(k,\omega) - \bar{x}(k)\|] + L_4 E[r(k,\omega)], \quad k \geq k_0, \tag{3.2.30}$$

where $r(k,\omega)$ is as given in (3.2.29) with $p = 1$. A similar discussion for the $p = 2$ case can be given analogously.

Example 3.2.4. We now consider Example 1.1.1. In this case,

$$\psi(k, k_0, x, y)\psi^{-1}(s, k_0, \bar{x}, \bar{y}) = \Phi(k, s) \quad \text{and}$$
$$R(s, y, \omega) = (A(s, \omega) - \hat{A}(s))y.$$

Again, we assume that $V(k, x) = x$ and

$$\|\Phi(k, s)\| \leq L, \quad k \geq s. \tag{3.2.31}$$

From these conditions, it is obvious that conditions (3.2.27) and (3.2.28) are valid with $\lambda_1(s, \omega) = \|A(s, \omega) - \hat{A}(s)\|$ and $\lambda_2(s, \|z\|, \omega) = \|A(s, \omega) - \hat{A}(s)\|\|z\|$. Furthermore, $\mathcal{D}^*W(k, s, y, z, \omega) = \Phi(k, s)$ $R(s, y, \omega)$ satisfies condition (3.2.15). Now, by following the argument of Example 3.2.1, a relation similar to (3.2.30) can be obtained analogously.

The following result provides another sufficient condition for error estimates on the pth moment deviation of a solution process of (1.1.1) with the solution process of either (1.1.2) or (1.1.3).

Theorem 3.2.3. *Let the hypotheses of Theorem 3.2.2 be satisfied, except that (ii) is replaced by*

(ii*)

$$\begin{aligned}
&\sum_{i=1}^{m} |\mathcal{D}^*W_i(k, s, y, z, \omega)| \\
&\quad \leq \eta_1(k-s)\bar{a}(\|x(k, \omega) - \bar{x}(k)\|) + \lambda_1(s, \omega)C_1(\|y - z\|) \\
&\quad\quad + \eta_2(k-s) \sum_{r=k_0+1}^{k} C_2(\|y(r-1, \omega) - \bar{x}(r-1)\| \\
&\quad\quad + \nu_2(r, \|\bar{x}(r-1), \omega))\lambda_2(r, \omega) + \nu_1(s, \|z\|, \omega),
\end{aligned} \tag{3.2.32}$$

where $\bar{a} \in \mathcal{CK}$, $C_1, C_2 \in \mathcal{K}$, $\nu_1, \nu_2 \in M[I(k_0) \times R_+, R[\Omega, R_+]]$, η_1 *and* η_2 *are summable sequences on* $I(k_0)$ *whose sums are less than or equal to* $\bar{\eta} : \lambda_1, \lambda_2 \in M[I(k_0), R[\Omega, R_+]]$.

Then,

$$b(\|y(k,\omega)-\overline{x}(k)\|^p) \le a^*(\|x(k,\omega)-\overline{x}(k)\|)+r(k,\omega), \quad k \ge k_0, \tag{3.2.33}$$

and

$$\begin{aligned} &b(E[\|y(k,\omega)-\overline{x}(k)\|^p]) \\ &\quad \le a^*(E\|x(k,\omega)-\overline{x}(k)\|)+E[r(k,\omega)], \quad k \ge k_0, \end{aligned} \tag{3.2.34}$$

where $a^*(\nu) = a(\nu^p)+\bar{\eta}\bar{a}(\nu)$, $\nu \in R_+$ *and* $r(k,\omega)$ *is the solution process of* (3.2.17) *with*

$$\begin{aligned} G(k,u,\omega) &= \left(\sum_{i=1}^{2} C_i \left(b^{-1}(a^*(\|x(k,\omega)-\overline{x}(k)\|)+u)\right)^{1/p}\right) g_1(k,\omega) \\ &\quad + g_2(k,\|\overline{x}(k-1)\|,\omega), \end{aligned} \tag{3.2.35}$$

$g_1 = \max\{\lambda_1, \bar{\eta}\lambda_2\}$ *and* $g_2 = \nu_1 + \bar{\eta}\lambda_2\nu_2$.

Proof. By following the proof of Theorem 1.5.2 and using (1.5.32), we have

$$\begin{aligned} &b(\|y(k,\omega)-\bar{x}(k)\|^p) \\ &\quad \le a(\|x(k,\omega)-\bar{x}(k)\|^p)+\bar{\eta}\bar{a}(\|x(k,\omega)-\bar{x}(k)\|) \\ &\qquad + \sum_{s=k_0+1}^{k} \lambda_1(s,\omega)C_1(\|y(s-1,\omega)-\bar{x}(s-1)\|) \\ &\qquad + \sum_{s=k_0+1}^{k} \bar{\eta}\lambda_2(s,\omega)C_2(\|y(s-1,\omega)-\bar{x}(s-1)\|) \\ &\qquad + \sum_{s=k_0+1}^{k} \bar{\eta}\lambda_2(s,\omega)\nu_2(s,\|\bar{x}(s-1)\|,\omega) \\ &\qquad + \sum_{s=k_0+1}^{k} \nu_1(s,\|\bar{x}(s-1)\|,\omega). \end{aligned}$$

Hence, it is written as

$$
\begin{aligned}
& b(\|y(k,\omega)-\bar{x}(k)\|^p) \\
& \quad \le a^*(\|x(k,\omega)-\bar{x}(k)\|) \\
& \qquad + \sum_{s=k_0+1}^{k}\left(\sum_{i=1}^{2} C_i(\|y(s-1,\omega)-\bar{x}(s)\|)\right) g_1(s,\omega) \\
& \qquad + \sum_{s=k_0+1}^{k} g_2(s,\|\bar{x}(s-1),\omega\|), \quad k \ge k_0+1, \qquad (3.2.36)
\end{aligned}
$$

where g_1, g_2 and a^* are as defined in the theorem. Now, by imitating the rest of the proof of Theorem 3.2.2, the conclusions of the theorem follow immediately. The details are left to the reader. The proof is complete.

Now, we present an example to illustrate the applicability of Theorem 3.2.3.

Example 3.2.5. We consider systems (1.1.55), (1.1.56) and (1.1.57). We take $V(k,x) = \|x\|^2$ and assume that the fundamental solution $\Phi(k,k_0)$ associated with (1.1.55) satisfies

$$
|\Phi(k,k_0)| \le \eta(k-k_0), \quad k \ge k_0, \qquad (3.2.37)
$$

where η is a sequence of positive terms and $\sum_{s=k_0}^{\infty}\eta(s)$ exists and is equal to some $\bar{\eta} > 0$. To verify the validity of hypothesis (ii*) of Theorem 3.2.3, we need to compute $\mathcal{D}^*W(k,s,y,z,\omega)$ and find an estimate for it. We note that $\mathcal{D}^*W(k,s,y,z,\omega)$ is as given in (1.1.49), that is,

$$
\begin{aligned}
& \mathcal{D}^*W(k,s,y,z,\omega) \\
& \quad = \left(\sum_{r=k_0+1}^{k} \Phi(r,k_0)R(r,y(r-1,\omega),\omega)\right) \\
& \qquad + 2(y_0(\omega)-z_0)^T\Phi^T(k,k_0)\Phi(k,s)R(s,y,\omega) \\
& \quad = 2(\Phi(k,k_0)(y_0(\omega)-z_0))^T\Phi(k,s)R(s,y,\omega) \\
& \qquad + \left(\sum_{r=k_0+1}^{k} \Phi(k,r)R(r,y(r-1,\omega),\omega)\right)^T \Phi(k,s)R(s,y,\omega)
\end{aligned}
$$

$$= 2(\Phi(k,k_0)(y_0 - z_0))^T\Phi(k,s)\,[R(s,y,\omega) - R(s,z,\omega)]$$
$$+ 2(\Phi(k,k_0)(y_0 - z_0))^T\Phi(k,s)R(s,z,\omega)$$
$$+ \left(\sum_{r=k_0+1}^{k}\Phi(k,r)[R(r,y(r-1,\omega),\omega) - R(r,\overline{x}(r-1),\omega)]\right)^T$$
$$\times \Phi(k,s)[R(s,y,\omega) - R(s,z,\omega)]$$
$$+ \left(\sum_{r=k_0+1}^{k}\Phi(k,r)[R(r,y(r-1,\omega),\omega) - R(r,\overline{x}(r-1),\omega)]\right)^T$$
$$\times \Phi(k,s)R(s,z,\omega)$$
$$+ \left(\sum_{r=k_0+1}^{k}\Phi(k,r)R(r,\bar{x},\omega)\right)^T\Phi(k,s)[R(s,y,\omega) - R(s,z,\omega)]$$
$$+ \left(\sum_{r=k_0+1}^{k}\Phi(k,r)R(r,\bar{x}(r-1),\omega)\right)^T\Phi(k,s)R(s,z,\omega).$$

From this, together with (3.2.37) and algebraic computations, one obtains an upper bound similar to (3.2.32) on $\mathcal{D}^*W(k,s,y,z,\omega)$ with $a(u) = b(u) = u$, $p = 2$, $\eta_1 = \eta_2 = \eta$ in (3.2.37), $c_1(u) = c_2(u) = u^2$, $\lambda_1 = \lambda\bar{\eta}$, $\lambda_1 = \bar{\lambda}\bar{\eta}$, $\nu_2(s,u,\omega) = u^2$, $\nu_1(s,u,\omega) = \lambda_1(s,\omega)\|z\|^2$, $\lambda(s,\omega) = \bar{\lambda}(s,\omega)(1+2\bar{\eta})$, $\bar{\lambda}(s,\omega) = \|A(s,\omega) - \hat{A}(s)\|^2$, $\bar{a} = (1+2\bar{\eta})a$, and $g(k,u,\omega) = 2((1+2\bar{\eta})\|x(k,\omega) - \bar{x}(k)\|^2 + u)g_1(k,\omega) + g_2(k,\|\bar{x}(k-1)\|,\omega)$, where g_1 and g_2 are as defined in Theorem 3.2.3 in the context of the above functions. The estimates (3.2.33) and (3.2.34) can be obtained by the application of Theorem 3.2.3. In this case, $r(k,\omega)$ has the following representation:

$$r(k,\omega) = \begin{cases} u_0(\omega), & \text{for } k = k_0,\\ u_0(\omega)\prod_{s=k_0+1}^{k}(1+2g_1(s,\omega)) \\ \quad + \sum_{s=k_0+1}^{k}P(s,\omega)\prod_{\tau=s+1}^{k} \\ \quad (1+2g_1(\tau,\omega)), & \text{for } k \geq k_0+1, \end{cases} \tag{3.2.38}$$

where

$$P(s,\omega) = g_2(s, \|\bar{x}(s-1)\|, \omega) + 2(1+2\bar{\eta})\|x(s,\omega) - \bar{x}(s)\|^2 g_1(s,\omega).$$

Now, we present an error estimate result in the context of Theorem 1.1.5.

Theorem 3.2.4. *Let us assume that the hypotheses of Theorem* 1.1.5 *is satisfied. Further, assume that the functions* V *and* LV *satisfy the following inequalities:*

$$b(\|x\|^q) \leq \sum_{i=1}^{m} V_i(k,x,\omega) \tag{3.2.39}$$

and

$$LV(p, m(k,p,\phi(p),\omega)) \leq g(p, V(p, m(k,p+1,\phi(p),\omega)), \tag{3.2.40}$$

where $b \in \nu k$, $q \geq 1$; $m(k,p+1,\phi(p))$ *is the solution process of* (1.1.2) *through* $(p+1,\phi(p))$ *and* ϕ *is a sequence in* R^n; $g(p,x)$ *is a Borel measurable sequence that satisfies the following condition:*

$$g(p,x) - g(p,y) \geq 0 \quad \text{for } x \geq y. \tag{3.2.41}$$

Then,

$$\begin{aligned} &b\left(E\left[\|y(k,\omega) - w(k,\omega)\|^q\right]\right) \\ &\quad \leq \sum_{i=1}^{m} E\left[r_i(k,k_0,V(k_0,y_0-w_0,\omega))\right], \quad k \geq k_0, \end{aligned} \tag{3.2.42}$$

where $r(k,k_0,V(k_0,y_0-w_0,\omega))$ *is the solution process of the following comparison system of difference equations:*

$$\Delta u(k,\omega) = g(k-1, u(k-1,\omega)), \quad v(k_0, y_0 - w_0, \omega) \leq u_0. \tag{3.2.43}$$

Proof. From Theorem 1.1.5, we have

$$\begin{aligned} &V(k, y(k,k_0,y_0) - w(k,k_0,w_0,\omega)) \\ &\quad = V(k_0, m(k,k_0,y_0-w_0,\omega)) \\ &\qquad + \sum_{p=k_0}^{k-1} LV(p, m(k,p,y(p)-w(p),\omega)). \end{aligned} \tag{3.2.44}$$

Using (3.2.40), (3.2.44) reduces to

$$
\begin{aligned}
&V(k, y(k, k_0, y_0) - w(k, k_0, w_0), \omega) \\
&\quad \leq V(k_0, m(k, k_0, y_0 - w_0), \omega) \\
&\qquad + \sum_{p=k_0}^{k-1} g(p, V(p, m(k, p+1, y(p) - w(p)), \omega)).
\end{aligned} \tag{3.2.45}
$$

We set

$$
\begin{aligned}
n(k) &= V(k_0, m(k, k_0, y_0 - w_0), \omega) \\
&\quad + \sum_{p=k_0}^{k-1} g(p, V(p, m(k, p+1, y(p) - w(p)), \omega))
\end{aligned} \tag{3.2.46}
$$

and

$$
n(k_0) = V(k_0, y_0 - w_0, \omega). \tag{3.2.47}
$$

From (3.2.45) and (3.2.46), we have

$$
V(k, y(k, k_0, y_0) - w(k, k_0, w_0), \omega) \leq n(k) \tag{3.2.48}
$$

and

$$
\begin{aligned}
\Delta n(k) = n(k) - n(k-1) = g(k-1, V(k-1, y(k-1, k_0, y_0) \\
- \ w(k-1, k_0, w_0), \omega)).
\end{aligned} \tag{3.2.49}
$$

From (3.2.41), (3.2.47) and (3.2.48), (3.2.49) reduces to

$$
\Delta n(k) \leq g(k-1, n(k-1)), \quad \text{for } k \geq k_0, \tag{3.2.50}
$$

and

$$
V(k_0, y_0 - w, \omega) = n(k_0). \tag{3.2.51}
$$

From (3.2.50), (3.2.50) and (3.2.49), we apply Corollary 1.2.2 and obtain

$$
n(k) \leq r(k, k_0, n(k_0)), \text{ for } k \geq k_0.
$$

This, together with (3.2.48) and (3.2.39), establishes

$$b(\|y(k,k_0,y_0 - w(k,k_0,w_0)\|^q) \leq \sum_{i=1}^{m} r_i(k,k_0,n(k_0)). \tag{3.2.52}$$

Now, using the property of b, we have

$$b(E[\|y(k,k_0,y_0) - w(k,k_0,w_0)\|^q] \leq \sum_{i=1}^{m} E[r_i(k,k_0,n(k_0))]. \tag{3.2.53}$$

This completes the proof of the theorem.

Example 3.2.6. From Corollary 1.1.6(vi), we have

$$\begin{aligned}
&\|y(k,k_0,y_0) - w(k,k_0,w_0)\|^2 \\
&\quad = \|\Phi(k,k_0)(y_0 - w_0)\|^2 \\
&\qquad + \sum_{p=k_0}^{k-1} LV(p,m(k,p,y(p) - w(p)),\omega),
\end{aligned}$$

where

$$\begin{aligned}
&LV(p,m(k,p,y(p) - w(p)),\omega) \\
&\quad = 2\alpha m^T(k,p,y(p) - w(p))\Phi(k,p+1) \\
&\qquad \times [f(p,y(p),\omega) - h(p,w(p),\omega) - A(p)(y(p) - w(p))] \\
&\qquad + \alpha\,[f(p,y(p),\omega) - h(p,w(p),\omega) - A(p)(y(p) - w(p)]^T \\
&\qquad \times \Phi^T(k,p+1)\Phi(k,p+1) \\
&\qquad \times \cdot [f(p,y(p),\omega) - h(p,w(p),\omega) - A(p)(y(p) - w(p)] \\
&\quad = 2\alpha m^T(k,p,y(p) - w(p))\Phi(k,p+1) \\
&\qquad \times (f(p,y(p),\omega) - h(p,w(p),\omega) \\
&\qquad - 2\alpha m^T(k,p,y(p) - w(p))\Phi(k,p+1)A(p)(y(p) - w(p)) \\
&\qquad + \alpha(f(p),y(p),\omega) - h(p,w(p),\omega))^T\Phi^T(k,p+1)\Phi(k,p+1) \\
&\qquad \times (f(p,y(p),\omega) - h(p,w(p),\omega) \\
&\qquad - 2\alpha(A(p)(y(p) - w(p))^T\Phi^T(k,p+1)\Phi(k,p+1) \\
&\qquad \times (f(p,y(p),\omega) - h(p,w(p),\omega))
\end{aligned}$$

$$
\begin{aligned}
&+\alpha(A(p)(y(p)-w(p))^T\Phi^T(k,p+1)\Phi(k,p+1)\\
&\times(A(p)(y(p)-w(p)). \qquad (3.2.54)
\end{aligned}
$$

Using $m(k,p+1,u)=\Phi(k,p+1)u$ and assuming the commutative property of $A(p)$ and $\Phi(k,p+1)$, we rewrite the terms in (3.2.54) as follows:

$$
\begin{aligned}
&\alpha(f(p,y(p),\omega)-h(p,w(p),\omega))\Phi^T(k,p+1)\Phi(k,p+1)\\
&\quad\times(f(p,y(p),\omega)-h(p,w(p),\omega))\\
&\quad=\alpha m^T(k,p+1,f(p,y(p),\omega)-h(p,w(p),\omega))\\
&\quad\times m(k,p+1,f(p,y(p),\omega)-h(p,w(p),\omega)),\\
&\quad-2\alpha(A(p)(y(p)-w(p)))^T\Phi^T(k,p+1)\Phi(k,p+1)\\
&\quad\times(f(p,y(p),\omega)-h(p,w(p),\omega)) \qquad (3.2.55)\\
&\quad=-2\alpha(y(p)-w(p))^T(\Phi(k,p+1)A(p))^T\Phi(k,p+1)\\
&\quad\times(f(p,y(p),\omega)-h(p,w(p),\omega))\\
&\quad=-2\alpha(y(p)-w(p))^T(\Phi(k,p+1)(I+A(p)-I)^T\Phi(k,p+1))\\
&\quad\cdot(f(p,y(p),\omega)-h(p,w(p),\omega))\\
&\quad=-2\alpha(y(p)-w(p))^T(\Phi(k,p)-\Phi(k,p+1))^T\Phi(k,p+1)\\
&\quad\cdot(f(p,y(p),\omega)-h(p,w(p),\omega))\\
&\quad=-2\alpha(y(p)-w(p))^T\Phi^T(k,p)\Phi(k,p+1)\\
&\quad\times(f(p,y(p),\omega)-h(p,w(p),\omega))\\
&\quad+2\alpha(y(p)-w(p))^T\Phi^T(k,p+1)\Phi(k,p+1)\\
&\quad\times(f(p,y(p),\omega)-h(p,w(p),\omega))\\
&\quad=-2\alpha m^T(k,p,y(p)-w(p))\Phi(k,p+1)\\
&\quad\times(f(p,y(p),\omega)-h(p,w(p),\omega))\\
&\quad+2\alpha m^T(k,p+1,y(p)-w(p))m(k,p+1,f(p,y(p),\omega)\\
&\quad-h(p,w(p),\omega)) \qquad (3.2.56)
\end{aligned}
$$

and

$$\begin{aligned}
&\alpha(A(p)(y(p)-w(p))^T\Phi^T(k,p+1)\Phi(k,p+1)A(p)(y(p)-w(p))\\
&\quad=\alpha(y(p)-w(p))^T(\Phi(k,p+1)A(p))^T\Phi(k,p+1)\\
&\qquad\times A(p)(y(p)-w(p))\\
&\quad=\alpha(y(p)-w(p))^T(\Phi(k,p)-\Phi(k,p+1))^T\Phi(k,p+1)\\
&\qquad\times A(p)(y(p)-w(p))\\
&\quad=\alpha(y(p)-w(p))^T\Phi^T(k,p)\Phi(k,p+1)A(p)(y(p)-w(p))\\
&\qquad-\alpha(y(p)-w(p))^T\Phi^T(k,p+1)\Phi(k,p+1)A(p)(y(p)-w(p))\\
&\quad=\alpha m^T(k,p,y(p)-w(p))\Phi(k,p+1)A(p)(y(p)-w(p))\\
&\qquad-\alpha m^T(k,p+1,y(p)-w(p))\Phi(k,p+1)A(p)(y(p)-w(p)).
\end{aligned} \tag{3.2.57}$$

From (3.2.55), (3.2.56) and (3.2.57), (3.2.54) reduces to

$$\begin{aligned}
&LV(p,m(k,p,y(p)-w(p),\omega))\\
&\quad=\alpha m^T(k,p+1,f(p,y(p),\omega)-h(p,w(p),\omega))\\
&\qquad\times m(k,p+1,f(p,y(p),\omega)-h(p,w(p),\omega))\\
&\qquad+2\alpha m^T(k,p+1,y(p)-w(p))m(k,p+1,f(p,y(p),\omega)\\
&\qquad-h(p,w(p),\omega))-\alpha m^T(k,p,y(p)-w(p))\\
&\qquad\times\Phi(k,p+1)A(p)(y(p)-w(p))\\
&\qquad-\alpha m^T(k,p+1,y(p)-w(p))\Phi(k,p+1)A(p)(y(p)-w(p)).
\end{aligned} \tag{3.2.58}$$

We rewrite the last two terms in (3.2.58) as follows:

$$\begin{aligned}
&-\alpha m^T(k,p,y(p)-w(p))\Phi(k,p+1)A(p)(y(p)-w(p))\\
&-\alpha m^T(k,p+1,y(p)-w(p))\Phi(k,p+1)A(p)(y(p)-w(p))\\
&\quad=-\alpha m^T(k,p,y(p)-w(p))A(p)\Phi(k,p+1)(y(p)-w(p))\\
&\qquad-\alpha(y(p)-w(p))^T(\Phi(k,p+1)A(p))^Tm(k,p+1,y(p)-w(p))\\
&\quad=-\alpha[m^T(k,p,y(p)-w(p))A(p)+(y(p)-w(p))^T\\
&\qquad\times(A(p)\Phi(k,p+1))^T]
\end{aligned}$$

$$
\begin{aligned}
&\cdot m(k,p+1,y(p)-w(p))\\
&\quad = -\alpha[(y(p)-w(p))^T\Phi^T(k,p)A(p) + (y(p)-w(p))^T\\
&\qquad\qquad \times (\Phi^T(k,p+1)A^T(p))]\\
&\cdot m(k,p+1,y(p)-w(p))\\
&\quad = -\alpha(y(p)-w(p))^T[(\Phi(k,p+1)(I+A(p)))^T A(p)\\
&\qquad + \Phi^T(k,p+1)A^T(p)]\\
&\cdot m(k,p+1,y(p)-w(p))\\
&\quad = -\alpha(y(p)-w(p))^T[((I+A(p))\Phi(k,p+1))^T A(p)\\
&\qquad + \Phi^T(k,p+1)A^T(p)]\\
&\cdot m(k,p+1,y(p)-w(p))\\
&= -\alpha(y(p)-w(p))^T[\Phi^T(k,p+1)(I+A(p))^T A(p)\\
&\qquad + \Phi^T(k,p+1)A^T(p)]\\
&\cdot m(k,p+1,y(p)-w(p))\\
&\quad = -\alpha(y(p)-w(p))^T\Phi^T(k,p+1)[(I+A(p))^T A(p) + A^T(p)]\\
&\qquad \times m(k,p+1,y(p)-w(p))\\
&\quad = -\alpha m^T(k,p+1,y(p)-w(p))[A(p) + A^T(p)A(p)\\
&\qquad + A^T(p) + I - I]m(k,p+1,y(p)-w(p))\\
&\quad = -\alpha m^T(k,p+1,y(p)-w(p))\\
&\qquad \times (I+A(p))^T(I+A(p))m(k,p+1,y(p)-w(p))\\
&\qquad + \alpha m^T(k,p+1,y(p)-w(p))m(k,p+1,y(p)-w(p)).
\end{aligned}
\tag{3.2.59}
$$

From (3.2.59), (3.2.58) reduces to

$$
\begin{aligned}
&LV(p,m(k,p,y(p)-w(p)))\\
&\quad = \alpha m^T(k,p+1,f(p,y(p),w)-h(p,w(p),w))\\
&\qquad \cdot m(k,p+1,f(p,y(p),w)-h(p,w(p),w))\\
&\qquad + 2\alpha m^T(k,p+1,y(p)-w(p))\\
&\qquad \times m(k,p+1,f(p,y(p)-h(p,w(p),\omega))
\end{aligned}
$$

$$\begin{aligned}&- \alpha m^T(k, p+1, y(p) - w(p))(I + A(p))^T\\&\times (I + A(p))m(k, p+1, y(p) - w(p))\\&+ \alpha m^T(k, p+1, y(p) - w(p))m(k, p+1, y(p) - w(p)).\end{aligned} \tag{3.2.60}$$

We assume that

$$\begin{aligned}&m^T(k, p+1, f(p, y(p), \omega) - h(p, w(p), \omega))\\&\quad \times m(k, p+1, f(p, y(p), \omega) - h(p, y(p), \omega))\\&\quad \le \lambda_1(p, \omega)\|m(k, p+1, y(p) - w(p))\|^2,\\&m^T(k, p+1, y(p) - w(p))m(k, p+1, f(p, y(p), \omega) - h(p, w(p), \omega))\\&\quad \le \lambda_2(p, \omega)\|m(k, p+1, y(p) - w(p))\|^2,\\&m^T(k, p+1, y(p) - w(p))(I + A(p))^T\\&\quad \times (I + A(p))m(k, p+1, y(p) - w(p))\\&\quad \ge \lambda_3(p, \omega)\|m(k, p+1, y(p) - w(p))\|^2,\end{aligned} \tag{3.2.61}$$

where λ_1, λ_2 and λ_3 are discrete-time nonnegative stochastic processes and λ is the smallest eigenvalue of the symmetric matrix $(I + A(p))^T(I + A(p))$. From the choice of the Lyapunov function $V(x) = \alpha\|x\|^2$, (3.2.60) and (3.2.61), we obtain the following inequality:

$$LV(p, m(k, p, y(p) - w(p))) \le g(p, V(p, m(k, p+1, y(p) - w(p)))), \tag{3.2.62}$$

where g is defined by

$$g(p, u, \omega) = (\lambda_1(p, \omega) + 2\lambda_2(p, \omega) - \lambda_3(p, \omega) + 1)u. \tag{3.2.63}$$

We further assume that

$$\mu(p, \omega) = \lambda_1 + 2\lambda_2 - \lambda + 1 \ge 0 \text{ almost surely.} \tag{3.2.64}$$

Repeating the argument used in Theorem 3.2.4, we have

$$E[V(y(k, k_0, y_0) - w(k, k_0, w_0))] \le E[r(k, k_0, r_0)] \tag{3.2.65}$$

whenever

$$V(y_0 - w_0) \leq r_0. \tag{3.2.66}$$

Moreover,

$$\begin{aligned} &E[\|y(k, k_0, y_0) - w(k, k_0, w_0)\|^2] \\ &\leq E\left[\|y_0 - w_0\|^2 \prod_{p=k_0}^{k-1} (1 + \mu(p, w))\right] \\ &\leq E\left[\|y_0 - w_0\|^2 \exp\left[\sum_{p=k_0}^{k-1} \mu(p, w)\right]\right]. \end{aligned}$$

Remark 3.2.1. The feasibility of assumption (3.2.61) is assured by the following special cases:

(a) For h in (1.1.60), and f in (1.1.1) satisfies the hypothesis $(\mathrm{H}_{(1.1.2)})$ with $f(p, 0, \omega) \equiv 0$. In this case, $w(p, k_0, w_0, \omega) = w_0$ and

$$\begin{aligned} f(p, y, \omega) - f(p, 0, \omega) &= \int_0^1 \frac{\partial f}{\partial y}(p, \theta(y - w_0), \omega) d\theta (y - w_0) \\ &= A(p, y - w_0, \omega)(y - w_0). \end{aligned} \tag{3.2.67}$$

(b) For $h(p, w, \omega) = f(p, w, \omega)$ in (1.1.60), and f in (1.1.1) satisfies the hypothesis $(\mathrm{H}_{(1.1.2)})$. In this case, $w(p, k_0, w_0) = y(p, k_0, w_0)$, and it is the solution process of (1.1.1) through (k_0, w_0). Moreover,

$$\begin{aligned} f(p, y, \omega) - f(p, w, \omega) &= \int_0^1 \frac{\partial f}{\partial y}(p, w + \theta(y - w), \omega) d\theta (y - w) \\ &= A(p, y - w, \omega)(y - w), \end{aligned} \tag{3.2.68}$$

where

$$A(p, w, y - w, \omega) = \int_0^1 \frac{\partial f}{\partial y}(p, w + \theta(y - w), \omega) d\theta. \tag{3.2.69}$$

Example 3.2.7. We consider Example 1.1.1, and its equivalent representation is as follows:

$$\begin{aligned}\Delta y(k,w) &= (A(k,w)-I)y(k,w), \quad y(k_0,w)=y_0(w),\\ \Delta w(k) &= (A(k,w)-I)w(k,w), \quad w(k_0,w)=w_0(w),\\ \Delta m(k) &= (\hat{A}(k)-I)m(k), \quad m(k_0)=m_0,\\ \Phi(k,p) &= \prod_{i=p}^{k-1}[\hat{A}(i)-I+I] = \prod_{i=p}^{k-1}\hat{A}(i),\end{aligned}$$

$$m(k,p,y(p)-w(p)) = \Phi(k,p)(y(p)-w(p)). \tag{3.2.70}$$

From (3.2.60) and (3.2.70), we obtain

$$\begin{aligned}&LV(p,m(k,p,y(p)-w(p)))\\ &\quad= \alpha m^T(k,p+1,(A-I)y(p)-(A-I)w(p))\\ &\qquad\times m(k,p+1,(A-I)y(p)-(A-I)w(p))\\ &\qquad+2\alpha m^T(k,p+1,y(p)-w(p))\\ &\qquad\times m(k,p+1,(A-I)y(p)-(A-I)w(p))\\ &\qquad-\alpha m^T(k,p+1,y(p)-w(p))(I+A(p))^T\\ &\qquad\times(I+A(p))m(k,p+1,y(p)-w(p))\\ &\qquad+\alpha m^T(k,p+1,y(p)-w(p))m(k,p+1,y(p)-w(p)).\end{aligned} \tag{3.2.71}$$

By an application of (3.2.61), (3.2.71) reduces to

$$\begin{aligned}&LV(p,m(k,p,y(p)-w(p)))\\ &\quad\le \alpha\lambda_1(p,w)\|m(k,p+1,y(p)-w(p))\|^2\\ &\qquad+2\alpha\lambda_2(p,w)\|m(k,p+1,y(p)-w(p))\|^2\\ &\qquad-2\alpha\lambda_3(p,w)\|m(k,p+1,y(p)-w(p))\|^2\\ &\qquad+\alpha\|m(k,p+1,y(p)-w(p))\|^2\\ &\quad=[\alpha\lambda_1(p,w)+2\alpha\lambda_2(p,w)-\alpha\lambda_3(p,w)\\ &\qquad+\alpha]\|m(k,p+1,y(p)-w(p))\|^2\\ &\quad=\mu(p,w)\|m(k,p+1,y(p)-w(p))\|^2,\end{aligned} \tag{3.2.72}$$

where $\mu(p,w) = \alpha\lambda_1(p,w) + 2\alpha\lambda_2(p,w) - \alpha\lambda_3(p,w) + \alpha$. From the choice of $V(x) = |x|^2$, (3.2.72) reduces to

$$\begin{aligned} &LV(p, m(k,p,y(p-w(p))) \\ &\quad \leq g(p, V(p, m(k,p+1, y(p)-w(p))), \omega), \end{aligned} \tag{3.2.73}$$

where

$$g(p,u,w) = \mu(p,w)u$$

and

$$\Delta u(k,w) = g(k-1, u(k-1,w)). \tag{3.2.74}$$

From the choice of the Lyapunov function $V(x) = \alpha\|x\|^2$, equations (3.2.70) and (3.2.65) reduce to

$$E\|y(k,k_0,y_0) - w(k,k_0,w_0)\|^2 \leq E\left[\|y_0 - w_0\|^2 \prod_{p=k_0}^{k-1}(1+\mu(p,w))\right]. \tag{3.2.75}$$

3.2.2 *Error estimates: Ito–Doob-type stochastic iterative system — II*

In this section, we develop error estimates of the pth moment derivation of a solution process of (2.1.1) with the solution processes of (2.2.71), (2.1.2) or (2.1.3). We employ the comparison method and the variation of constants method to derive the error estimates.

Now, we present a few error estimate results by employing the comparison theorems.

Theorem 3.2.5. *Let the hypothesis of Theorem* 2.2.3 *be satisfied. Further, assume that*

$$b(\|x\|^p) \leq \sum_{i=1}^m V_i(k,x,\omega), \tag{3.2.76}$$

where $b \in \mathcal{CK}$, and it is a convex function with $p \geq 1$. Then,

$$\begin{aligned} &b(E(\|y(k,\omega) - w(k)\|^p) \\ &\quad \leq \sum_{i=1}^m E(r_i(k,k_0,V(k_0,m(k,k_0,y_0-w_0),\omega),\omega)), \quad k \geq k_0, \end{aligned}$$

where $w(k) \equiv w(k,k_0,w_0)$ is the solution process of (1.1.60).

3.3 Relative Stability

We present a few relative stability results that assure the relative stability of the pth moment of the system of difference equations (1.1.1) with respect to (1.1.60), in particular, (1.1.2) and (1.1.3). These results provide sharper error estimates of the pth moment derivation of a solution process of (1.1.1) relative to either system (1.1.2) or (1.1.3). Further, results are illustrated by suitable examples.

In the following, we present relative stability results in the context of the comparison method.

Theorem 3.3.1. *Let the hypotheses of Theorem* 1.2.5 *be satisfied. Further, assume that*

$$b(\|x\|^p) \leq \sum_{i=1}^{m} V_i(k, x, \omega) \leq a(k, \|x\|^p), \tag{3.3.1}$$

where $b \in \mathcal{VK}$, $a \in \mathcal{CK}$ *and* $p \geq 1$. *Then,*

(i) (DJR_1) *of* (1.2.9) *and* (1.1.2) *implies* (DRM_1) *of* (1.1.1) *and* (1.1.60);
(ii) (DJR_2) *of* (1.2.9) *and* (1.1.2) *implies* (DRM_1) *of* (1.1.1) *and* (1.1.60).

Proof. Let $\epsilon > 0$, $k_0 \in I(k_0)$ be given. Assume that (DJR_1) of (1.2.9) and (1.1.60) holds. Then, for $b(\epsilon^p) > 0$ and $k_0 \in I(k_0)$, there exists a $\delta = \delta(\epsilon, k_0)$ such that

$$\|y_0(\omega) - z_0\|_p \leq \delta \tag{3.3.2}$$

implies

$$\sum_{i=1}^{m} E[\nu_i(k, k_0, y_0(\omega) - w_0, \omega)] < b(\epsilon^p), \quad k \geq k_0, \tag{3.3.3}$$

and $r(k, k_0, u_0(\omega), \omega)$ is a solution process of (1.2.20). Now, we claim that if (3.3.2) is valid, then $\|y(k, \omega) - w(k)\|_p < \epsilon$, for all $k \geq k_0$. Suppose that this is false. Then, there would exist solution processes $y(k, \omega)$ and $w(k)$ satisfying (3.3.2) such that

$$\|y(k_1, \omega) - w(k_1)\|_p = \epsilon, \quad \text{for some } k_1 > k_0, \tag{3.3.4}$$

and

$$\|y(k, \omega) - w(k, \omega)\|_p \leq \epsilon, \quad \text{for } k_0 \leq k \leq k_1. \tag{3.3.5}$$

From the conclusion of Theorem 1.2.5, the convexity of b and relations (3.3.1) and (3.3.4), we have

$$b(E[\|y(k_1,\omega)-w(k_1,\omega)\|^p]) \le \sum_{i=1}^{m} E[V_i(k_1,y(k_1,\omega)-w(k_1,\omega),\omega)]$$
$$\le \sum_{i=1}^{m} E[r_1(k_1,k_0,u_0(\omega),\omega)].$$

This, together with (3.3.3) and (3.3.4), leads to the contradiction

$$b(\epsilon^p) \le \sum_{i=1}^{m} E[V_i(k_1,y(k_1,\omega)-w(k_1,\omega),\omega)]$$
$$\le \sum_{i=1}^{m} E[r_i(k_1,k_0,u_0(\omega),\omega)] < b(\epsilon^p).$$

This completes the proof of (i). The proof of (ii) can be given analogously. Thus, the proof of the theorem is complete.

Example 3.3.1. Consider Example 1.2.6. In this case, $p = 1$, $a(r) = b(r) = r$ and the comparison equation is as described in (1.2.30). To apply Theorem 3.3.1, it is enough to show that (1.2.30), (1.2.65) or (1.2.93) is jointly relatively stable in the mean. For this purpose, we assume that

$$E\left[\exp\left[\sum_{s=k_0+1}^{k}(\alpha_p(s)+\mu(s,\omega)+\nu(s,\omega)-1)\right]\right]$$

is bounded for $k \ge k_0$. This, together with the right-hand expression of (3.2.6) and the Hölder inequality, establishes property (DJR_1) of (1.2.30) and (1.2.65) (or (1.2.93)). Thus, property (DRM_1) of (1.1.52), (1.2.65) or (1.2.93) follows from the application of Theorem 3.3.1.

Example 3.3.2. Consider the initial value problems (1.1.55), (1.1.56) and (1.1.57). From Example 3.2.7, for a linear system, we arrive at (3.2.10). To conclude that the comparison equations (1.2.30) and (1.1.56) (or (1.1.57)) possess property (DJR_1), we assume that

$$E\left[\exp\left[\sum_{s=k_0+1}^{k}(\alpha_p(s)+\mu(s,\omega)+\nu(s,\omega)-1)\right]\right] \tag{3.3.6}$$

is bounded for $k \geq k_0$. Moreover, (3.3.6) can be rewritten as

$$\exp\left(\sum_{s=k_0+1}^{k}(\alpha_p(s)-1)\right)E\left[\exp\left(\sum_{s=k_0+1}^{k}(\mu(s,\omega)+\nu(s,\omega))\right)\right]. \tag{3.3.7}$$

If $E[\alpha(s,\omega)] = 0$ and $E[(\alpha(s,\omega)-1)(\alpha(\tau,\omega)-1)] = C(s-\tau)$, then (3.3.7) can be rewritten as

$$\exp\left(\sum_{s=k_0+1}^{k}(\alpha_p(s)-1)\right)\exp\left(\frac{1}{2}\sum_{u=k_0+1}^{k}\sum_{s=k_0+1}^{k}C(u-s)\right). \tag{3.3.8}$$

By an application of Theorem 3.3.1, we conclude that (1.1.55) and (1.1.56) (or (1.1.57)) have property (DRM_1).

Next, we employ the method of variation of constants to investigate the pth moment relative stability of (1.1.1) with respect to either system (1.1.2) or (1.1.3).

Theorem 3.3.2. *Suppose that all the hypotheses of Theorem* 1.1.4 *hold. Moreover, assume that*

(i) $\hat{E}(k,0) \equiv 0$, *for* $k \in I(k_0)$;
(ii) $b(\|x\|^p) \leq \sum_{i=1}^m |V_i(k,x,\omega)| \leq a(\|x\|^p)$, *where* $b \in \mathcal{VK}$, $a \in \mathcal{K}$ *and* $p \geq 1$;
(iii) $\sum_{i=1}^m |\mathcal{D}^*W_i(k,x(k,s,y)-x(k,s,z),\omega)|$
$\leq \eta(k-s)\Big[\sum_{i=1}^m |V_i(k_0,x(k,\omega)-\bar{x}(k),\omega)|$
$+\lambda_1(s,\omega)\sum_{i=1}^m |V_i(s-1,y(k,\omega)-\bar{z}(k),\omega)|$
$+\sum_{r=k_0+1}^k \eta(k-r)\lambda_2(s,\omega)\Big(\sum_{i=1}^m |V_i(r-1,y(k,\omega)-\bar{x}(k),\omega)|$
$+\lambda_3(r,\omega)\sum_{i=1}^m |V_i(r-1,\bar{x},\omega)|\Big) + \lambda_3(s,\omega)\sum_{i=1}^m |V_i(s,z,\omega)|\Big]$,
for $k_0 \leq s \leq k$ *and* $\|y-z\|^p < p$, $\|z\|^p < p$ *and some* $p > 0$;

(iv) $\eta, \lambda_i, i = 1, 2, \ldots, 5, \nu_i, i = 1, 2, 3$ *satisfy*

$$\sum_{\tau=k_0}^{\infty} \eta(\tau) = \bar{\eta}, \quad \bar{\eta} \in R^+,$$

$$\beta(k-k_0)\left(1 + \sum_{s=k_0+1}^{k} \eta(k-s)\right) \le \nu_1(k-k_0),$$

$$\left(1 + \sum_{s=k_0+1}^{k} \eta(k-s)\lambda_2(s,\omega)\right) \sum_{s=k_0+1}^{k} \eta(k-s)\lambda_3(s,\omega)$$
$$\times\, \beta(s-1-k_0) \le \nu_2(k-k_0)\lambda_4(k,\omega),$$

$$\eta(k-s)(\lambda_1(s,\omega) + \sum_{r=k_0+1}^{k} \eta(k-r)\lambda_2(r,\omega))$$
$$\le \nu_3(k-s)\lambda_5(s,\omega);$$

(v) $\sum_{i=1}^{m} |V_i(k_0, x(k,\omega) - \bar{x}(k), \omega)| \le \alpha(\|y_0(\omega) - z_0\|^p)\beta(k-k_0)$ *whenever* $\|y_0(\omega) - z_0\|^p \le \rho$ *and* $\|z_0\|^p < \rho, \alpha \in \mathcal{CK}$. *Then,*

(1) *The relative boundedness of the pth moment iterative process is as follows:*

$$\nu_1(k-k_0), \quad \nu_2(k-k_0)E[\lambda_4(k,\omega)],$$
$$\times \sum_{s=k_0+1}^{k} E^{1/2}\Bigg[\nu_1(s-k_0)\nu_3(k-s)\lambda_5(s,\omega)$$
$$\times \exp\left(\sum_{\tau=s+1}^{k} \nu_3(k-\tau)\lambda_5(\tau,\omega)\right)\Bigg]^2$$

and

$$\sum_{s=k_0+1}^{k} E^{1/2}\Bigg[\nu_2(s-k_0)\lambda_4(s,\omega)\nu_3(k-s)\lambda_5(s,\omega)$$
$$\times \exp\left(\sum_{\tau=s+1}^{k} \nu_3(k-\tau)\lambda_5(\tau,\omega)\right)\Bigg]^2$$

imply (DRM_1) *of* (1.1.1) *and* (1.1.2);

(2) *The pth moment relative asymptotic stability of the iterative process is as follows:*

$$\lim_{k\to\infty} \nu_1(k-k_0) = 0,$$

$$\lim_{k\to\infty} \nu_2(k-k_0)E[\lambda_4(k,\omega)] = 0,$$

$$\lim_{k\to\infty} \sum_{s=k_0+1}^{k} E^{1/2}\Bigg[\nu_1(s-k_0)\nu_3(k-s)\lambda_5(s,\omega),$$
$$\times \exp\left(\sum_{\tau=s+1}^{k} \nu_3(k-\tau)\lambda_5(\tau,\omega)\right)\Bigg]^2 = 0$$

and

$$\lim_{k\to\infty} \sum_{s=k_0+1}^{k} E^{1/2}\Bigg[\nu_2(s-k_0)\lambda_4(s,\omega)\nu_3(k-s)\lambda_5(s,\omega)$$
$$\times \exp\left(\sum_{\tau=s+1}^{k} \nu_3(k-\tau)\lambda_5(\tau,\omega)\right)\Bigg]^2 = 0$$

imply (DRM_2) *of* (1.1.1) *and* (1.1.2).

Proof. Let $y(k,\omega) = y(k,k_0,y_0(\omega),\omega)$, $x(k,\omega) = x(k,k_0,y_0,\omega)$ and $\bar{x}(k) = x(k,k_0,z_0)$ be the solution processes as defined in Theorem 1.1.4. From hypothesis (iii) and Theorem 1.1.4, we have

$$\sum_{i=1}^{m} |V_i(k, y(k,\omega)-\bar{x}(k),\omega)| \le \sum_{i=1}^{m} |V_i(k_0, x(k,\omega)-\bar{x}(k),\omega)|$$
$$+ \sum_{s=k_0+1}^{k} \left(\sum_{i=1}^{m} \mathcal{D}^* W(k, x(k,s,y(s,\omega)) - x(k,s,\bar{x}(s)),\omega)\right) \tag{3.3.9}$$

as long as $E\|y(k,\omega)-z(k)\|^p < \rho$, $E\|z(k)\|^p < \rho$. By setting

$$m(k,\omega) = \sum_{i=1}^{m} |V_i(k, y(k,\omega)-\bar{x}(k),\omega)|$$

and using hypotheses (ii), (iii) and (iv), relation (3.3.9) reduces to

$$
\begin{aligned}
m(k,\omega) &\le \alpha(\|y_0(\omega)-z_0\|^p)\beta(k-k_0)\\
&\quad+\sum_{s=k_0+1}^{k}\eta(k-s)\{\alpha(\|y_0(\omega)-z_0\|^p)\\
&\quad\times\beta(k-k_0)+\lambda_1(s,\omega)m(s-1,\omega)\\
&\quad+\sum_{r=k_0+1}^{k}\eta(k-r)\lambda_2(s,\omega)\\
&\quad\times(m(r-1,\omega)+\lambda_3(r,\omega)\alpha\|z_0\|^p\beta(r-1-k_0))\\
&\quad+\lambda_3(s,\omega)\alpha\|z_0\|^p\beta(s-1-k_0)\}\\
&\le \alpha\|y_0(\omega)-z_0\|^p\beta(k-k_0)\left(1+\sum_{s=k_0+1}^{k}\eta(k-s)\right)\\
&\quad+\sum_{s=k_0+1}^{k}\eta(k-s)\lambda_1(s,\omega)m(s-1,\omega)\\
&\quad+\left(\sum_{s=k_0+1}^{k}\eta(k-s)\lambda_2(s,\omega)\right)\sum_{r=k_0+1}^{k}\eta(k-r)m(r-1-k_0)\\
&\quad+\alpha\|z_0\|^p\left(\sum_{s=k_0+1}^{k}\eta(k-s)\lambda_2(s,\omega)\right)\\
&\quad\times\sum_{r=k_0+1}^{k}\eta(k-r)\lambda_3(r,\omega)\beta(r-1-k_0)\\
&\quad+\alpha\|z_0\|^p\sum_{s=k_0+1}^{k}\eta(k-s)\lambda_3(s,\omega)\beta(s-1-k_0)\\
&\le \alpha\|y_0(\omega)-z_0\|^p\left(1+\sum_{s=k_0+1}^{k}\eta(k-s)\right)\beta(k-k_0)\\
&\quad+\alpha\|z_0\|^p\left(1+\sum_{s=k_0+1}^{k}\eta(k-s)\lambda_2(s,\omega)\right)
\end{aligned}
$$

$$\times \sum_{s=k_0+1}^{k} \eta(k-s)\lambda_3(s,\omega)\beta(s-1-k_0)$$

$$+ \sum_{s=k_0+1}^{k} \eta(k-s)(\lambda_1(s,\omega)$$

$$+ \sum_{r=k_0+1}^{k} \eta(k-r)\lambda_2(r,\omega))m(s-1,\omega)$$

$$\leq \alpha\|y_0(\omega)-z_0\|^p\nu_1(k-k_0) + \alpha\|z_0\|^p\nu_2(k-k_0)\lambda_4(k,\omega)$$

$$+ \sum_{s=k_0+1}^{k} \nu_3(k-s)\lambda_5(s,\omega)m(s-1,\omega).$$

Therefore, we can rewrite the above inequality as follows:

$$m(k,\omega) \leq n(k,\omega) + \sum_{s=k_0+1}^{k} \nu_3(k-s)\lambda_5(s,\omega)m(s-1,\omega), \quad (3.3.10)$$

where

$$n(k,\omega) = \alpha\|y_0(\omega)-z_0\|^p\nu_1(k-k_0) + \alpha\|z_0\|^p\nu_2(k-k_0)\lambda_4(k,\omega)$$

as long as $E\|y(k,\omega)\|^p \leq \rho$, $E\|\bar{x}(k)\|^p \leq \rho$. (3.3.10) gives

$$m(k,\omega) \leq n(k,\omega) + \sum_{s=k_0+1}^{k} n(s,\omega)\nu_3(k-s)\lambda_5(s,\omega)$$

$$\times \exp\left(\sum_{\tau=s+1}^{k} \nu_3(k-\tau)\lambda_5(\tau,\omega)\right) \quad (3.3.11)$$

as long as $E\|y(k,\omega)\|^p \leq \rho$, $E\|\bar{x}(k)\|^p \leq \rho$. Equation (3.3.11), together with the definition of $m(k,\omega)$, reduces to

$$\sum_{i=1}^{m} |V_i(k,y(k,\omega)-\bar{x}(k),\omega)| \leq n(k,\omega)$$

$$+ \sum_{s=k_0+1}^{k} n(s,\omega)\nu_3(k-s)\lambda_5(s,\omega)\exp\left(\sum_{\tau=s+1}^{k} \nu_3(k-\tau)\lambda_5(\tau,\omega)\right). \quad (3.3.12)$$

By an application of hypothesis (ii) and (3.3.12), we obtain

$$
\begin{aligned}
b(E[\|y(k,\omega) - \bar{x}(k)\|^p]) &\le E[n(k,\omega)] \\
&+ \sum_{s=k_0+1}^{k} E\Bigg[n(s,\omega)\nu_3(k-s)\lambda_5(s,\omega) \\
&\qquad \times \exp\left(\sum_{\tau=s+1}^{k} \nu_3(k-\tau)\lambda_5(\tau,\omega)\right)\Bigg] \qquad (3.3.13)
\end{aligned}
$$

as long as $E[\|y(k,\omega)\|^p] \le \rho$ and $E[\|\bar{x}(k)\|^p] \le \rho$. It is obvious that properties (DRM_1) and (DRM_2) of (1.1.1) and (1.1.2) (or (1.1.3)) follow from the given assumptions of β, λ_i, $i = 1, 2, ..., 5$, ν_i, $i = 1, 2, 3$, η, and by following the argument used in the proof of Theorem 3.1.3. This completes the proof of the theorem.

Remark 3.3.1. We remark that Theorems 3.3.1 and 3.3.2 also provide sufficient conditions for the relative stability properties of (1.1.1) and (1.1.2) (or (1.1.3)) with a probability of one.

Example 3.3.3. We consider the initial value problems (1.1.55), (1.1.56) and (1.1.57). We take $V(k, x, \omega) = \|x\|^2$. Assume that

$$
\begin{aligned}
\|\Phi(k,s)\| &\le \nu(k-s), \\
\|\bar{x}(s-1)\| &\le \eta(s-1-k_0)\|z_0\|^2, \\
\bar{\eta} &= \sum_{s=k_0+1}^{\infty} \eta(k-s) \text{ is finite}
\end{aligned}
$$

and

$$
\eta(k-s)\eta^2(s-1-k_0) \le \eta(k-s)\eta(s-k_0) \le K_1\eta(k-k_0), \quad K_1 \in R^+.
$$

Set

$$
\begin{aligned}
\bar{\lambda}(s,\omega) &= \|A(s,\omega) - \hat{A}(s)\|, \\
\lambda_1(s,\omega) &= \bar{\lambda}^2(s,\omega)\left(1 + \frac{1}{2}\sum_{r=k_0+1}^{k} \eta(k-r)\right), \\
\lambda_1^*(s,\omega) &= 2\lambda_1(s,\omega), \quad \hat{\lambda}(s,\omega) = \lambda_1^*(s,\omega) + \bar{\lambda}^2(s,\omega)\bar{\eta}
\end{aligned}
$$

and

$$V(k_0, x(k,\omega) - \bar{x}(k), \omega) \leq \alpha(\|y_0(\omega) - z_0\|^2)\eta^2(k - k_0).$$

Now, by an application of hypothesis (iii) of Theorems 3.3.2 and 1.1.4, we have

$$V(k, y(k,\omega) - \bar{x}(k), \omega) \leq V(k_0, x(k,\omega) - \bar{x}(k), \omega)$$
$$+ \sum_{s=k_0+1}^{k} \mathcal{D}^* W(k, x(k, s, y(s,\omega)) - x(k, s, \bar{x}(s)), \omega), \tag{3.3.14}$$

where

$$\|\mathcal{D}^* W(k, x(k, s, y(s,\omega)) - x(k, s, \bar{x}(s)), \omega)\|$$
$$\leq \|2(\Phi(k, k_0)(y_0, (\omega) - z_0)^T \Phi(k, s) R(s, y, \omega)\|$$
$$+ \| \sum_{r=k_0+1}^{k} (\Phi(k, r) R(r, y, \omega))^T \Phi(k, s) R(s, y, \omega)\|$$
$$\leq \eta(k - s)[2\|x(k, k_0, y_0(\omega)) - x(k, k_0, z_0)\| \|R(s, y, \omega)\|$$
$$+ \frac{1}{2} \sum_{r=k_0+1}^{k} \|\Phi(k, r)\|(\|R(r, y, \omega)\|^2 + \|R(s, y, \omega)\|^2)]$$
$$\leq \eta(k - s)[\|x(k, k_0, y_0(\omega)) - x(k, k_0, z_0)\|^2 + \|R(s, y, \omega)\|^2$$
$$+ \frac{1}{2} \sum_{r=k_0+1}^{k} \|\Phi(k, r)\|(\|R(r, y, \omega)\|^2 + \|R(s, y, \omega)\|^2)]$$
$$\leq \eta(k - s)[V(k, x(k,\omega) - \bar{x}(k), \omega) + \bar{\lambda}^2(s,\omega)\|y(s - 1, \omega)\|^2$$
$$+ \frac{1}{2} \sum_{r=k_0+1}^{k} \eta(k - r)(\bar{\lambda}^2(r,\omega)\|y(r - 1, \omega)\|^2$$
$$+ \bar{\lambda}^2(s,\omega)\|y(s - 1, \omega)\|^2)]$$
$$\leq \eta(k - s)[V(k, x(k,\omega) - \bar{x}(k), \omega)$$
$$+ \bar{\lambda}^2(s,\omega)\|y(s - 1, \omega)\|^2 \left(1 + \frac{1}{2} \sum_{r=k_0+1}^{k} \eta(k - r)\right)$$

$$
\begin{aligned}
&+\frac{1}{2}\sum_{r=k_0+1}^{k}\eta(k-r)\bar{\lambda}^2(r,\omega)\|y(r-1,\omega)\|^2]\\
&\le \eta(k-s)[V(k,x(k,\omega)-\bar{x}(k),\omega)+\lambda_1(s,\omega)\|y(s-1,\omega)\|^2\\
&+\frac{1}{2}\sum_{r=k_0+1}^{k}\eta(k-r)\bar{\lambda}^2(r,\omega)\|y(r-1,\omega)\|^2]\\
&\le \eta(k-s)[V(k,x(k,\omega)-\bar{x}(k),\omega)+2\lambda_1(s,\omega)\\
&\times\|y(s-1,\omega)-z(s-1)\|^2+2\lambda_1(s,\omega)\|z(s-1)\|^2\\
&+\sum_{r=k_0+1}^{k}\eta(k-r)\bar{\lambda}^2(r,\omega)(\|y(r-1,\omega)-x(r-1)\|^2\\
&+\|\bar{x}(r-1)\|^2)]\\
&\le \eta(k-s)[V(k,x(k,\omega)-\bar{x}(k))+\lambda_1^*(s,\omega)\\
&\times V(s,y(s-1,\omega)-z(s-1),\omega)\\
&+\sum_{r=k_0+1}^{k}\eta(k-r)\bar{\lambda}^2(r,\omega)V(r,y(r-1,\omega)-x(r-1),\omega)\\
&+\sum_{r=k_0+1}^{k}\eta(k-r)\bar{\lambda}^2(r,\omega)V(r,\bar{x}(r-1),\omega)\\
&+\lambda_1^*(s,\omega)V(s,z(s-1),\omega)]. \qquad (3.3.15)
\end{aligned}
$$

This verifies hypothesis (iii) of Theorem 3.3.2. Let $m(k,\omega) = V(k,y(k,\omega)-\bar{x}(k),\omega)$. We note that

$$
\begin{aligned}
&V(k_0,x(k,\omega)-\overline{x}(k),\omega)+\sum_{s=k_0+1}^{k}\eta(k-s)V(k,x(k,\omega)-\bar{x}(k),\omega)\\
&+\sum_{s=k_0+1}^{k}\eta(k-s)\left(\sum_{r=k_0+1}^{k}\eta(k-r)\bar{\lambda}^2(r,\omega)V(r,\bar{x}(r-1),\omega\right)
\end{aligned}
$$

$$
\begin{aligned}
&+ \sum_{s=k_0+1}^{k} \eta(k-s)\lambda^*(s,\omega)V(s,z(s-1),\omega) \\
&\le \alpha\|y_0(\omega)-z_0\|^2\eta^2(k-k_0) \\
&\quad + \bar{\eta}\alpha\|y_0(\omega)-z_0\|^2\eta^2(k-k_0) \\
&\quad + \bar{\eta}\sum_{r=k_0+1}^{k} \eta(k-s)\bar{\lambda}^2(r,\omega)\eta^2(r-1-k_0)\|z_0\|^2 \\
&\quad + \sum_{r=k_0+1}^{k} \eta(k-s)\lambda_1^*(s,\omega)\eta^2(s-1-k_0)\|z_0\|^2 \\
&\le \alpha\|y_0(\omega)-z_0\|^2(1+\bar{\eta})\eta(k-k_0) \\
&\quad + K_1\|z_0\|^2\left(\sum_{r=k_0+1}^{k}(\bar{\lambda}^2(r,\omega)+\lambda_1^*(r,\omega))\right)\eta(k-k_0)
\end{aligned}
$$

and

$$
\begin{aligned}
&\sum_{s=k_0+1}^{k} \eta(k-s)\lambda_1^*(s,\omega)V(s,y(s-1,\omega)-z(s-1),\omega) \\
&\qquad + \sum_{s=k_0+1}^{k}\eta(k-s)\left(\sum_{r=k_0+1}^{k}\eta(k-r)\bar{\lambda}^2(r,\omega)\right. \\
&\qquad \left.\times\ V(r,y(r-1,\omega)-\bar{x}(r-1)\vphantom{\sum_{r=k_0+1}^{k}}\right) \\
&\quad \le \sum_{s=k_0+1}^{k}\eta(k-s)(\lambda_1^*(s,\omega)m(s-1,\omega)) \\
&\qquad + \bar{\eta}\sum_{r=k_0+1}^{k}\eta(k-r)\bar{\lambda}^2(r,\omega)m(r-1,\omega)
\end{aligned}
$$

$$\leq \sum_{s=k_0+1}^{k} \eta(k-s)(\lambda_1^*(s,\omega) + \bar{\lambda}^2(s,\omega)\bar{\eta})m(s-1,\omega)$$

$$\leq \sum_{s=k_0+1}^{k} \eta(k-s)\hat{\lambda}(s,\omega)m(s-1,\omega).$$

From these considerations, together with (3.3.14) and (3.3.15), we obtain

$$\begin{aligned} m(k,\omega) \leq (\alpha\|y_0(\omega) - z_0\|^2)(1+\bar{\eta}) + K_1\|z_0\|^2 \\ \times \left(\sum_{s=k_0+1}^{k} \bar{\lambda}^2(s,\omega)(3+\bar{\eta}) \right) \eta(k-k_0) \\ + \sum_{s=k_0+1}^{k} \eta(k-s)\hat{\lambda}(s,\omega)m(s-1,\omega). \end{aligned}$$

That is,

$$m(k,\omega) = n(k,\omega) + \sum_{s=k_0+1}^{k} \eta(k-s)\hat{\lambda}(s,\omega)m(s-1,\omega), \qquad (3.3.16)$$

where

$$\begin{aligned} n(k,\omega) &= \left\{ \alpha\|y_0(\omega) - z_0\|^2(1+\bar{\eta}) + K_1\|z_0\|^2 \right. \\ &\quad \left. \times \left(\sum_{s=k_0+1}^{k} \bar{\lambda}^2(s,\omega)(3+\bar{\eta}) \right) \right\} \eta(k-k_0). \end{aligned}$$

Equation (3.3.16) reduces to

$$\begin{aligned} m(k,\omega) \leq n(k,\omega) + \sum_{s=k_0+1}^{k} n(s,\omega)\eta(k-s)\hat{\lambda}(s,\omega) \\ \times \exp\left(\sum_{\tau=s+1}^{k} \eta(k-\tau)\hat{\lambda}(\tau,\omega) \right). \end{aligned} \qquad (3.3.17)$$

From the definition of $m(k,\omega)$, we obtain

$$V(k, y(k,\omega) - \bar{x}(k), \omega) \le n(k,\omega) + \sum_{s=k_0+1}^{k} n(s,\omega)\eta(k-s)\hat{\lambda}(s,\omega)\exp\left(\sum_{\tau=s+1}^{k}\eta(k-\tau)\hat{\lambda}(\tau,\omega)\right). \tag{3.3.18}$$

If we assume

$$\lim_{k\to\infty} \alpha(1+\bar{\eta})\eta(k-k_0) = 0,$$

$$\lim_{k\to\infty} \alpha K_1 \sum_{s=k_0+1}^{k} E[\bar{\lambda}^2(s,\omega)](3+\bar{\eta})\eta(k-k_0) = 0,$$

$$\lim_{k\to\infty} \sum_{s=k_0+1}^{k} E^{1/2}\left[\alpha(1+\bar{\eta})\eta(s-k_0)\eta(k-s)\hat{\lambda}(s,\omega) \times \exp\left(\sum_{\tau=s+1}^{k}\eta(k-\tau)\hat{\lambda}(\tau,\omega)\right)\right]^2 = 0,$$

$$\lim_{k\to\infty} \sum_{s=k_0+1}^{k} E^{1/2}\left[K_1\bar{\lambda}^2(s,\omega)(3+\bar{\eta})\eta(s-k_0)\eta(k-s)\hat{\lambda}(s,\omega) \times \exp\left(\sum_{\tau=s+1}^{k}\eta(k-\tau)\hat{\lambda}(\tau,\omega)\right)\right]^2 = 0,$$

then from (3.3.18), the relative asymptotic stability of (1.1.55) and (1.1.56) (or (1.1.57)) follows by an application of Theorem 3.3.2.

3.4 Notes and Comments

The various qualitative and quantitative properties are outlined in this chapter. In particular, the definitions of stability, error estimates and relative stability are introduced, and sufficient conditions are given in terms of stochastic systems of iterative processes in Sections 3.1, 3.2 and 3.3, respectively. The method of generalized variation of parameters and comparison theorems developed in Chapters 1 and 2 in the context of energy/Lyapunov function are systematically used in a coherent manner. Section 3.1.1 deals with Theorem 3.1.1, which is based on the comparison Theorem 1.2.3, and Theorems 3.1.2 and 3.1.3 for the qth moment stability and asymptotic stability of systems of iterative processes under random parametric processes, which are based on the generalized variation of parameter Theorem 1.1.3. These results stemmed from the results of Ladde and Sambandham [33, 34]. These are also generalizations and extend the results of Lakshmikantham and Trigiante [39]. Section 3.1.2 contains Theorem 3.1.4, which is based on the comparison Theorem 2.2.1, and Theorems 3.1.5 and 3.1.6 for the qth moment stability and asymptotic stability of systems of iterative processes under Gaussian process perturbations, which are based on the generalized variation of parameter Theorem 2.1.2. The presented results are drawn from Ladde and Siljak [36, 37]. Section 3.2.1 consists of Theorem 2.2.1 based on the comparison Theorem 1.2.5 and the generalized variation of parameter Theorem 1.1.4-based Theorems 3.2.2 and 3.2.3, coupled with Theorem 3.2.2, and Theorem 1.1.5-based Theorem 3.2.4 for the qth moment error estimation characterizing stochastic effects ("stochastic versus deterministic" issues) on systems of iterative processes under random parametric processes. These results stemmed from the results of Ladde and Sambandham [28–34]. These also extend the results of Lakshmikantham and Trigiante [39] and Ladde and Sambandham [33, 34]. Section 3.2.2 consists of the comparison Theorem 2.2.3-based Theorem 3.2.4 and the generalized variation of parameter Theorem 2.1.2-based Theorem 3.2.5 for the qth moment error estimates of systems of iterative processes under Gaussian process perturbations. Section 3.3.1 deals with the comparison Theorem 1.2.5-based Theorem 3.3.1 and the generalized variation of parameter Theorem 1.1.4-based Theorem 3.3.2 for the qth moment relative stability and asymptotic stability of systems of iterative processes under random parametric processes. The results parallel to the above for the qth moment relative stability and asymptotic stability of systems of iterative processes under Gaussian process perturbations are left as exercises. These results stemmed from the results of

Ladde and Sambandham [31, 33, 34] and Ladde and Siljak [37]. These are also generalizations and extend the results of Lakshmikantham and Trigiante [39]. Further, several corollaries, remarks and examples are given to exhibit the scope of the roles of these very general results in all sections in a systematic and unified way.

Chapter 4

Random Polynomials

4.0 Introduction

In mathematical and engineering applications, it is well known that the knowledge of solutions to algebraic equations is important. For earlier studies on the theory of random polynomial.

In this chapter, we investigate numerical solutions of random algebraic equations. For this purpose, we utilize comparison theorems to study the convergence and order of convergence of the iterates of solutions of random equations. As special cases of such theorems, the convergences of the secant method and Muller's numerical scheme are analyzed. We demonstrate how these methods can be used to develop other higher-degree methods. This chapter provides a systematic and unified approach to these problems. Moreover, the results concerning error estimates between the solution process of algebraic equations and iterates are discussed. Finally, we provide examples to illustrate our results. The presented analysis is directly applicable to the study of simple roots of random polynomials. For related results, we refer to [5, 32].

4.1 Numerical Schemes

First, we formulate an iteration process by employing Newton's divided difference formula. Let

$$F(x, a_0(\omega), \ldots, a_n(\omega)) = 0 \tag{4.1.1}$$

be a random algebraic equation, where the coefficients a_i, $i = 0, \ldots, n$, are random parameters and $F \in C[D, R]$, where D is an open subset of R^{n+2}. Let us denote $f(x, \omega) = F(x, a_0(\omega), \ldots, a_n(\omega))$.

Using Newton's divided difference form of interpolating a random polynomial,

$$\begin{aligned} p(x, \omega) = f(x_k, \omega) &+ (x - x_k) f[x_k, x_{k-1}, \omega] \\ &+ (x - x_k)(x - x_{k-1}) f[x_k, x_{k-1}, x_{k-2}, \omega] \\ &+ \cdots + (x - x_k)(x - x_{k-1}) \cdots (x - x_{k-m+1}) \\ &\times f[x_k, x_{k-1}, \ldots, x_{k-m}, \omega], \end{aligned} \tag{4.1.2}$$

we obtain

$$\begin{aligned} p(x, \omega) = f(x_k, \omega) &+ (x - x_k) w_m + (x - x_k)^2 w_{m-1} \\ &+ \cdots + (x - x_k)^m w_1, \end{aligned} \tag{4.1.3}$$

where

$$\begin{aligned} w_m &= f[x_k, x_{k-1}, \omega] + (x_k - x_{k-1}) f[x_k, x_{k-1}, x_{k-2}, \omega] \\ &= f[x_k, x_{k-1}, \omega] + f[x_k, x_{k-2}, \omega] - f[x_{k-1}, x_{k-2}, \omega], \\ w_{m-1} &= f[x_k, x_{k-1}, \omega] + f[x_k, x_{k-2}, \omega] + f[x_k, x_{k-3}, \omega], \\ &\quad - f[x_{k-1}, x_{k-2}, \omega] - f[x_{k-1}, x_{k-3}, \omega] \\ &\quad - (x_k - x_{k-2}) f[x_{k-1}, x_{k-2}, x_{k-3}, \omega], \\ w_{m-2} &= f[x_k, x_{k-1}, \omega] + f[x_k, x_{k-2}, \omega] \\ &\quad + f[x_k, x_{k-3}, \omega] + f[x_k, x_{k-4}, \omega] \\ &\quad - f[x_{k-1}, x_{k-2}, \omega] - f[x_{k-1}, x_{k-3}, \omega] - f[x_{k-3}, x_{k-4}, \omega] \\ &\quad - (x_k - x_{k-2}) f[x_{k-1}, x_{k-2}, x_{k-3}, \omega] \\ &\quad - (x_k - x_{k-3}) f[x_{k-2}, x_{k-3}, x_{k-4}, \omega] \\ &\quad + (x_k - x_{k-2})(x_k - x_{k-3}) f[x_{k-1}, \ldots, x_{k-4}, \omega], \\ &\ \ \vdots \\ w_1 &= f[x_k, x_{k-1}, \ldots, x_{k-m}, \omega]. \end{aligned}$$

If $p(x, \omega) = 0$, then (4.1.3) is an mth-degree polynomial equation in $x - x_k$. Therefore, by solving (4.1.3), when $p(x, \omega) = 0$, we obtain

$$x - x_k = g(x_k, x_{k-1}, \ldots, x_{k-m}, \omega).$$

If we denote x by x_{k+1}, then this equation can be written as

$$x_{k+1} = x_k + g(x_k, x_{k-1}, \ldots, x_{k-m}, \omega). \tag{4.1.4}$$

By taking $(m+1)$ initial conditions, $x_0(\omega), \ldots, x_m(\omega)$, (4.1.3) provides an iterative scheme $\{x_k(\omega), k \geq m+1\}$ for the solution of (4.1.1).

Let $\alpha(\omega)$ be a simple root of $f(x, \omega)$. To discuss the convergence and order of convergence of $\alpha(\omega) - x_{k+1}(\omega)$, we derive an analytic expression for this using Newton's divided difference formula. By setting $x = \alpha(\omega)$, (4.1.2) reduces to

$$\begin{aligned}
p(\alpha(\omega), \omega) = {} & f(x_k, \omega) + (\alpha(\omega) - x_k) f[x_k, x_{k-1}, \omega] \\
& + (\alpha(\omega) - x_k)(\alpha(\omega) - x_{k-1}) f[x_k, x_{k-1}, x_{k-2}, \omega] \\
& + \cdots \\
& + (\alpha(\omega) - x_k)(\alpha(\omega) - x_{k-1}) \cdots (\alpha(\omega) - x_{k-m+1}) \\
& \times f[x_k, x_{k-1}, \ldots, x_{k-m}, \omega],
\end{aligned}$$

where x_k's are determined using (4.1.4). From (4.1.4), we have

$$\begin{aligned}
p(x_{k+1}, \omega) = 0 = {} & f(x_k, \omega) + (x_{k+1} - x_k) f[x_k, x_{k-1}, \omega] \\
& + (x_{k+1} - x_k)(x_{k+1} - x_{k-1}) f[x_k, x_{k-1}, x_{k-2}, \omega] \\
& + \cdots \\
& + (x_{k+1} - x_k)(x_{k+1} - x_{k-1}) \cdots (x_{k+1} - x_{k-m+1}) \\
& \times f[x_k, \ldots, x_{k-m}, \omega].
\end{aligned}$$

From these equations and letting

$$\begin{aligned}
R(\alpha, \omega) = {} & \big[(\alpha(\omega) - x_k) + (\alpha(\omega) - x_{k-1}) - (\alpha(\omega) - x_{k+1})\big] \\
& \times f[x_k, x_{k-1}, x_{k-2}, \omega] \\
& + \big[(\alpha(\omega) - x_k)(\alpha(\omega) - x_{k-1})
\end{aligned}$$

$$
\begin{aligned}
&+ (\alpha(\omega) - x_{k-1})(\alpha(\omega) - x_{k-2}) \\
&+ (\alpha(\omega) - x_{k-2})(\alpha(\omega) - x_k) \\
&- \{(\alpha(\omega) - x_k) + (\alpha(\omega) - x_{k-1}) \\
&+ (\alpha(\omega) - x_{k-2})\}(\alpha(\omega) - x_{k+1}) + (\alpha(\omega) - x_{k+1})^2] \\
&\times f[x_k, \ldots, x_{k-3}, \omega] \\
&+ \cdots \\
&+ \big[(\alpha(\omega) - x_k) \cdots (\alpha(\omega) - x_{k-m+1}) + \cdots \\
&+ (\alpha(\omega) - x_{k-2}) \cdots (\alpha(\omega) - x_{k-m})(\alpha(\omega) - x_k) \\
&- \{(\alpha(\omega) - x_k) \cdots (\alpha(\omega) - x_{k-m+2}) + \cdots \\
&+ (\alpha(\omega) - x_{k-2}) \cdots (\alpha(\omega) - x_{k-m})\}(\alpha(\omega) - x_{k+1}) \\
&+ \cdots \\
&+ (-1)^{m-2}\{(\alpha(\omega) - x_k) + (\alpha(\omega) - x_{k-1}) + \cdots \\
&+ (\alpha(\omega) - x_{k-m+1})\}(\alpha(\omega) - x_{k+1})^{m-2} \\
&(-1)^{m-1}(\alpha(\omega) - x_{k+1})^{m-1}\big] f[x_k, \ldots, x_{k-m}, \omega], \qquad (4.1.5)
\end{aligned}
$$

followed by algebraic manipulation, we obtain

$$p(\alpha(\omega), \omega) = (\alpha(\omega) - x_{k+1})[f[x_k, x_{k-1}, \omega] + R(\alpha(\omega), \omega)]. \qquad (4.1.6)$$

From (4.1.6), we have

$$
\begin{aligned}
\alpha(\omega) - x_{k+1} &= \frac{p(\alpha(\omega), \omega)}{f[x_k, x_{k-1}, \omega] + R(\alpha, \omega)} \\
&= -(\alpha(\omega) - x_k)(\alpha(\omega) - x_{k-1}) \cdots (\alpha(\omega) - x_{k-m}) \\
&\quad \times \frac{f[x_k, x_{k-1}, \ldots, x_{k-m}, \omega]}{f[x_k, x_{k-1}, \omega] + R(\alpha, \omega)}. \qquad (4.1.7)
\end{aligned}
$$

Illustration 4.1.1. We note that when $m = 1$, (4.1.3) reduces to

$$p(x, \omega) = f(x_k, \omega) + (x(\omega) - x_k) f[x_k, x_{k-1}, \omega]. \qquad (4.1.8)$$

If $p(x, \omega) = 0$, then

$$x(\omega) - x_k(\omega) = -\frac{f(x_k, \omega)}{f[x_k, x_{k-1}, \omega]}.$$

By denoting $x(\omega)$ by $x_{k+1}(\omega)$, we obtain

$$x_{k+1}(\omega) = x_k(\omega) - \frac{f(x_k, \omega)}{f[x_k, x_{k-1}, \omega]}.$$

That is,

$$x_{k+1}(\omega) = x_k(\omega) - (x_k(\omega) - x_{k-1}(\omega)) \frac{f(x_k, \omega)}{f(x_k, \omega) - f(x_{k-1}, \omega)}. \quad (4.1.9)$$

Illustration 4.1.2. For $m = 2$, (4.1.3) can be written as

$$p(x, \omega) = f(x_k, \omega) + w_2(x(\omega) - x_k(\omega)) + w_1(x(\omega) - x_k(\omega))^2, \quad (4.1.10)$$

where

$$\begin{aligned} w_2 &= f[x_k, x_{k-1}, \omega] + f[x_k, x_{k-2}, \omega] - f[x_{k-1}, x_{k-2}, \omega], \\ w_1 &= f[x_k, x_{k-1}, x_{k-2}, \omega]. \end{aligned}$$

If $p(x, \omega) = 0$, then by solving the resulting quadratic equation, we obtain

$$x(\omega) - x_k(\omega) = \frac{-w_2 \pm \sqrt{w_2^2 - 4f(x_k, \omega) f[x_k, x_{k-1}, x_{k-2}, \omega]}}{2f[x_k, x_{k-1}, x_{k-2}, \omega]}.$$

Denoting $x(\omega)$ by $x_{k+1}(\omega)$ in this equality and simplifying, we obtain

$$x_{k+1}(\omega) = \frac{2f(x_k, \omega)}{w_2 \pm \sqrt{w_2^2 - 4f(x_k, \omega) f[x_k, x_{k-1}, x_{k-2}, \omega]}}, \quad (4.1.11)$$

with the sign chosen to maximize the magnitude of the denominator.

4.2 Formulation of Difference Inequality and Comparison Theorem

In this section, we formulate the difference inequality and outline the comparison theorems.

Theorem 4.2.1. *Let*

$$M \equiv M(\omega) = \max_I \left| \frac{f^{(m+1)}(\eta_k, \omega)}{(m+1)![f'(\xi_k, \omega) + R(\zeta_k, \omega)]} \right|, \tag{4.2.1}$$

where η_k *is between* $x_k, \dots, x_{k-m}, \alpha$, ξ_k *is between* x_k *and* x_{k-1} *and* ζ_k *is between* x_i*'s in each divided difference in* $R(\alpha, \omega)$. *Let* $e_k = M|\alpha(\omega) - x_k|$. *Then,* (4.1.7) *can be rewritten as*

$$M^m e_{k+1} = e_k e_{k-1} \cdots e_{k-m} \left| \frac{f^{(m+1)}(\eta_k, \omega)}{(m+1)![f'(\xi_k, \omega) + R(\zeta_k, \omega)]} \right|. \tag{4.2.2}$$

From (4.2.1), (4.2.2) *reduces to*

$$M^m e_{k+1} \le e_k e_{k-1} \cdots e_{k-m}. \tag{4.2.3}$$

The details are outlined in Corollary 1.2.4. Inequality (4.2.3) is a difference inequality for $M|\alpha(\omega) - x_{k+1}|$ in terms of $M|\alpha(\omega) - x_i(\omega)|$, $i = k - m, \dots, k$.

We also remark that (4.1.9) exhibits the well-known secant method of iteration. This is presented as a corollary.

Corollary 4.2.1. *From* (4.1.2) *and* (4.1.9), (4.2.2) *and* (4.2.3) *reduce to*

$$M e_{k+1} = e_k e_{k-1} \left| \frac{f''(\eta_k, \omega)}{2f'(\xi_k, \omega)} \right| \tag{4.2.4}$$

and

$$e_{k+1} \le e_k e_{k-1}, \tag{4.2.5}$$

respectively.

It is obvious that (4.2.5) is a useful inequality to study the convergence and order of convergence of the iterates of (4.1.1) using the secant method.

As before, we remark that (4.1.11) is the well-known Muller's method of iteration. Furthermore, we present the difference inequality corresponding to (4.1.10) and (4.1.11) as a corollary.

Corollary 4.2.2. *From* (4.2.2) *inequality,* (4.2.3) *reduces to*

$$M^2 e_{k+1} = e_k e_{k-1} e_{k-2} \frac{|f^{(3)}(\eta_k, \omega)|}{|6[f'(\xi_k, \omega) + R(\zeta_k, \omega)]|} \tag{4.2.6}$$

and

$$M e_{k+1} \leq e_k e_{k-1} e_{k-2}, \tag{4.2.7}$$

respectively.

It is obvious that (4.2.7) is a useful inequality to study the convergence and order of convergence of the iterates of (4.1.1) using Muller's method.

4.3 Stability and Convergence Analysis

To discuss the convergence and order of convergence of $\alpha(\omega) - x_{k+1}(\omega)$ defined by (4.1.7), we employ the comparison theorem developed in Section 4.2. First, we state a few definitions.

Definition 4.3.1. A sequence of iterates $\{x_k(\omega), k \geq 0\}$ converges with a probability of one (a.s., a.e.) to the random variable $\alpha(\omega)$ if

$$P(\omega : x_k(\omega) \to \alpha(\omega) \text{ as } k \to \infty) = 1,$$

whenever

$$|\alpha(\omega) - x_0(\omega)| \leq \delta,$$

for some $\delta > 0$.

Definition 4.3.2. A sequence of iterates $\{x_k(\omega), k \geq 0\}$ is said to converge with a probability of one and with an order of q (a.s., a.e.) to the random variable $\alpha(\omega)$ if there exists a nonnegative random number c such that

$$P\left(\omega : \frac{|\alpha(\omega) - x_{k+1}(\omega)|}{|\alpha(\omega) - x_k(\omega)|^q} \to c, \text{ as } k \to \infty\right) = 1,$$

whenever

$$|\alpha(\omega) - x_0(\omega)| \leq \delta,$$

for some real number $\delta > 0$.

Definition 4.3.3. From Definition 4.3.2, if $q = 1$, the sequence is said to converge linearly to $\alpha(\omega)$ w.p. 1.

In this case, we require $c < 1$ w.p. 1, where c is called the rate of linear convergence of $x_k(\omega)$ to $\alpha(\omega)$ w.p. 1.

Definition 4.3.4. From Definition 4.3.2, if $q = 2$, the convergence is called quadratic convergence, etc.

Next, we state and prove a theorem which establishes the convergence and order of convergence of (4.1.4).

Theorem 4.3.1. *Let $f(x,\omega)$ be defined on the complete probability space $(\Omega, \mathcal{B}, P)$. Let $\Omega_0 \subset \Omega$ be such that $P(\Omega_0) = 1$. For each $\omega \in \Omega_0$, let $f(\alpha,\omega) = 0$, $f'(\alpha,\omega) \neq 0$ and $f(x,\omega), f'(x,\omega), f''(x,\omega), \ldots, f^{(m+1)}(x,\omega)$ be sample continuous for all values of x in the closed finite interval I. Let*

$$M \equiv M(\omega) = \max_I \left| \frac{f^{(m+1)}(x,\omega)}{(m+1)!\,[f'(x,\omega) + R(x,\omega)]} \right| < \infty, \tag{4.3.1}$$

where R is as defined in (4.1.5). Let $x_0(\omega), \ldots, x_m(\omega)$ be random variables such that

$$M|\alpha(\omega) - x_i| < 1, \qquad i = 0, 1, \ldots, m. \tag{4.3.2}$$

Then, the sequence $\{x_k(\omega)\}$ defined by (4.1.4) converges w.p. 1 to $\alpha(\omega)$, and the order of convergence is q ($q < 2$). Moreover,

$$|\alpha(\omega) - x_k(\omega)| \leq \beta^{q_k^0} \delta^{q_k+1}, \tag{4.3.3}$$

where $0 < \delta_q < 1$ such that $M|\alpha(\omega) - x_i(\omega)| \leq \delta$, $i = 0, 1, \ldots, m$, $\beta = M^{1-m}$ and q_{k+1} is the solution to the difference equation

$$\begin{aligned} q_{k+1} &= q_k + q_{k-1} + \cdots + q_{k-m}, \\ q_0 &= q_1 = \cdots = q_m = 1. \end{aligned} \tag{4.3.4}$$

q_k^0 is the solution process of (1.2.19).

Proof. It is obvious that (4.2.3) can be rewritten as

$$z_{k+1}(\omega) \le G(k, z_k(\omega), \omega), \tag{4.3.5}$$

where G is as defined in (1.2.16) with $z_k^{(i)}(\omega) = e_{k-m+j-1}$, for $j = 1, 2, \ldots, m+1$, and $z_k(\omega) = [e_0, e_1, \ldots, e_m]^T$. By applying Corollary 1.2.4, we get

$$z_k(\omega) \le r_k(\omega), \quad k \ge m, \text{ w.p. } 1, \tag{4.3.6}$$

where $r_k(\omega)$ is the solution process of (1.2.17) with $z_m = [e_0, e_1, \ldots, e_m]^T$, and it is expressed in (1.2.18). As in the proof of Corollary 1.2.4, in order to prove the convergence of $\{x_k(\omega)\}$ w.p. 1, it is enough if we show that the solution process $r_k(\omega)$ of the comparison system (1.2.17) converges to zero w.p. 1. For any given $0 < \delta < 1$, we assume that

$$e_i \le \delta, \quad i = 0, 1, \ldots, m. \text{ w.p. } 1. \tag{4.3.7}$$

Using this identity, one can find the following estimates of $r_k(\omega)$. From (1.2.18), we have

$$r_k(\omega) = \begin{pmatrix} \beta^{q_{k-m}^0} e_m^{q_{k-m}^m} e_{m-1}^{q_{k-m}^{m-1}} \cdots e_0^{q_{k-m}^0} \\ \beta^{q_{k-m+1}^0} e_m^{q_{k-m+1}^m} e_{m-1}^{q_{k-m+1}^{m-1}} \cdots e_0^{q_{k-m}^0} \\ \cdots\cdots\cdots \\ \beta^{q_k^0} e_m^{q_k^m} e_{m-1}^{q_k^{m-1}} \cdots e_0^{q_k^0} \end{pmatrix}$$

$$\le \begin{pmatrix} \beta^{q_{k-m}^0} \delta^{q_{k-m+1}} \\ \beta^{q_{k-m+1}^0} \delta^{q_{k-m+2}} \\ \cdots\cdots\cdots \\ \beta^{q_k^0} \delta^{q_{k+1}} \end{pmatrix}, \tag{4.3.8}$$

where q_k and q_k^i, $i = 0, 1, \ldots, m$, are as defined in (1.2.19) and (4.3.4), and we note that $q_k \to \infty$ as $k \to \infty$. From (4.3.8), we obtain

$$e_k \le \beta^{q_k^0} \delta^{q_{k+1}} \le (M^{-m+1})^{q_k^0} \delta^{q_{k+1}}. \tag{4.3.9}$$

This shows that $x_k(\omega) \to \alpha(\omega)$ as $k \to \infty$ w.p. 1.

To prove the order of convergence, we proceed as follows. Let $\epsilon = \frac{1}{q_s}$, where q_s is given by (4.3.4), for some $s > 0$, such that $\bar{M}^{-}\epsilon > 0$ and

$$\bar{M} \equiv \bar{M}(\omega) = \left| \frac{f^{(m+1)}(\alpha, \omega)}{(m+1)! f'(\alpha, \omega)} \right|. \tag{4.3.10}$$

Then, based on continuity, we can find an interval $I = (\alpha - \delta(\epsilon), \alpha + \delta(\epsilon))$ such that

$$M_{-\epsilon} \equiv M - \epsilon < \left| \frac{f^{(m+1)}(x, \omega)}{(m+1)![f'(x, \omega) + R(x, \omega)]} \right| < M + \epsilon \equiv M_\epsilon, \tag{4.3.11}$$

for $x \in I$. Then, from (4.1.7), we obtain

$$\bar{M}_\epsilon^{m-1} e_{k+1} \le e_k e_{k-1} \cdots e_{k-m}, \tag{4.3.12}$$

$$e_k e_{k-1} \cdots e_{k-m} \le \bar{M}_{-\epsilon}^{m-1} e_{k+1}. \tag{4.3.13}$$

Let $u_i = e_i$, $y_i = e_i$, $i = 0, 1, \ldots, m$. Then, from (4.3.12) and an application of Corollary 1.2.4, we have

$$e_{k+1} \le \bar{M}_\epsilon^{-(m-1)q_k^0} u_m^{q_k^m} u_{m-1}^{q_k^{m-1}} \cdots u_0^{q_k^0}. \tag{4.3.14}$$

From Remark 1.2.2 and (4.3.14), we obtain

$$\bar{M}_{-\epsilon}^{-(m-1)q_k^0} y_m^{q_k^m} y_{m-1}^{q_k^{m-1}} \cdots y_0^{q_{k-m}^0} \le e_{k+1}. \tag{4.3.15}$$

By applying (4.3.14) and (4.3.15), using the fact that $q_k^j - q q_{k-1}^j \le 1$, for $j = 0, 1, 2, \ldots, m$ and the nature of $M_{-\epsilon}$, M_ϵ, we obtain

$$e_{k+1} \le c e_k^q, \qquad \text{for } k \ge m+1, \tag{4.3.16}$$

where q is the positive root of

$$x^m - x^{m-1} - \cdots - 1 = 0, \tag{4.3.17}$$

which is the auxiliary equation of (4.3.4). This implies that q is the order of convergence of $\alpha(\omega) - x_k(\omega)$ w.p. 1. This concludes the proof of the theorem.

Next, we state two corollaries which are special cases of Theorem 4.3.1.

Corollary 4.3.1. *Let $f(x,\omega)$ satisfy all the hypotheses of Theorem 4.3.1 with $m = 1$ and Corollary 4.2.1. Let $f(x,\omega)$, $f'(x,\omega)$ and $f''(x,\omega)$ be sample continuous for all values of x in the closed finite interval I. Let*

$$M \equiv M(\omega) = \max_I \left| \frac{f^{(2)}(x,\omega)}{2f'(x,\omega)} \right| < \infty. \tag{4.3.18}$$

Let $x_0(\omega)$ and $x_1(\omega)$ be random variables such that

$$M|\alpha(\omega) - x_i(\omega)| < 1, \qquad i = 0, 1.$$

Then, the sequence $\{x_k(\omega)\}$ defined by (4.1.9), which is the secant method of iteration, converges to $\alpha(\omega)$ w.p. 1, and the order of convergence is $q \approx 1.62$. Moreover,

$$|\alpha(\omega) - x_k(\omega)| \leq M^{-q_k^0} \delta^{q_{k+1}}, \tag{4.3.19}$$

where $0 < \delta < 1$ such that $M|\alpha(\omega) - x_i(\omega)| < \delta$, $i = 0, 1$, $q_{k+1} = q_k + q_{k-1}$, $q_0 = q_1 = 1$, $q_k^1 = q_{k-1}^1 + q_{k-1}^0$, $q_k^0 = q_{k-1}^1$, $[q_1^0, q_1^1] = [0, 1]$.

Corollary 4.3.2. *Let $f(x,\omega)$ satisfy all the hypotheses of Theorem 4.3.1 with $m = 2$ and Corollary 4.2.2. Let $f(x,\omega)$, $f'(x,\omega)$, $f''(x,\omega)$ and $f'''(x,\omega)$ be sample continuous for all values of x in the closed finite interval I. Let*

$$M \equiv M(\omega) = \max_I \left| \frac{f^{(3)}(\eta_k,\omega)}{6\,[f'(\xi_k(\omega)) + R(\eta_k,\omega)]} \right| < \infty, \tag{4.3.20}$$

where $R(\eta_k(\omega)) = [(\alpha - x_k) + (\alpha - x_{k-1}) - (\alpha - x_{k+1})]\frac{f''(\eta_k,\omega)}{2}$. Let $x_0(\omega)$, $x_1(\omega)$ and $x_2(\omega)$ be random variables such that

$$M|\alpha(\omega) - x_i(\omega)| < 1, \quad i = 0, 1, 2. \tag{4.3.21}$$

Then, the sequence $\{x_k(\omega)\}$ defined by (4.1.11), which is Muller's method of iteration, converges to $\alpha(\omega)$ w.p. 1, and the order of convergence is $q \approx 1.84$. Moreover,

$$|\alpha(\omega) - x_k(\omega)| \leq M^{-q_k^0} \delta^{q_{k+1}},$$

where $0 < \delta < 1$ such that $M|\alpha(\omega) - x_i(\omega)| \leq \delta$, $i = 0, 1, 2$, q_{k+1} is the solution process of $q_{k+1} = q_k + q_{k-1} + q_{k-2}$, $q_0 = q_1 = q_2 = 1$ and q_k^0 is *the solution process of* $q_k^2 = q_{k-1}^2 + q_{k-1}^1, q_k^1 = q_{k-1}^2 - q_{k-1}^0, q_k^0 = q_{k-1}^2,$ $[q_1^0, q_1^1, q_1^2] = [0, 1, 0]$.

Remark 4.3.1. We remark that in the secant method, real choices of $x_0(\omega)$ and $x_1(\omega)$ will lead to a real value of $x_2(\omega)$. But in Muller's method, real choices of $x_0(\omega)$, $x_1(\omega)$ and $x_2(\omega)$ may lead to complex roots of $f(x, \omega)$. This is an important aspect of Muller's method.

4.4 Error Estimates and Relative Stability

In this section, we obtain the error between the theoretical and approximate solution $x_n(\omega)$ of $f(x, \omega) = 0$. Suppose that the random equation $f(x, \omega) = 0$ derives its randomness from variables $a_0(\omega)$, $a_1(\omega), \ldots, a_n(\omega)$. Then, the approximate solution $x_k(\omega)$ obtained from the secant and Muller's methods can be considered a function of $a_0(\omega), \ldots, a_n(\omega)$, that is, $x_k(a_0, \ldots, a_n)$. Let $\mu(a_0, a_1, \ldots, a_n)$ be the joint probability density function of $a_0(\omega), \ldots, a_n(\omega)$. Then, the mean of $x_k(\omega)$ is given by

$$E(x_k(\omega)) = \int_{-\infty}^{\infty} \cdots \int_{-\infty}^{\infty} x_k(a_0, \ldots, a_n)\mu(a_0, \ldots, a_n)da_0 \cdots da_n, \tag{4.4.1}$$

and this is evaluated using numerical integration. The numerical approximation of integral (4.4.1) would be of the form

$$A_k = \sum_{j_0=1}^{k_0} \cdots \sum_{j_n=1}^{k_n} a_{j_0 \cdots j_n} x_k(a_{0,j_0}, a_{1,j_1}, \ldots, a_{n,j_n})\mu(a_{0,j_0}, \ldots, a_{n,j_n}), \tag{4.4.2}$$

where k_0, $k_1, \ldots, k_n$, $a_{j_0 \cdots j_n}^k$ are finite and positive numbers. In the following theorem, we state the error between (4.4.1) and (4.4.2) for Muller's method. A similar theorem for the secant method is left as an exercise.

Theorem 4.4.1. *Let* $f(\alpha, \omega) = 0$, $f'(\alpha, \omega) \neq 0$ *and* $f(x, \omega)$, $f'(x, \omega), \ldots, f^{(m+1)}(x, \omega)$ *be sample continuous for all values of* x

in the closed interval I. Let

$$M \equiv M(\omega) = \max_I \frac{|f^{(m+1)}(\eta_k, \omega)|}{|(m+1)![f'(\xi_k, \omega) + R(\eta_k, \omega)]|} < \infty.$$

Let $M|\alpha(\omega) - x_i(\omega)| < 1$, $i = 0, \ldots, m$, and A_k be a numerical approximation of the integral (4.4.1), where $\{x_k\}$ are iterates described by (4.1.4). Then,

$$|E\alpha(\omega) - A_k| \leq M^{-q_k^0} \delta^{q_{k+1}} + \delta_1, \tag{4.4.3}$$

where $M|\alpha(\omega) - x_i(\omega)| \leq \delta$, $i = 0, 1, \ldots, m$, and δ_1 is the absolute error of $E(x_k)$.

Proof. The approximation of $E[\alpha(\omega)]$ is subject to error from numerical approximation and numerical integration. Therefore,

$$\begin{aligned}
|E[\alpha(\omega)] - A_k| &\leq |E[\alpha(\omega)] - E[x_k(\omega)] + E[x_k(\omega)] - A_k| \\
&\leq |E[\alpha(\omega)] - E[x_k(\omega)]| + |E[x_k(\omega)] - A_k| \\
&\leq E|\alpha(\omega) - x_k(\omega)| + \delta_1 \\
&\leq M^{-q_k^0} \delta^{q_{k+1}} + \delta_1,
\end{aligned}$$

where q_k^0 and q_{k+1} are defined in (4.3.4). This completes the proof of the theorem.

Remark 4.4.1. We remark that for $m = 1, 2$, Theorem 4.4.1 will give errors due to the secant method and Muller's method, respectively.

The concept of relative stability is closely associated with the concept of the error estimate of a non-smooth iterative process relative to a smooth iterative one. Therefore, to avoid repetition, we simply formulate a problem for the non-smooth roots relative to smooth roots of polynomial equations. The details can be reformulated based on the general techniques described in Section 3.3. Therefore, it is enough to formulate a problem of relative stability in the framework of polynomial equations. The iterative process based on Newton's divided difference formula described in (4.1.4) or (4.1.7) represents a non-smooth iterative process. We then find a corresponding smooth iterative process, namely, that based on Newton's divided difference formula, as

$$\begin{cases} \alpha(\omega) - x_{k+1} = -(\alpha(\omega) - x_k)(\alpha(\omega) - x_{k-1}) \cdots (\alpha(\omega) - x_{k-m}) \\ \quad \frac{f[x_k, x_{k-1}, \ldots, x_{k-m}, \omega]}{f[x_k, x_{k-1}, \omega] + R(\alpha, \omega)}, & (4.1.7) \\ x_{k+1} = x_k + g(x_k, x_{k-1}, \ldots, x_{k-m}, \omega). & (4.1.4) \end{cases}$$

$$\begin{cases} \alpha - m_{k+1} = -(\alpha - m_k)(\alpha - m_{k-1}) \cdots (\alpha - m_{k-m}) \\ \quad \frac{h[m_k, m_{k-1}, \ldots, m_{k-m}]}{h[m_k, m_{k-1}] + R(\alpha)}, & (4.1.7) \\ m_{k+1} = m_k + H(m_k, m_{k-1}, \ldots, m_{k-m}). & (4.1.4) \end{cases}$$

$$\begin{cases} \alpha(\omega) - x_{k+1} = -(\alpha(\omega) - x_k)(\alpha(\omega) - x_{k-1}) \cdots (\alpha(\omega) - x_{k-m}) \\ \quad \frac{h[x_k, x_{k-1}, \ldots, x_{k-m}, \omega]}{h[x_k, x_{k-1}, \omega] + R(\alpha, \omega)}, & (4.1.7) \\ x_{k+1} = x_k + H(x_k, x_{k-1}, \ldots, x_{k-m}, \omega). & (4.1.4) \end{cases}$$

4.5 Numerical Examples

To illustrate the scope and usefulness of these methods, we consider an example. We consider a random algebraic equation. We choose the secant and Muller's methods to solve it.

Example 4.5.1. Let us find the real roots of $x^2 + Bx - C = 0$, where B and C are independent random variables with respective probability functions

$$f_B(b) = \begin{cases} 1, & 1 \leq b \leq 2, \\ 0, & \text{elsewhere}; \end{cases}$$

$$f_C(c) = \begin{cases} 0.5, & 2 \leq c \leq 4, \\ 0, & \text{elsewhere}. \end{cases}$$

In Table 4.1, we present estimates for the positive root of $x^2 + Bx + C = 0$.

The results in Table 4.1 are for a sample size of 100. We note that for the secant method, the number of iterations never exceeded 6, and the average number of iterations was 5.3. For Muller's method, the number of iterations never exceeded 2; therefore, the average number of iterations was also 2. In the last column of Table 4.1, for the secant method, the result given is for the fifth iteration of the sample, and

Table 4.1: Estimates for the positive root of $x^2 + Bx + C = 0$.

	$E(x_k)$	$\text{Var}(x_k)$	$E(\alpha)$	$E\|\alpha - x_k\|$	$E\|x_k - A_k\|$	$E\|\alpha - A_k\|$
Secant method	1.134302	0.003188	1.134302	0.0	0.051618	0.051618
Muller's method	1.133889	0.003175	1.133889	0.0	0.0007	0.0007

for Muller's method, it is the second iteration. Furthermore, Table 4.1 gives the errors in the numerical approximation and integration, and hence the error between the actual value and numerical approximation of the zero of the random algebraic polynomial of degree 2. The foregoing analysis provides a systematic formulation of convergence, the rate of convergence and error estimates. The numerical examples illustrate the usefulness of these methods for random modeling and its applications.

4.6 Notes and Comments

The random iterative process formulation by employing Newton's divided difference form of interpolation polynomial is from Bharuch-Reid and Sambandham [5] and Atkinson [2]. The convergence, error estimate and stability analyses are based on the results of Ladde and Sambandham [28, 32, 34] and are presented in Chapter 3 in a systematic and unified way. The contents of Sections 4.1 and 4.2 are due to Ladde and Sambandham [32]. Corollaries 4.2.1 and 4.2.2 are well-known results of Muller's iteration [2, 5, 32]. See also Bharucha-Reid and Sambandham [5]. The stability, convergence, error estimate and relative stability results in Sections 4.3 and 4.4 are based on the results of Ladde and Sambandham [32] and Lax [40].

Chapter 5

Numerical Schemes for Systems of Differential Equations with Random Parameters

5.0 Introduction

In this chapter, variational comparison theorems for a class of stochastic hybrid systems in the context of a vector Lyapunov-like function and a system of differential inequalities are developed. These comparison results are applied to study the convergence and relative stability analysis of stochastic large-scale approximation schemes for initial value problems in the framework of a stochastic hybrid system. Examples are given to illustrate the usefulness of the results.

We first formulate variational comparison theorems in the context of a vector Lyapunov-like function and a system of differential inequalities for a stochastic hybrid system. We then present a procedure for developing convergence and stability analysis for a stochastic numerical scheme for approximating solutions of stochastic large-scale systems of ordinary differential equations. The procedure is based on comparison and decomposition-aggregation methods in the framework of hybrid systems. Both the system of stochastic differential equations and its theoretical iterative algorithm (numerical scheme) are decomposed into lower-order subsystems. Each stochastic iterative subsystem is assigned to a subprocessor for parallel distributed computation of a cost/Lyapunov-like function that is assigned to each subprocessor. The individual Lyapunov functions are stacked up to form a vector Lyapunov function, and comparison theorems are derived to obtain error estimates.

In Section 5.1, the formulation of the problem is presented. In Section 5.2, a variational comparison theorem in the context of vector Lyapunov-like functions and a system of differential inequalities are established. In addition, a few auxiliary comparison results are formulated. In Section 5.3, in the context of the obtained variational comparison theorems, the sufficient conditions that provide the local discretization/truncation error, local analytic error and total error estimates are outlined. Furthermore, these results are applied to establish sufficient conditions for Dahlquist-type convergence and relative stability. In Section 5.3, variational comparison results of Section 5.2 are extended for studying stochastic large-scale hybrid systems. Section 5.5 deals with the error estimates, convergence and relative stability in moment and probability results for stochastic large-scale numerical schemes. These results are parallel to the results of Section 5.3. In Sections 5.6 to 5.9, local convergence analysis for large-scale algorithms are outlined. Examples are given to illustrate the fruitfulness of the results.

5.1 Theoretical Large-Scale Algorithms

The error estimation problem for stochastic numerical schemes with regard to a system of large-scale stochastic differential equations is proposed for large-scale stochastic hybrid systems. For this purpose, we assume the following: Let

$$\begin{cases} dx = F(t,x,\omega)dt, \quad x(t_0,\omega) = x_0(\omega), \\ \Delta y(k,\omega) = B(t_k, y(k,\omega),\omega), \quad y(k_0,\omega) = y_0(\omega), \\ y(t_0,\omega) = y_0(\omega) = x(t_0,\omega) = x_0(\omega), \end{cases} \tag{5.1.1}$$

where $F \in M[[t_0, t_0+\alpha] \times R^n \times \Omega, R^n]$ is continuous in x and satisfies the Carathéodory-type condition, that is, it is continuous in x for each (t,ω). Moreover, $\|F^i(k,x,\omega)\| \leq \lambda_i(k,\omega)$, $\sum_{i=1}^{p} \lambda_i(k,\omega) < \infty$, $I(0,N) = \{0,1,\ldots,N\}$, $B \in M[J \times R^n \times \Omega, R^n]$, $J = [t_0, t_0+\alpha]$, $t_0 \in [0,\infty)$, $t \geq t_0$, $k \geq k_0$, and

$$\Delta y(k,\omega) = y(k+1,\omega) - y(k,\omega), \quad k \in I[0,N] \text{ and } i \in I[1,p].$$

We assume that $x(t,\omega) = x(t,t_0,x_0,\omega)$ and $y(k,\omega) = y(k,k_0,\omega)$ exist and are solution processes of (5.1.1).

The system (5.1.1) is a large-scale system. It is studied here in a "piece by piece" manner, and conclusions about the pieces are utilized to form conclusions about the complex system as a whole. For this purpose, we decompose the stochastic system (5.1.1) into p interconnected (perturbed) subsystems of the following form:

$$\begin{cases} dx_i = f^i(t, x_i, \omega)dt + R^i(t, x, \omega)dt, \quad x_i(t_0, \omega) = x_i(t_0), \\ \Delta y_i(k, \omega) = b^i(t_k, y_i(k, \omega), \omega) + d_i(t_k, y(k, \omega), \omega), \\ y_i(k, \omega) = y_i^0(\omega), \\ y_i(t_0, \omega) = x_i(t_0, \omega) = x_i^0(\omega), \quad k \in [0, N], \ i \in [1, p], \end{cases} \tag{5.1.2}$$

where

$$f^i = (f_0^{iT}, f_1^{iT}, \ldots, f_j^{iT}, \ldots, f_m^{iT})^T,$$

and

$$R^i = (R_0^{iT}, R_1^{iT}, \ldots, R_j^{iT}, \ldots, C_m^{iT})^T$$

$$f_j^i \in (J \times R^{n_i}, R^{n_i}) \quad \text{and} \quad R^j \in C[J \times R^n, R^{n_i}],$$

for $i \in I(0, m)$;

$$b^i = (b_0^{iT}, b_1^{iT}, \ldots, b_m^{iT}, \ldots, b_\ell^{iT}, \ldots, b_{\ell_q}^{iT})^T$$

and

$$d^i = (d_0^{iT}, d_1^{iT}, \ldots, d_m^{iT}, \ldots, d_\ell^{iT}, \ldots, d_{\ell_q}^{iT})^T,$$

$$b_\ell^i \in M[J \times R^{n_i}, R^{n_i}] \quad \text{and } d_\ell^i \in M[J \times R^n, R^{n_i}],$$

for $\ell \in I(1, q)$; $x_0 = (x_1^{0T}, x_2^{0T}, \ldots, x_p^{0T})^T$ and $x = (x_1^T, x_2^T, \ldots, x_p^T)$, $y = (y_1^T, y_2^T, \ldots, y_p^T)^T$ and $\sum_{i=1}^p n_i = n$, for each $i \in I(1, p)$.

In (5.1.2), the interactions among p subsystems of the continuous part are described by c^i and those of the discrete part by d^i. The isolated subsystems corresponding to (5.1.2) are given by

$$\begin{cases} dx_i = f^i(t, x_i, \omega)dt, \quad x_i(t_0, \omega) = x_i(t_0), \\ \Delta y_i(k, \omega) = b^i(t_k, y_i(k, \omega), \omega), \quad y_i(k_0, \omega) = y_i^0(\omega), \\ y_i(t_0, \omega) = x_i(t_0, n) = x_i^0(\omega), \end{cases} \tag{5.1.3}$$

respectively, for $i \in I[1, p]$ and $k \in (0, N)$.

Example 5.1.1. In this example, we present the exact algorithm as a part of equation (5.1.1). Let $h = t_k - t_{k-1}$.

(i) For a given function $F(t, y, \omega)$, the Ito–Taylor series approximation can be written as

$$\begin{aligned} y(k+1,\omega) &= y(k,\omega) + F(t_k, y(k,\omega),\omega)(t_{k+1} - t_k) \\ &\quad + LF(t_k, y(k,\omega),\omega)\frac{(t_{k+1} - t_k)^2}{2}. \end{aligned}$$

This simplifies to

$$\begin{aligned} \Delta y(k,\omega) &= hF(t_k, y(k,\omega),\omega) + L\frac{h^2}{2}F(t_k, y(k,\omega),\omega) \\ &= B(t_k, y(k,\omega),\omega). \end{aligned}$$

(ii) Using integral approximation, we get

$$\begin{cases} dx = F(t, x_i, \omega)dt, \\ \Delta y(k,\omega) = \displaystyle\int_{t_k}^{t_{k+1}} F(t, y(k,\omega),\omega)dt. \end{cases} \tag{5.1.4}$$

Using the midpoint rule in (5.1.4), we obtain

$$\begin{aligned} \Delta y(k,\omega) &= (t_{k+1} - t_k)F\left(\frac{t_k + t_{k+1}}{2}, y(k,\omega),\omega\right) \\ &= hF\left(\frac{t_k + t_{k+1}}{2}, y(k,\omega),\omega\right) = B(t_k, y(k,\omega),\omega). \end{aligned}$$

Similarly, using the left-hand end point rule in (5.1.4), we obtain

$$\begin{aligned} \Delta y_i(k,\omega) &= (t_{k+1} - t_k)f^i(t_k, y_i(k,\omega),\omega) = hf^i(t_k, y_i(k,\omega),\omega) \\ &= b^i(t_k, y_i(k,\omega),\omega). \end{aligned}$$

From (i) and (ii), we obtain by Taylor approximation and integral approximation,

$$\Delta y_i(k,\omega) = b^i(t_k, y_i(k,\omega),\omega). \tag{5.1.5}$$

Therefore, equation (5.1.3), based on (5.1.5), can be rewritten as

$$\begin{cases} dx_i = f^i(t, x_i, \omega)dt, \quad x_i(t_0, \omega) = x_i^0(\omega), \\ \Delta y_i(k, \omega) = b^i(t_k, y_i(k, \omega), \omega), \quad y_i(t_0, \omega) = y_i^0(\omega), \\ y_i(t_0, \omega) = x_i(t_0, \omega). \end{cases} \tag{5.1.6}$$

If $b^i(t_k, y_i(k, \omega), \omega) = f^i(t_k, y_i(k, \omega), \omega)$, then using (5.1.5), the Euler approximation or the integral approximation can be written as

$$\begin{cases} dx_i = f^i(t_k, x_i, \omega)dt, \quad x_i(t_0, \omega) = x_i^0(\omega), \\ \Delta y_i(t_k, \omega) = hf^i(t_k, y_i(t_k, \omega), \omega), \quad y_i(t_0, \omega) = y_i^0(\omega), \\ y_i(t_0, \omega) = x_i(t_0, \omega). \end{cases} \tag{5.1.7}$$

5.2 Local Theoretical Discretization Error Estimates for Isolated Subsystems

A comparison principle provides a very general framework for studying the qualitative properties of dynamic systems. In this section, an attempt is made to formulate a variational comparison theorem for the isolated hybrid subsystems (7.6.6) in Section 7.6. These results connect the solution processes of stochastic hybrid subsystems, the maximal solution of comparison and the auxiliary systems of differential equations.

By setting $\bar{y}_i = y_i + b^i(t_k, y_i(k, \omega), \omega)$, the ith isolated hybrid subsystem (7.6.6) can be written as

$$\begin{cases} dx_i = f^i(t, x, \omega)dt, \quad x_i(t_k, \omega) = x_i^k(\omega), \\ d\bar{y}_i(t) = b^i(t_k, y_k, \omega)dt, \quad y_i(t_k, \omega) = y_i^k(\omega), \quad t \in J_k, \\ y_i(t_0, \omega) = y_i^0(\omega), \quad x_i(t_0, \omega) = x_i^0(\omega), \end{cases} \tag{5.2.1}$$

for $k \in I(0, N)$ and $i \in I[1, p]$. We note that $\bar{y}_i(t_{k+1}, \omega) = y_i(t_{k+1}, \omega), \bar{y}_i(t_k, \omega) = y_i(t_k, \omega)$. For further application, we need the following auxiliary system:

$$dz_i = G^i(t, z_i, \omega)dt, \quad z_i(t_0, \omega) = z_i^0(\omega), \tag{5.2.2}$$

where $G^i \in M[J \times R^{n_i} \times \Omega, R^{n_i}]$, and it is twice continuously differentiable with respect to z_i; the solution process $z_i(t, t_0, z_i^0, \omega)$ of

(5.2.2) exists for $t \geq t_0$; and it is unique and continuous with respect to the initial conditions $(t_0, z_i^0, (\omega))$ and locally Lipschitzian in z_i^0, for $i \in I[1,p]$.

For $q_i \geq 1$ and $V^i \in C[J \times R^{n_i} \times \Omega, R^{q_i}]$, $\frac{\partial}{\partial t}V^i(t, z_i, \omega)$, $\frac{\partial}{\partial x}V^i(t, z_i, \omega)$ and $\frac{\partial^2}{\partial x^2}V_i(t, z_i, \omega)$ are continuous in $(t, z_i) \in (R_+ \times R^n)$, and $\frac{\partial^2}{\partial x^2}V^i(t, z_i, \omega)$ is locally Lipschitzian in z_i, for each $t_0 \leq t_k \leq s \leq t \leq t_{k+1} \leq t_0 + \alpha$, $\sum_{i=1}^{p} q_i = q$, and $k \in I(0, N)$. We define the operator $D^+V^i_{(5.2.1)}$ as follows:

$$
\begin{aligned}
D^+_{(5.2.1)}&V^i(s, t_k, z_i(t, s, y_i - x_i, \omega), \omega) \\
&= \lim_{h \to 0^+} \sup \frac{1}{h}\big[V^i(s + h, z_i(t, s + h, y_i - x_i + F^i(s, t_k, y_i, x_i)), \omega) \\
&\quad - V^i(s, z_i(t, s, y_i - x_i, \omega), \omega)\big], \qquad (5.2.3)
\end{aligned}
$$

with $i \in I[1, p]$, for $F^i(s, t_k, y_i, x_i) = (-f^{iT}(s, x_i, \omega), b^{iT}(t_k, y_i, \omega))$.

Remark 5.2.1. Under the assumptions of G^i and V^i, for each $t_0 \leq t_k \leq s \leq t \leq t_{k+1} \leq t_0 + \alpha$ and $k \in I[0, N)$,

$$
\begin{aligned}
&D^+_{(5.2.1)}V^i(s, t_k, z_i(t, s, \Delta x_i, \omega), \omega) \\
&\quad \leq L_i^D V^i(s, t_k, y_i, z_i(t, s, \Delta x_i, \omega), \omega), \qquad (5.2.4)
\end{aligned}
$$

where L_i^D is defined by

$$
\begin{aligned}
L_i^D &V^i(s, t_k, y_i, z_i(t, s, \Delta x_i, \omega), \omega) \\
&= V_z^i(s, z_i(t, s, \Delta x_i, \omega), \omega) \\
&\quad + \frac{1}{2}\operatorname{tr}\frac{\partial^2}{\partial z_i^2}V^i(s, z_i(t, s, \Delta x_i, \omega), \omega)\theta_i^D(t, s, t_k, x_i, y_i) \\
&\quad + V_{zi}^i(s, z_i(t, s, \Delta x_i, \omega), \omega)\Bigg[\frac{1}{2}\operatorname{tr}\frac{\partial^2}{\partial x_i \partial x_i}z_i(t, s, \Delta x_i, \omega)F^{i^T}F^i \\
&\quad + \Phi^i(t, s, \Delta x_i)M_i^D(s, t_k, y_i, x_i)\Bigg],
\end{aligned}
$$

where

$$\begin{aligned}\Delta x_i &= y_i - x_i,\\ \theta_i^D(t, s, t_k, x_i, y_i) &= \Phi^{i^T}(t, s, \Delta x_i) F^{i^T}(s, t_k, y_i, x_i)\\ &\quad \times F^i(s, t_i, y_i, x_i)\Phi^i(t, s, \Delta x_i),\end{aligned}$$

and

$$M_i^D(s, t_k, y_i, x_i) = \lim_{h\to 0} \frac{1}{h}\left[F^i(s, t_k, y_i, x_i) - hG^i(s, \Delta x_i)\right].$$

Remark 5.2.2. D^+V^i is analogous to (5.2.3) with respect to the continuous part of (5.1.6), and its corresponding $Li^D V^i$ is defined by

$$\begin{aligned}&L_i^D V^i(s, z_i(t, s, \Delta x_i, \omega), \omega)\\ &= V_s^i(s, z_i(t, s, \Delta x_i, \omega), \omega) + \frac{1}{2}\operatorname{tr}\frac{\partial^2}{\partial z_i^2} V^i(s, z_i(t, s, \Delta x_i, \omega), \omega)\\ &\quad \times \Delta\theta_i^D(t, s, x_i) + V_{zi}^i(s, z_i(t, s, \Delta x_i, \omega), \omega)\\ &\quad + \left[\frac{1}{2}\operatorname{tr}\frac{\partial^2}{\partial x_i \partial x_i} z_i(t, s, \Delta x_i, \omega)\Delta\lambda_i^D(s, x)\right.\\ &\quad \left. + \Phi^i(t, s, \Delta x_i)\Delta M_i^D(s, x_i)\right],\end{aligned} \tag{5.2.5}$$

where

$$\begin{aligned}\Delta M_i^D(s, x) &= \lim_{h\to 0}\frac{1}{h}\left[\Delta f^i(s, x_i, \omega) - hG^i(s, \Delta x_i, \omega)\right],\\ \Delta\lambda_i^D(s, x_i) &= f^i(s, y_i) - f^i(s, x_i),\\ \Delta\theta_i^D(t, s, x_i) &= \Phi^i(t, s, \Delta x_i)\Delta\lambda_i^D(s, x)\Phi^{iT}(t, s, \Delta x_i),\\ \Delta x_i &= y_i - x_i.\end{aligned}$$

In the following, we provide local theoretical/discretization and local analytic error estimates for the solution process of the ith isolated subsystem (5.1.3) in the framework of the ith isolated hybrid subsystem (7.6.6).

Theorem 5.2.1. *Assume that*

(i) $g^i \in M[J \times J \times R^{n_i} \times R^{q_i} \times \Omega, R_+^{q_i}]$, $g_i(t, t_k, y_i, u_i, \omega)$ *is concave and quasimonotone nondecreasing in* u_i *for each* $(t, t_k, y_i) \in J \times J \times R^{n_i}$.

(ii) $r_i^k(t, \omega) = g^i(t, t_k, y_i, u_i, \omega)$ *is the maximal solution of the following comparison system:*

$$du_i = g^i(t, t_k, y_i, u_i, \omega)dt, \quad u_i^k(t_k, \omega) = u_i^k(\omega), \quad t \in J_k. \tag{5.2.6}$$

(iii) $z_i(t, s, x_i, \omega)$ *is the solution process of the auxiliary system* (5.2.2) *through* (s, x_i), $t_0 \le s \le t \le t_0 + \alpha$.

(iv) $V^i \in C[J \times R^{n_i} \times \Omega, R_+^{q_i}]$; $\frac{\partial}{\partial t} V^i(s, z_i, \omega)$, $\frac{\partial}{\partial x} V^i(s, z_i, \omega)$ *and* $\frac{\partial^2}{\partial x^2} V^i(s, z_i, \omega)$ *exist and are continuous on* $J \times R^{n_i}$; $\frac{\partial^2}{\partial x^2} V^i(s, z_i, \omega)$ *is locally Lipschitzian in* z_i, *for each* $t_0 \le t_k \le s \le t \le t_{k+1} < t_0 + \alpha$; *and* $L_i^D V^i(s, t_k, y_i, z_i(t, s, \Delta x_i))$ *defined in* (5.2.4) *satisfies the following relation:*

$$\begin{aligned} &L_i^D V^i(s, t_k, y_i, z_i(t, s, \Delta x_i), \omega) \\ &\quad \le g^i(s, t_k, y_i, V^i(s, z_i(t, s, \Delta x_i)), \omega). \end{aligned} \tag{5.2.7}$$

(v) $\hat{x}_i^k(t, \omega) = \hat{x}_i(t, t_k, x_i^k, \omega)$ *and* $\bar{y}_i^k(t, \omega) = \bar{y}_i(t, t_k, y_i^k, \omega)$ *are solutions of* (5.2.1) *through* (t_k, x_i^k) *and* (t_k, y_i^k), *respectively, where* $\{y^k\}_0^N$ *is a sequence in* $R[(\Omega, F, P), R^n]$ *and*

$$y^k = (y_1^{kT}, y_2^{kT}, \ldots, y_i^{kT}, \ldots, y_p^{kT})^T.$$

$y_i^k \in R[(\Omega, F, P), R^{n_i}]$, *and* $V^i(s, z_i(t, s, \bar{y}_i^k(s, \omega)) - \hat{x}_i^k(s, \omega))$ *exists for* $s \in [t_k, t]$, $t \in J_k$ *and* $k \in I(0, N)$. *Then,*

$$V^i(t, z_i(t, t_k, \bar{y}_i^k - \hat{x}_i^k, \omega), \omega) \le r_i(t, t_k, u_i^k, \omega) \tag{5.2.8}$$

whenever

$$V^i(t_k, z_i(t, t_k, y_i^k - x_i^k, \omega), \omega) \le u_i^k(\omega), \tag{5.2.9}$$

for $k \in I(0, N)$, $t \in J_k$ *and each* $i \in I(1, p)$.

Proof. Let $\hat{x}_i^k(t, \omega)$ and $\bar{y}_i^k(t)$ be the solution processes of (5.2.1) through (t_k, x_i^k) and (t_k, y_i^k), respectively. We set

$$m^i(s, \omega) = V^i(s, z_i(t, s, \bar{y}_i^k(s, \omega) - \hat{x}_i^k(s, \omega)), \omega),$$

for $t_k \le s \le t \le t_{k+1}$. We note that $m^i(t_k, \omega) = V^i(t_k, z_i(t, t_k, y_i^k - x_i^k, \omega), \omega)$.

Corollary 5.2.1. *Assume that all the hypotheses of Theorem* 5.2.1 *are satisfied, except that* (5.2.9) *and* (*v*) *are replaced by* $V^i(t_k, z(t, t_k, y_i^k - x_i^k, \omega), \omega) = u^k$ *and* $(v)'$, *respectively.* $\hat{x}_i^k(t, \omega)$ *and* $\bar{y}_i^k(t, \omega)$ *are the solution processes of the ith isolated hybrid subsystem* (5.1.6) *described by* (5.2.1) *through* (t_k, y_i^k) *and* (t_k, y_i^k), *respectively. Then,*

$$V^i(t_{k+1}, y_i^{k+1} - \hat{x}_i^k(t_{k+1}), \omega) \\ \leq v_i(t_{k+1}, t_k, V^i(t_k, z_i(t_{k+1}, t_k, 0, \omega), \omega)), \qquad (5.2.10)$$

for all $i \in I(1, p)$ *and* $k \in I(0, N)$.

Example 5.2.1. In this example, we illustrate the result in Theorem 5.2.1 for special values of V. Let the hybrid system be

$$\begin{cases} dx = f(t, x, \omega)dt, \\ \Delta y_k = hf(k, y, \omega), \text{ and} \\ dz = F(t, z, \omega)dt. \end{cases}$$

Let us assume that $V(t, x) = x^2$. Then, $V(s, z(t, s, x)) = z^2(t, s, x)$. This gives

$$\frac{\partial V}{\partial x} = 2x, \\ \frac{\partial V}{\partial t} = 0, \\ \frac{\partial^2 V}{\partial x^2} = 2,$$

$$d_s z(t, s, x) = \frac{\partial z}{\partial s} ds + \frac{\partial z}{\partial x} dx = -\Phi(t, s, x)F(s, x, \omega)ds \\ + \Phi(t, s, x)[F(s, x, \omega) - F(s, x, \omega) + f(s, x, \omega)]ds \\ = [-\Phi(t, s, x)F(s, x, \omega) + \Phi(t, s, x)f(s, x, \omega)]ds,$$

$$dV = \frac{\partial V}{\partial s} + \frac{\partial V}{\partial x} d_s z + \frac{1}{2}\frac{\partial^2 V}{\partial x^2}(dz)^2,$$

$$\begin{aligned}
dV &= 0 + 2z(t,s,x)d_s z + \left(\frac{1}{2}\right) 2(dz)^2 \\
&= 2z(t,s,x)[-\Phi(t,s,x)F(s,x,\omega)ds + \Phi(t,s,x)f(s,x,\omega)ds] \\
&= 2z(t,s,x)\Phi(t,s,x)[f(s,x,\omega) - F(s,x,\omega)]ds.
\end{aligned}$$

Let

$$\begin{aligned}
LV(s,z(t,s,x)) &= 2z(t,s,x)\Phi(t,s,x)[f(s,x,\omega) - F(s,x,\omega)] \\
&\leq \alpha(t,s,\omega)V(s,z(t,s,x)).
\end{aligned}$$

Therefore,

$$dV = LV(s,z(t,s,x))ds.$$

Let

$$\begin{aligned}
dm(s) &= dV(s,z(t,s,\bar{y}(s) - x(s))) \\
&= dz^2(t,s,\bar{y}(s) - x(s)), \\
dm(s) &= 2z(t,s,\bar{y}(s) - x(s))[\Phi(t,s,\bar{y}(s) - x(s)) \\
&\quad \times (-F(s,x,\omega) + f(s,x,\omega))]ds \\
&= 2z(t,s,\bar{y}(s) - x(s))\Phi(t,s,\bar{y}(s) - x(s))(f - F)ds.
\end{aligned}$$

Let $2z(t,s,\bar{y}(s) - x(s))\Phi(t,s,\bar{y}(s) - x(s))(f - F)(1 - h) \leq \lambda(s,\omega)z^2(t,s,\bar{y}(s) - x(s))$. Then,

$$D^+(m(s)) \leq \lambda(s,\omega)m(s).$$

This gives

$$m(s) \leq m(t_k)e^{\int_{t_k}^{s} \lambda(s,\omega)ds}, \quad t_k \leq t \leq t_{k+1}.$$

Therefore, if $z(t,s,\bar{y}(s) - x(s)) = \bar{y}(s) - x(s)$, then

$$(\bar{y}(k) - x(t))^2 = m(t) \leq m(t_k)e^{\int_{t_k}^{t} \lambda(s,\omega)ds}.$$

That is,

$$\begin{aligned}
\|\bar{y}(t_{k+1}) - x(t_{k+1})\|^2 &\leq z(t_{k+1}, t_k, (\bar{y}(t_k) - x(t_k))^2)e^{\int_{t_k}^{t} \lambda(s,\omega)ds} \\
&= \left(\prod_{i=0}^{k} \bar{y}(t_i) - x(t_i)\right) e^{\sum_{i=0}^{k} \int_{t_i}^{t_{i+1}} \lambda(s,\omega)ds}.
\end{aligned}$$

5.3 Local Analytical Error Estimates for Isolated Subsystems

In this section, we state an auxiliary comparison theorem that provides an estimate of local analytic error for the solution process of the ith isolated subsystem (5.1.3) by using the solution process of the ith isolated subsystem (7.6.6).

Theorem 5.3.1. *Assume that*

(i) $g^i \in M[J \times R^{q_i} \times \Omega, R_+^{q_i}]$, $g^i(t, u_i, \Omega)$ *is concave and quasimonotone nondecreasing in* u_i, *for each* $t \in J$, $i \in I[1,p]$.

(ii) $r_i^k(t,\omega) = r_i(t, t_k, u_i^k, \omega)$ *is the maximal solution of the following comparison system:*

$$du_i = g^i(t, u_i, \omega), \quad u_i(t_k, \omega) = u_i^k(\omega), \tag{5.3.1}$$

$i \in I[1,p]$.

(iii) $z_i(t, s, x_i, \omega)$ *is the solution process of the auxiliary system* (5.2.2) *through* (s, x_i), $t_0 \leq s \leq t \leq t_0 + \alpha$.

(iv) $V^i \in C[J \times R^{n_i} \times \Omega, R_+^{q_i}]$; $\frac{\partial}{\partial t}V^i(s, z_i, \omega)$, $\frac{\partial}{\partial x}V^i(s, z_i, \omega)$ *and* $\frac{\partial^2}{\partial x^2}V^i(s, z_i, \omega)$ *exist and are continuous on* $J \times R^{n_i} \times \Omega$; $\frac{\partial^2}{\partial x^2}V^i(s, z_i)$ *is locally Lipschitzian in* z_i, *for each* $t_0 \leq t_k \leq s \leq t \leq t_{k+1} < t_0 + \alpha$; *and* $L_i^D V^i(s, z_i(t, s, \Delta x_i, \omega), \omega)$ *defined in* (5.2.5) *satisfies the following relation:*

$$\begin{aligned} &L_i^D V^i(s, z_i(t, s, \Delta x_i, \omega), \omega) \\ &\quad \leq g^i(s, V^i(s, z_i(t, s, x_i - \bar{x}_i, \omega), \omega)). \end{aligned} \tag{5.3.2}$$

(v) $\hat{x}_{ix_i^0}(t, \omega)$ *and* $\hat{x}_i^k(t, \omega)$ *are solution processes of the continuous part of the ith isolated subsystem* (5.1.3) *and the hybrid subsystem* (7.6.6) *through* (t_0, x_i^0) *and* (t_k, y_i^k), *respectively, when* $\{y^k\}_0^N$ *is a sequence in* $R[(\Omega, F, P), R^n]$ *and*

$$y^k = ({y_1}^{kT}, {y_2}^{kT}, \ldots, y_i^{kT}, \ldots, y_p^{kT})^T,$$

${y_i}^k \in R[(\Omega, F, P), R^{n_i}]$, *and* $V^i(s, z_i(t, s, \hat{x}_i^k(s) - \hat{x}_{ix_i^0}(s), \omega), \omega)$ *exists for* $s \in [t_k, t]$, $t \in J_k$ *and* $k \in I(0, N)$. *Then,*

$$V^i(t, \hat{x}_i^k(t) - \hat{x}_{ix_i^0}(t), \omega) \leq r_i(t, t_k, u_i^k, \omega) \tag{5.3.3}$$

whenever

$$V^i(t_k, z_i(t, t_k, y_i^k - \hat{x}_{ix_i^0}(t_k), \omega), \omega) \leq u_i^k, \tag{5.3.4}$$

for all $i \in I[1,p]$, $t \in J_k$ *and* $k \in I(0,N)$.

Proof. Let $\hat{x}_{ix_i^0}(t)$ and $\hat{x}_i^k(t)$ be the solution processes of (5.1.3) and (7.6.6) through (t_0, x_i^0) and (t_k, y_i^k), respectively. We define

$$m^i(s) = V^i(s, z_i(t, s, \hat{x}_i^k(s) - \hat{x}_{ix_i^0}(s))), \quad \text{for } t_k \leq s \leq t \leq t_{k+1}. \tag{5.3.5}$$

By using the fundamental properties of the solution process of (5.2.2), we note that

$$m^i(t) = V^i(t, z_i(t, t, \hat{x}_i^k(t) - \hat{x}_{ix_i^0}(t)), \omega) = V^i(t, \hat{x}_i^k(t) - \hat{x}_{ix_i^0}(t), \omega)$$

and

$$m^i(t_k) = V^i(t_k, z_i(t, t_k, y_i^k - \hat{x}_{ix_i^0}(t_k)), \omega). \tag{5.3.6}$$

The rest of the proof of the theorem follows by imitating the standard argument.

Corollary 5.3.1. *Assume that all the hypotheses of Theorem* 5.3.1 *are satisfied, except that* (5.3.4) *is replaced by*

$$V^i(t_k, z_i(t_{k+1}, t_k, y_i^k - \hat{x}_{ix_i^0}(t_{k+1})), \omega) = u_i^k, \tag{5.3.7}$$

for all $i \in I[1,p]$ *and* $k \in I(0,N)$. *Then,*

$$\begin{aligned} &V^i(t_{k+1}, \hat{x}_i^k(t_{k+1}) - \hat{x}_{ix_i^0}(t_{k+1}), \omega) \\ &\quad \leq r_i(t_{k+1}, t_k, V^i(t_k, z_i(t_{k+1}, t_k, y_i^k - \hat{x}_{ix_i^0}(t_k))), \omega), \end{aligned} \tag{5.3.8}$$

where $r_i(t_{k+1}, t_k, u_i^k, \omega)$ *is the solution process of* (5.3.1) *through* (t_k, u_i^k), *for all* $i \in I[1,p]$ *and* $k \in I(0,N)$.

Remark 5.3.1. If $f^i \equiv 0$, then Theorems 5.2.1 and 5.3.1 reduce to the usual comparison theorem as special cases.

Example 5.3.1. Consider the following:

(i) For $i \in I[1,p]$, the comparison functions $g^i(t,u_i) \equiv 0$ are admissible in Theorem 5.3.1. In this case, (5.3.5) can be reduced to

$$\begin{aligned} V^i(t_{k+1}, \hat{x}_i^k(t_{k+1}) - \hat{x}_{ix_i^0}(t_{k+1}), \omega) \\ \leq V^i(t_k, z_i(t_{k+1}, t_k, y_i^k - \hat{x}_{ix_i^0}(t_k)), \omega). \end{aligned} \tag{5.3.9}$$

(ii) Let $G^i(t)$ be $q_i \times q_i$ continuous matrices whose off-diagonal entries are nonnegative, for each $i \in I(1,p)$. The comparison functions $g_i(t,u_i) = G^i(t)u_i$ are also admissible in Theorem 5.3.1. Here, $r_i(t,t_0,u_i^0) = \Phi_c^i(t,s)u_i^0$, where $\Phi_c^i(t,s)$ is the fundamental matrix solution of $du_i = G^i(t)u_i$, $\Phi_c^i(s,s) = I_{q_i \times q_i}$ (an identity matrix).

Example 5.3.2. For each $i \in I(1,p)$, the functions $V^i(s,z_i) = \|z_i\|$ and $V^i(s,z_i) = z_i^T P_i z_i$ are admissible in Theorems 5.2.1 and 5.3.1 with $q_i = 1$. Here, P_i is a nonnegative definite $n_i \times n_i$ matrix, for each $i \in I[1,p]$.

Remark 5.3.2. From various combinations of functions in Theorems 5.2.1 and 5.3.1, Corollaries 5.2.1 and 5.3.1, Remarks 5.2.1, 5.2.2 and 5.3.1 and Examples 5.3.1 and 5.3.2, one can obtain various possible special estimates of type (5.2.8), (5.2.10), (5.3.3) and (5.3.8). These particular estimates provide a basis for comparison with the existing results in the literature. This is illustrated in the following.

5.4 Global Theoretical Discretization Error Estimates for Isolated Subsystems

In this section, we propose to study the error estimates between the continuous- and discrete-time subsystems of isolated hybrid subsystems. The obtained results are applied to study error estimates in the numerical analysis of stochastic initial value problems. For each ith isolated hybrid subsystem (7.6.6), we associate a vector Lyapunov-like function. As a byproduct of this study, we also analyze the relative stability and convergence of this problem.

We are ready to present a very general result that establishes a global theoretical/discretization error estimate for the solution process of the ith subsystem (5.1.3).

Theorem 5.4.1. *Assume that all the hypotheses of Theorem 5.3.1 are satisfied. Further, assume that*

(a) *the sequence $\{y^k\}_0^N$ satisfies the following condition:*

$$\|y_i^k - \hat{x}^{k-1}(t_k)\| \leq \chi_i^{k-1}(N), \tag{5.4.1}$$

where $\chi_i^{k-1}(N)$ is a discrete process defined by

$$\chi_i^{k-1}(N) = \beta_i^{k-1}(h_N^k, N), \tag{5.4.2}$$

$\beta_i^{k-1} \in C[[0, h^0], R_+], h^0 > 0$, *for each* $i \in I(1,p)$, *and for all* $k \in I(1,N), N \in I(1,\infty)$;

(b) $w_i(t_k) = w_i(t_k, t_{k-1}, \omega_i^0)$ *is the solution process of an iterative process:*

$$w_i^k(N) = W^i(t_k, t_{k-1}, w_i^{k-1}(N)), \quad w_i^0(N) = w_i^0, \tag{5.4.3}$$

where

$$W^i(t_k, t_{k-1}, w_i^{k-1}(N)) = r_i(t_k, t_{k-1}, w_i^{k-1}(N)) + \mathcal{T}_i^k \chi_i^{k-1}(N),$$

where $r_i(t_k)$ is defined in Theorem 5.3.1 and $\mathcal{T}_i^k = \mathcal{M}^i(t_{k+1}, t_k)$ is a q_i-dimensional local Lipschitz constant vector relative to both $V^i(s, z_i)$ and $z_i(t, s, x_i)$ with respect to x_i, z_i and t, for each $i \in I(1,p)$ and all $k \in I(1,N)$.

Then,

$$V^i(t_k, z_i(t_{k+1}, t_k, y_i^k - \hat{x}_{ix_i^0}(t_k)), \omega) \leq w_i(t_k), \quad k \in I(1,N), \tag{5.4.4}$$

whenever

$$V^i(t_0, z_i(t_1, t_0, y_i^0 - x_i^0), \omega) \leq w^0, \tag{5.4.5}$$

for each $i \in I(1,p)$.

Proof. The immediate goal is to establish a system of difference inequality with respect to

$$V^i(t_k, z_i(t_{k+1}, t_k, y_i^k - \hat{x}_{ix_i^0}(t_k))).$$

For this purpose, we set $V^i(t_{k-1}, z_i(t_k, t_{k-1}, y_i^{k-1} - \hat{x}_{ix_i^0}(t_{k-1}))) = u_i^{k-1}$ and apply Corollary 5.3.1 to obtain an estimate (5.3.8) which can be written as

$$V^i(t_i, \hat{x}_i^{k-1}(t_k) - \hat{x}_{ix_i^0}(t_k), \omega)$$
$$\leq r_i(t_k, t_{k-1}, V^i(t_{k-1}, z_i(t_k, t_{k-1}, y_i^{k-1} - \bar{x}_{ix_i^0}(t_{k-1}))), \omega). \tag{5.4.6}$$

From (5.4.1), by noting the fact about the uniqueness of the solution of (5.2.2),

$$V^i(t_k, \hat{x}_i^{k-1}(t_k) - \hat{x}_{ix_i^0}(t_k)) = V^i(t_k, z_i(t_k, t_k, \hat{x}_i^{k-1}(tk) - \hat{x}_{ix_i^0}(t_{k-1}))),$$

and using the Lipschitzian property of $V^i(s, z_i)$ and $z_i(t, s, x_i)$ in z_i coupled with algebraic manipulations, we arrive at

$$V^i(t_k, z_i(t_{k+1}, t_k, y_i^k - \hat{x}_{ix_i^0}(t_k)), \omega)$$
$$\leq V^i(t_k, \hat{x}_i^{k-1}(t_k) - \hat{x}_{ix_i^0}(t_k), \omega) + \mathcal{M}^i(t_{k+1}, t_k)\chi_i^{k-1}(N). \tag{5.4.7}$$

From (5.4.6) and (5.4.7), we obtain the desired system of difference inequality:

$$V^i(t_k, z_i(t_{k+1}, t_k, y_i^k - \hat{x}_{ix_i^0}(t_k)), \omega)$$
$$\leq r_i(t_k, t_{k-1}, V^i(t_{k-1}, z_i(t_k, t_{k-1}, y_i^{k-1} - \hat{x}_{ix_i^0}(t_{k-1}))), \omega)$$
$$+ \mathcal{T}_i^k \chi_i^{k-1}(N). \tag{5.4.8}$$

Now, we set

$$m_i(k) = V^i(t_k, z_i(t_{k+1}, t_k, y_i^k - \hat{x}_{ix_i^0}(t_k)), \omega)$$

and note that $m_i(0) = V^i(t_0, z_i(t_1, t_0, y_i^0 - x_i^0))$ and

$$m_i(k-1) = V^i(t_{k-1}, z_i(t_k, t_{k-1}, y_i^{k-1} - \hat{x}_{ix_i^0}(t_{k-1})), \omega).$$

From these considerations and the definition of W^i in (5.4.3), inequalities (5.4.8) and (5.4.5) reduce to

$$m_i(k) \leq W^i(t_k, t_{k-1}, m_i(k-1)), \quad k \in I(1, N), \tag{5.4.9}$$

and

$$m_i(0) \leq w_i^0, \tag{5.4.10}$$

respectively, for each $i \in I[1, p]$. Now, by the application of Lemma 3.1 [7], the system of difference inequalities and the comparison theorem, we obtain the desired inequality

$$V^i(t_k, z_i(t_{k+1}, t_k, y_i^k - \hat{x}_{ix_i^0}(t_k)), \omega) \le w_i(t_k),$$
$$k \in I(0, N) \quad \text{and} \quad i \in I[1, p].$$

This completes the proof of the theorem.

Remark 5.4.1. It is important to note that the hypotheses of Theorem 5.4.1 provide a crucial iterative scheme to estimate a global error for the solution process of (5.1.3) in the framework of the hybrid subsystems (7.6.6). Of course, this is achieved by inequality (5.4.8).

Remark 5.4.1 poses a question: under less demanding and more feasible conditions on functions, is there a better way to obtain a system of difference inequalities similar to (5.4.8)? In the following, we make a few attempts to provide solutions to this problem.

Theorem 5.4.2. *Assume that all the hypotheses of Theorem* 5.3.1 *are satisfied. Further, assume that*

(i) *the solution process* $z_i(t, s, x_i)$ *of* (5.2.2) *satisfies the following relation:*

$$\|z_i(t, s, x_i)\|^{p_i} \le \lambda_i(t_k, t_{k-1})\gamma_i(\|x_i\|^{p_i}), \tag{5.4.11}$$

where $p_i \ge 1$, $\lambda_i \in C[R_+ \times R_+, R_+]$ *and* $\gamma_i \in C[R_+, R_+]$, *with* $\gamma_i \in \mathcal{CK}$;

(ii) *for each* $t \in J$, $V^i(t, z_i, \omega)$ *satisfies the following relations:*

$$V^i(t, z_i) \le \phi^i(\|z_i\|^{p_i}) \tag{5.4.12}$$

and

$$\theta_i(\|z_i\|^{p_i}) \le \sum_{j=1}^{q_i} V_j^i(t, z_i, \omega), \tag{5.4.13}$$

where $\phi^i \in C[R_+, R_+^{q_i}]$, $\theta_i \in C[R_+, R_+]$, $\theta_i \in \mathcal{VK}$ *and all components of* $\phi^i \in \mathcal{CK}$;

(iii) $w_i(t_k) = w_i(t_k, t_{k-1}, w_i^0)$ *is the solution process of a scalar iterative process:*

$$w_i^k(N) = W^i(t_k, t_{k-1}, w_i^{k-1}(N)), \quad w_i^0(N) = w_i^0, \tag{5.4.14}$$

where

$$\begin{aligned} &W^i(t_k, t_{k-1}, w_i^{k-1}(N)) \\ &= 2^{p_i}\left[(\theta_i)^{-1}\left(\sum_{j=1}^{q_i} r_{ij}(t_k, t_{k-1}, \phi^i(\lambda_i^k \gamma_i(w_i^{k-1}(N))))\right)\right. \\ &\qquad \left. + \left[\chi_i^{k-1}(N)\right]^{pi}\right], \end{aligned}$$

$\lambda_i^k = \lambda_i(t_k, t_{k-1})$, *for each* $i \in I(1,p)$ *and all* $k \in I(1,N)$. *Then,*

$$\|y_i^k - \hat{x}_{ix_i^0}(t_k)\|^{p_i} \le w_i(t_k), \quad k \in I(1,N), \tag{5.4.15}$$

whenever

$$\|y_i^0 - x_i^0\|^{p_i} \le w_i^0, \tag{5.4.16}$$

for each $i \in I[1,p]$.

Proof. Again, the immediate goal is to establish a scalar difference inequality with respect to $\|y_i^k - \hat{x}_{ix_i^0}(t_k)\|^{p_i}$. As before, by following the proof of Theorem 5.4.1, we arrive at (5.4.6). Now, from (5.4.11), (5.4.12) and Lemma 3.1 [7], we get

$$\begin{aligned} &r_i(t_k, t_{k-1}, V^i(t_{k-1}, z_i(t_k, t_{k-1}, y_i^{k-1} - \hat{x}_{ix_i^0}(t_{k-1}))), \omega) \\ &\quad \le r_i(t_k, t_{k-1}, \phi^i(\lambda_i^k \gamma_i(\|y_i^{k-1} - \hat{x}_{ix_i^0}(t_{k-1}))\|^{p_i})). \end{aligned} \tag{5.4.17}$$

From (5.4.13) and the property of θ_i, we have

$$\begin{aligned} &\|\hat{x}_i^{k-1}(t_k) - \hat{x}_{ix_i^0}(t_k)\|^{p_i} \\ &\quad \le (\theta_i)^{-1}\left(\sum_{j=1}^{q_i} V_j^i(t_k, \hat{x}_i^{k-1}(t_k) - \hat{x}_{ix_i^0}(t_k), \omega)\right). \end{aligned} \tag{5.4.18}$$

We note that

$$\begin{aligned}
&\|y_i^k - \hat{x}_{ix_i^0}(t_k)\|^{p_i} \\
&\leq 2^{p_i}\left[\|y_i^k - \hat{x}_i^{k-1}(t_k)\|^{p_i} + \|\hat{x}_i^{k-1}(t_k) - \hat{x}_{ix_i^0}(t_k)\|^{p_i}\right].
\end{aligned} \tag{5.4.19}$$

From (5.4.1), (5.4.6), (5.4.17), (5.4.18) and (5.4.19), one can obtain the following scalar difference inequality:

$$\begin{aligned}
&\|y_i^k - \hat{x}_{ix_i^0}(t_k)\|^{p_i} \\
&\leq 2^{p_i}\left[(\theta_i)^{-1}\left(\sum_{j=1}^{q_i} r_{ij}(t_k, t_{k-1}, \phi^i(\lambda_i^k \gamma_i(\|y^{k-1} - \hat{x}_{ix_i^0}(t_{k-1}))\|^{p_i}))\right)\right. \\
&\qquad \left. + \left[\chi_i^{k-1}(N)\right]^{p_i}\right].
\end{aligned} \tag{5.4.20}$$

This, together with the definition of W^i in (5.4.14), leads to the following scalar difference inequality:

$$m_i(k) \leq W^i(t_k, t_{k-1}, m_i(k-1)),$$

with $m_i(0) \leq x_i^0$.

Now, by imitating the proof of Theorem 5.4.1, the proof of the theorem follows immediately.

Corollary 5.4.1. *Assume that all the hypotheses of Theorem* 5.4.2 *are satisfied, except that the scalar comparison difference equation* (5.4.14) *is replaced by*

$$w_i^k(N) = W^i(t_k, t_{k-1}, w_i^{k-1}(N)), \quad w_i^0(N) = w_i^0, \tag{5.4.21}$$

where

$$\begin{aligned}
&W^i(t_k, t_{k-1}, w_i^{k-1}(N)) \\
&= 2^{p_i}\left[\lambda_i^k \gamma_i\left((\theta_i)^{-1}\left(\sum_{j=1}^{q_i} r_{ij}(t_k, t_{k-1}, \phi^i(w_i^{k-1}(N)))\right)\right)\right. \\
&\qquad \left. + \mathcal{L}_i^k[\chi_i^{k-1}(N)]^{p_i}\right],
\end{aligned}$$

$\mathcal{L}_i^k = [L_z^i(t_k, h_N^k)]^{p_i}$; $\mathcal{L}_z^i$ *is the local Lipschitz constant of* $z_i(t, s, x_i)$ *with respect to* x_i*, for each* (t, s) *and all* $i \in I(1, p)$ *and* $k \in I(1, N)$*. Then,*

$$\|z_i(t_k, t_{k-1}, y_i^{k-1} - \hat{x}_{ix_i^0}(t_{k-1}))\|^{p_i} \leq w_i(t_k) \quad k \in I(0, N). \quad (5.4.22)$$

Proof. Again, the goal is to establish a scalar difference inequality with respect to $\|z_i(t_{k+1}, t_k, y_i^k - \hat{x}_{ix_i^0}(t_k)\|^{p_i}$. As before, by following the proof of Theorem 5.4.1, we arrive at (5.4.6). From (5.4.12) and Lemma 3.1 [7], we have

$$\begin{aligned} &r_i(t_k, t_{k-1}, V^i(t_{k-1}, z_i(t_k, t_{k-1}, y_i^{k-1} - \hat{x}_{ix_i^0}(t_{k-1}))), \omega) \\ &\quad \leq r_i(t_k, t_{k-1}, \phi^i(\|z_i(t_k, t_{k-1}, y_i^{k-1} - \hat{x}_{ix_i^0}(t_{k-1}))\|^{p_i})). \quad (5.4.23) \end{aligned}$$

By imitating the argument in Theorem 5.4.2, using the Lipschitzian property of the solution process of (5.2.2), assumption (i) and (5.4.1), we have

$$\begin{aligned} &\|z_i(t_{k+1}, t_k, y_i^k - \hat{x}_{ix_i^0}(t_k))\|^{p_i} \\ &\quad \leq 2^{p_i} \left[\mathcal{L}_i^k(N)[\chi_i^{k-1}(N)]^{p_i} + \lambda_i^k \gamma_i(\|\hat{x}_i^{k-1}(t_k) - \hat{x}_{ix_i^0}(t_k)\|^{p_i})\right]. \\ &\qquad\qquad (5.4.24) \end{aligned}$$

From (5.4.6), (5.4.18), (5.4.23), (5.4.24) and the definition of W^i in (5.4.21), we arrive at the following scalar difference inequality:

$$\begin{aligned} &\|z_i(t_{k+1}, t_k, y_i^k - \hat{x}_{ix_i^0}(t_k)\|^{p_i} \\ &\quad \leq W(t_k, t_{k-1}, \|z_i(t_k, t_{k-1}, y_i^{k-1} - \hat{x}_{ix_i^0}(t_{k-1}))\|). \end{aligned}$$

This establishes the desired scalar difference inequality. Now, the rest of the proof can be completed by imitating the argument used in Theorem 5.4.1. The details are left to the reader.

The following example illustrates the scope of Theorems 5.4.1 and 5.4.2 and Corollary 5.4.1.

Example 5.4.1. Consider the following:

(i) Assume that the hypotheses of Theorem 5.4.2 are satisfied in the context of the following considerations: suppose that for each $i \in I(1,p)$, $f^i(t,z_i) = A^i(t)z_i$ in (5.2.2), where A^i is a continuous $n_i \times n_i$ matrix function. $\Phi^i(t,s)$ is a fundamental matrix solution of $z_i' = A^i(t)z_i$, $\Phi^i(s,s) = I_{n_i \times n_i}$ (an identity matrix). Further, assume that $g^i(t,u_i) \equiv 0$. In this case, $r_i(t,t_0,u_i^0) = u_i^0$, $z_i(t,s,x_i) = \Phi^i(t,s)x_i$. Then, (5.4.8) reduces to

$$\begin{aligned} &V_i(t_k, \hat{x}_i^{k-1}(t_k) - \hat{x}_{ix_i^0}(t_k)) \\ &\quad \leq V^i(t_{k-1}, \Phi^i(t_k,t_{k-1})(y_i^{k-1} - \hat{x}_{ix_i^0}(t_{k-1})), \omega), \end{aligned} \tag{5.4.25}$$

for each $i \in I[1,p]$ and all $k \in I(1,N)$. We note that $r_i(t,t_0,u_i^0)$ is nondecreasing in u_i^0. Furthermore, $\gamma_i(s) = s$, $\lambda_i^k = \|\Phi^i(t_k,t_{k-1})\|^{p_i}$. We assume that condition (ii) of Theorem 5.4.2 is satisfied. Then, the comparison functions $W^i(t_k,t_{k-1},w_i^{k-1}(N))$ in (5.4.3), (5.4.14) and (5.4.21) reduce to

$$\mathcal{T}_i^k \chi_i^{k-1}(N) + w_i^{k-1}(N), \tag{5.4.26}$$

$$2^{p_i}\left[(\theta_i)^{-1}\left(\sum_{j=1}^{q_i}\phi_j^i(\lambda_i^k w_i^{k-1}(N))\right) + [\chi_i^{k-1}(N)]^{p_i}\right] \tag{5.4.27}$$

and

$$2^{p_i}\left[\lambda_i^k(\theta_i)^{-1}\left(\sum_{j=1}^{q_i}\phi_j^i(w_i^{k-1}(N))\right) + \mathcal{L}_i^k[\chi_i^{k-1}(N)]^{p_i}\right], \tag{5.4.28}$$

respectively, for all $k \in I(1,N)$ and for each $i \in I[1,p]$.

(ii) By assuming the conditions as described in Example 5.4.1(i) and considering $g^i(t,u_i) = G^i(t)u_i$, where $G^i(t)$ is a $q_i \times q_i$ matrix as defined in Example 5.2.1(ii), we can see that in this case, $r_i(t,t_0,u_i^0) = \Phi_c^i(t,s)u_i^0$ is nondecreasing in u_i^0. Assuming that condition (ii) of Theorem 5.3.1 is satisfied, equation (5.4.18) reduces to

$$V^i(t_k, \hat{x}_i^{k-1}(t_k) - \hat{x}_{ix_i^0}(t_k))$$
$$\leq \Phi_c^i(t_k, t_{k-1})V^i(t_{k-1}, \Phi^i(t_k, t_{k-1})(y_i^{k-1} - x_{ix_i^0}(t_{k-1}), \omega)), \tag{5.4.29}$$

and in this case, the comparison functions $W^i(t_k, t_{k-1}, w_i^{k-1}(N))$ in (5.4.3), (5.4.14) and (5.4.21) become

$$\Phi_c^i(t_k, t_{k-1})w_i^{k-1}(N) + \mathcal{T}_i^k \chi^{k-1}(N), \tag{5.4.30}$$

$$2^{p_i}\left[(\theta_i)^{-1}\left(\sum_{j=1}^{q_i}\sum_{l=1}^{q_i}\Phi_{cjl}^i(t_k, t_{k-1})\phi_l^i(\lambda_i^k w_i^{k-1}(N))\right)\right.$$
$$\left. + \ [\chi_i^k(N)]^{p_i}\right] \tag{5.4.31}$$

and

$$2^{p_i}\left[\lambda_i^k(\theta_i)^{-1}\left(\sum_{j=1}^{q_i}\sum_{l=1}^{q_i}\Phi_{cjl}^i(t_k, t_{k-1})\phi_l^i(w_i^{k-1}(N))\right)\right.$$
$$\left. +\mathcal{L}_i^k[\chi_i^k(N)]^{p_i}\right], \tag{5.4.32}$$

respectively, for all $i \in I[1, p]$ and $k \in I(1, N)$.

(iii) We pick $V^i(s, z_i, \omega) = \|z_i\|^2$ and assume the conditions as described in Example 5.4.1(i). In this case, V^i and $z_i(t, s, x_i) = \Phi^i(t, s)x_i$ are continuously differentiable with respect to (s, x_i). Also, $L_i^D V^i(s, z_i(t, s, \Delta x_i), \omega)$, which is defined in (5.2.5), reduces to

$$2(\phi^i(t, s)\Delta x_i)^T \Phi^i(t, s)\Delta M^i(s, x_i)$$
$$+\mathrm{tr}(\Phi^i(t, s)\Delta\lambda^i(s, x_i)\Phi^i(t, s)), \tag{5.4.33}$$

where $\Delta M^i(s, x_i) = [\Delta a_0^i(s, x_i) - A^i(\Delta x_i)], \Delta\lambda^i(s, x_i) = \Delta\sigma^i(s, x_i)\Delta\sigma^{iT}(s, x_i)$, and $\sigma^i = [a_1^i, a_2^i, \ldots, a_m^i]$ is an $n \times m$

matrix whose elements are as defined in (5.1.2). It is assumed that

$$2(\phi^i(t,s)\Delta x_i)^T\Phi^i(t,s)\Delta M^i(s,x_i)$$
$$+\operatorname{tr}(\Phi^i(t,s)\Delta\lambda^i(s,x_i)\Phi^i(t,s)) \le \mu^i(s)\|\Phi^i(t,s)\Delta x_i\|^2, \tag{5.4.34}$$

where μ^i is an integrable function on the given interval J_k. Here, $g^i(s,u_i) = \mu^i(s)u_i$ with $\Phi_c^i(t_k,t_{k-1}) = \exp[\int_{t_{k-1}}^{t_k}\mu^i(s)ds]$ and $q_i = 1$. Here, $p_i = 2$, $\phi^i(s) = s = \theta_i(s)$, and $r_i(t,t_0,u_i^0) = \Phi_c^i(t_k,t_{k-1})u_i^0$ is nondecreasing u_i^0. In this case, (5.3.8) becomes

$$\|\hat{x}_i^{k-1}(t_k) - \hat{x}_{ix_i^0}(t_k)\|^2$$
$$\le \Phi_c^i(t_k,t_{k-1})\|\Phi^i(t_k,t_{k-1})(y_i^{k-1} - \hat{x}_{ix_i^0}(t_{k-1}))\|^2. \tag{5.4.35}$$

The comparison functions in (5.4.3), (5.4.14) and (5.4.21) are

$$\exp\left[\int_{t_{k-1}}^{t_k}\mu^i(s)ds\right]w_i^{k-1}(N) + \mathcal{T}_i^k\chi_i^{k-1}(N), \tag{5.4.36}$$

$$4\left[\exp\left[\int_{t_{k-1}}^{t_k}\mu^i(s)ds\right]\lambda_i^k w_i^{k-1}(N) + \left[\chi_i^{k-1}(N)\right]^2\right] \tag{5.4.37}$$

and

$$4\left[\lambda_i^k\exp\left[\int_{t_{k-1}}^{t_k}\mu^i(s)ds\right]w_i^{k-1}(N) + \mathcal{L}_i^k\left[\chi_i^{k-1}(N)\right]^2\right], \tag{5.4.38}$$

respectively. The solutions of (5.4.26), (5.4.30), (5.4.36), (5.4.37) and (5.4.38) have the following form:

$$w_i(t_k) = \prod_{j=1}^{k} H^i(t_j,t_{j-1})w_i^0 + \sum_{l=1}^{k}\prod_{j=l+1}^{k} H^i(t_j,t_{j-1})\nu_i^l(N), \tag{5.4.39}$$

where H_i and ν_i^l stand for coefficient functions with respect to homogeneous system/scalar difference equations and nonhomogeneous/perturbation functions in the respective comparison equations. Thus, (5.4.39) provides estimates corresponding to (5.4.4), (5.4.15) and (5.4.22), respectively. If in addition $f^i(t, z_i) \equiv 0$, then (5.4.35), (5.4.36), (5.4.37) and (5.4.38) lead to

$$\|\hat{x}^{k-1}(t_k) - \hat{x}_{ix_i^0}(t_k)\|^2 \leq \Phi_c^i(t_k, t_{k-1})\|(y_i^{k-1} - \hat{x}_{ix_i^0}(t_{k-1}))\|^2, \tag{5.4.40}$$

$$\exp\left[\int_{t_{k-1}}^{t_k} \mu^i(s)ds\right] w_i^{k-1}(N) + \chi_i^{k-1}(N), \tag{5.4.41}$$

$$4\left[\exp\left[\int_{t_{k-1}}^{t_k} \mu^i(s)ds\right] w_i^{k-1}(N) + [\chi_i^{k-1}(N)]^2\right] \tag{5.4.42}$$

and

$$4\left[\exp\left[\int_{t_{k-1}}^{t_k} \mu^i(s)ds\right] w_i^{k-1}(N) + [\chi_i^{k-1}(N)]^2\right], \tag{5.4.43}$$

respectively. The solutions corresponding to these comparison functions have the same form as in (5.4.39).

By employing comparison Theorem 5.2.1, the following provides a sufficient condition for the feasibility of hypothesis (a) of Theorem 5.4.1 in the context of the isolated hybrid subsystem (7.6.6).

Lemma 5.4.1. *Let the hypothesis of Theorem* 5.2.1 *and assumption* (ii) *of Theorem* 5.4.2 *be satisfied. Then,*

$$\|y_i^k - \hat{x}_i^{k-1}(t_k)\| \leq \chi_i^{k-1}(N), \tag{5.4.44}$$

where $y^k = (y_1^{kT}, y_2^{kT}, \ldots, y_i^{kT}, \ldots, y_p^{kT})^T, y_i^k \in R^{n_i}$, *as defined before, and* $\chi_i^{k-1}(N)$ *is defined by*

$$\begin{aligned}\chi_i^{k-1}(N) &= \beta_i^{k-1}(h_N^k, N)\\ &= (\theta_i)^{-1}\left(\sum_{j=1}^{q_i} r_{ij}(t_k, t_{k-1}, \phi^i(\|z_i(t_k, t_{k-1}, 0)\|))\right).\end{aligned} \tag{5.4.45}$$

Proof. Let $\hat{x}_i^{k-1}(t)$ and $\bar{y}_i^{k-1}(t)$ be the solution processes of the ith isolated hybrid subsystem (5.1.6) described by (5.2.1) through

(t_{k-1}, y_i^{k-1}) and (t_{k-1}, y_i^{k-1}), respectively, where $\{y^k\}_0^N$ is a sequence in R^n and $y^k = (y_1^{kT}, y_2^{kT}, \ldots, y_i^{kT}, \ldots, y_p^{kT})^T$, with $y_i^k \in R^{n_i}$. From (5.2.8) and (5.2.10), one can obtain

$$V^i(t_k, \bar{y}_i^{k-1}(t_k) - \hat{x}_i^{k-1}(t_k)) \leq r_i(t_k, t_{k-1}, V^i(t_{k-1}, z_i(t_k, t_{k-1}, 0))),$$

for all $i \in I[1, p]$ and $k \in I(1, N)$. This, together with the definition of $\bar{y}_i = y + (t - t_{k-1})b^i(t_{k-1}, y_i^{k-1}, h^{k-1})$, leads to

$$V^i(t_k, y_i^k - \hat{x}_i^{k-1}(t_k)) \leq r_i(t_k, t_{k-1}, V^i(t_{k-1}, z_i(t_k, t_{k-1}, 0))). \quad (5.4.46)$$

From (5.4.12), (5.4.13) and (5.4.46), we obtain the desired inequality

$$\|y_i^k - \hat{x}_i^{k-1}(t_k)\| \leq \chi_i^{k-1}(N),$$

where $\chi_i^{k-1}(N)$ is defined in (5.4.45). This completes the proof of the lemma.

5.5 Convergence Analysis

Now, we present a very general result that provides sufficient conditions under which Dahlquist-type convergence in moment and probability is assured.

Theorem 5.5.1. *Under the assumptions of Theorem* 5.4.1 *with* $V^i(t, 0) \equiv 0$ (*Theorem* 5.4.2 *or Corollary* 5.4.1), *the convergence of the solution of comparison equations* (5.4.3), ((5.4.14) *or* (5.4.21)) *implies convergence in the* p_i*th moment and in the probability of stochastic numerical schemes for the ith isolated subsystems* (5.1.3).

Proof. The proof of the theorem follows from relations (5.4.4) and (5.4.5) ((5.4.15) and (5.4.16) or (5.4.22) and (5.4.16)), the convergence of (5.4.3) ((5.4.14) or (5.4.21)) and by following the standard argument used in the qualitative study of systems of differential/difference equations.

The following example illustrates the scope of Theorem 5.5.1.

Example 5.5.1. Consider the following:

(i) The convergence of solutions of the comparison system of difference equations in (5.4.26), (5.4.30), (5.4.36) and (5.4.41) in Examples 5.4.1(i), 5.4.1(ii) and 5.4.1(iii), corresponding to comparison system (5.4.3) in Theorem 5.4.1, guarantees the convergence in moment of the stochastic numerical scheme in (5.1.3). The solutions of (5.4.26), (5.4.30), (5.4.36) and (5.4.41) have the form (5.4.39), that is,

$$w_i(t_k) = \sum_{j=1}^{k} H^i(t_j, t_{j-1})w_i^0 + \sum_{l=1}^{k} \prod_{j=l+1}^{k} H^i(t_j, t_{j-1})\nu_i^l(N). \tag{5.5.1}$$

Therefore, the convergence of the solutions of systems (5.4.26), (5.4.30), (5.4.36) and (5.4.41) is assured by the convergence of $\sum_{l=1}^{k} \prod_{j=l+1}^{k} H^i(t_j, t_{j-1})\nu_i^l(N)$. Of course, this expression depends on the maximal solutions of the comparison functions and the solutions of the auxiliary systems.

(ii) Similarly, the convergence of solutions of the comparison system of difference equations in (5.4.37) and (5.4.42) ((5.4.38) and (5.4.43)) in Examples 5.4.1(iii), corresponding to the comparison system (5.4.14) ((5.4.21)) in Theorem 5.4.2 (Corollary 5.4.1), guarantees the convergence in moment of the stochastic numerical scheme in (5.1.3). The solutions of (5.4.37) and (5.4.42) ((5.4.38) and (5.4.43)) have the form (5.4.39). Therefore, the convergence of the solutions of systems (5.4.26), (5.4.30), (5.4.36) and (5.4.41) is assumed by the convergence of $\sum_{l=1}^{k} \prod_{j=l+1}^{k} H^i(t_j, t_{j-1})\nu_i^l(N)$. In fact, the Lemma 5.4.1 assures the feasibility for hypothesis (a) in Theorem 5.4.1.

Finally, we present a very general result that provides sufficient conditions which guarantee the relative stability property of the stochastic numerical scheme (5.1.3). In addition, an example is given to illustrate the scope of the result.

Theorem 5.5.2. *Under the assumptions of Theorem* 5.4.1 *with* $V^i(t, 0) \equiv 0$ (*Theorem* 5.4.2 *or Corollary* 5.4.1), *the convergence of solution of the comparison equations* (5.4.3), ((5.4.14) *or* (5.4.21)) *implies relative stability in the p_ith moment and in the probability of the stochastic numerical scheme for the ith isolated subsystem* (5.1.3).

Example 5.5.2. Examples 5.5.1(i) and 5.5.1(ii) provide the basis for the relative stability property of the stochastic numerical scheme (5.1.3). From Example 5.5.1(ii) and Lemma 6.4.1, imply the comparison. The solution process (5.4.45) converges as $w_i^0 \to 0$, and $\sum_{l=1}^{k}\prod_{j=l+1}^{k} H^i(t_j, t_{j-1})\nu_i^l(N) \to 0$ as $N \to \infty$.

5.6 Local Convergence Analysis for Large-Scale Algorithms

In this section, we develop a few comparison results analogous to the results in Section 5.3. For this purpose and for the sake of simplicity, we associate a scalar Lyapunov-like function ($q_i = 1$) to each isolated hybrid subsystem (7.6.6).

By imitating the argument used to derive (5.2.1), one can derive the following form of large-scale hybrid system (7.6.5):

$$
\begin{aligned}
dx_i &= [\boldsymbol{a}^i(t, x_i, \omega) + \boldsymbol{c}^i(t, x_i, \omega)]dt, \quad x_i(t_k, \omega) = x_i^k,\\
d\bar{y}_i &= [\boldsymbol{b}^i(t_k, y_i, \omega) + \boldsymbol{d}^i(t_k, y_i, \omega)]dt, \quad y_i(t_k, \omega) = y_i^k, \quad t \in J_k,
\end{aligned}
\tag{5.6.1}
$$

$$
y_i(t_0, \omega) = x_i(t_0, \omega) = x_i^0, \quad k \in I(0, N) \quad \text{and} \quad i \in I[1, p].
$$

We use the same auxiliary system of differential equation (5.2.2). As stated before, we associate a scalar Lyapunov-like function to each isolated stochastic hybrid subsystem (5.1.6).

Similar to the definition $D^+_{(5.2.1)}V^i(s, t_k, z_i(t, s, \Delta x_i), \omega)$, one can define $D^+_{(5.6.1)}V^i(s, t_k, x, z_i(t, s, \Delta x_i), \omega)$. With respect to each subsystem of the large-scale system (5.6.1), operators similar to operators L_i^D(5.2.3) with regard to isolated subsystems can also be defined analogously, which are denoted by L_i^C, and

$$
\begin{aligned}
&D^+_{(5.6.1)}V^i(s, t_k, x, z_i(t, s, \Delta x_i), \omega)\\
&\quad \le L_i^C V^i(s, t_k, y, x, z_i(t, s, \Delta x_i), \omega),
\end{aligned}
\tag{5.6.2}
$$

where L_i^C is defined and can be represented as

$$
\begin{aligned}
&L_i^C V^i(s,t_k,y,x,z_i(t,s,\Delta x_i),\omega)\\
&\quad = L_i^D V^i(s,t_k,y_i,z_i(t,s,\Delta x_i,\omega)) + I^i(t,s,t_k,\mathbf{c}^i(s,x,\omega),\mathbf{d}^i(t_k,y,\omega)),
\end{aligned} \tag{5.6.3}
$$

$$
\begin{aligned}
&I^i(t,s,t_k,\mathbf{c}^i(t,x,\omega),\mathbf{d}^i(t_k,y,\omega))\\
&\quad = \frac{1}{2}\operatorname{tr}\left(\frac{\partial^2}{\partial z_i^2}V^i(s,z_i(t,s,\Delta x)\right)\Theta_i^C(t,s,t_k,x,y))\\
&\qquad + V_{z_i}^i(s,z_i(t,s,\Delta x_i))\left[\frac{1}{2}\operatorname{tr}\left(\frac{\partial^2}{\partial x_i \partial x_i}z_i(t,s,\Delta x_i)\Lambda_i^C(s,t_k,x,y)\right)\right.\\
&\qquad\qquad \left. + \Phi^i(t,s,\Delta x_i)M_i^C(s,t_k,y,x)\right],
\end{aligned}
$$

$$
\begin{aligned}
M_i^C(s,t_k,y,x) &= \lim_{h\to 0^+}\frac{1}{h}\mathbf{C}^i(s,t_k,y,x,\omega)\Delta\psi(s) - hf^i(s,\Delta x_i,\omega),\\
\Lambda_i^C(s,t_k,x,y) &= \lim_{h\to 0^+}\frac{1}{h}([\mathbf{F}^i(s,t_k,y_i,x_i,\omega)\\
&\quad + \mathbf{C}^i(s,t_k,y,x,\omega)]\Delta\psi(s))([\mathbf{F}^i(s,t_k,y_i,x_i,\omega)\\
&\quad + \mathbf{C}^i(s,t_k,y,x,\omega)]\Delta\psi(s))^T - \Lambda_i^D(s,t_k,x_i,y_i),\\
\Theta_i^C(t,s,t_k,y,x) &= \Phi^i(t,s,\Delta x_i)\Lambda_i^C(s,t_k,y,x)\Phi^{iT}(t,s,\Delta x_i),\\
\Delta x_i &= y_i - x_i,\\
\mathbf{C}^i(s,t_k,y,x,\omega) &= (-\mathbf{c}^{iT}(t,x,\omega),\mathbf{d}^{iT}(t_k,y,\omega))^T.
\end{aligned}
$$

Remark 5.6.1. Moreover, one can define D^+V^i analogously to (5.6.2) with respect to the continuous part of (7.6.5), and its corresponding $L_i^C V_i$ is defined by

$$
\begin{aligned}
&L_i^C V^i(s,x,z_i(t,s,\Delta x_i),\omega)\\
&\quad = L_i^D V^i(s,z_i(t,s,\Delta x_i),\omega) + I^i(t,s,t_k,\mathbf{c}^i(t,x,\omega),\omega),
\end{aligned} \tag{5.6.4}
$$

where

$$
\begin{aligned}
I^i(t,s,t_k,\mathbf{c}^i(t,x,\omega),\omega) &= \frac{1}{2}\mathrm{tr}\left(\frac{\partial^2}{\partial z_i^2}V^i(s,z_i(t,s,\Delta x_i),\omega)\right)\Delta\theta_i^C(t,s,x)\\
&\quad + V_{z_i}^i(s,z_i(t,s,\Delta x),\omega)\\
&\quad \times\left[\frac{1}{2}\,\mathrm{tr}\left(\frac{\partial^2}{\partial x_i\partial x_i}z_i(t,s,\Delta x_i)\Delta\lambda_i^C(s,x)\right)\right.\\
&\quad \left. + \Phi^i(t,s,\Delta x_i)\Delta M_i^C(s,x)\right],\\
M_i^C(s,x) &= \lim_{h\to 0^+}\frac{1}{h}\Delta\mathbf{c}^i(t,x,\omega)\Delta\xi(s) - hf^i(s,\Delta x_i,\omega),\\
\lambda_i^C(s,x) &= \lim_{h\to 0^+}\frac{1}{h}\left([\Delta\mathbf{a}^i(s,x_i,\omega)+\Delta\mathbf{c}^i(t,x,\omega)]i\Delta\xi(s)\right)\\
&\quad \times\left([\Delta\mathbf{a}^i(s,x_i,\omega)+\Delta\mathbf{c}^i(t,x,\omega)] - \lambda_i^D(s,x_i)\right),\\
\theta_i^C(t,s,x_i) &= \Phi^i(t,s,\Delta x_i)\lambda_i^C(s,x_i)\Phi^{iT}(t,s,\Delta x_i),\\
\Delta\mathbf{c}^i(t,x,\omega) &= [\mathbf{c}^i(t,y,\omega)-\mathbf{c}^i(t,x,\omega)],\\
\Delta\mathbf{a}^i(s,x_i,\omega) &= (\mathbf{a}^i(s,y_i,\omega)-\mathbf{a}^i(s,x_i,\omega))
\end{aligned}
$$

and $\Delta x_i = y_i - x_i$. The following theorem provides an estimate of the local theoretical/discretization error for the solution process of the complex stochastic system (5.1.2) in the framework of a solution process of the corresponding large-scale hybrid system (7.6.5). We simply state the theorem with proof.

Theorem 5.6.1. *Assume that*

(i) $G^i \in M[J\times J\times R^n\times R_+^p\times\Omega,R_+]$, $g^i\in M[J\times J\times R^{n_i}\times R_+\times\Omega,R_+]$, *for each* $(t,t_k,y_i^k)\in J\times J\times R^{n_i}$ *and* $(t,t_k,y^k)\in J\times J\times R^n$. $g^i(t,t_k,y_i^k,u_i,\omega)$ *and* $G^i(t,t_k,y^k,u,\omega)$ *are concave in* u_i *and* u, *respectively. Moreover,* $G^i(t,t_k,y^k,u,\omega)$ *is quasi-monotone nondecreasing in* u, *for each* $(t,t_k,y^k)\in J\times J\times R^n$ *and* $i\in I[1,p]$.

(ii) $r^k(t) = r(t, t_k, u^k)$ *is the maximal solution of the following comparison system:*

$$u_i' = g^i(t, t_k, y_i^k, u_i, \omega) + G^i(t, t_k, y^k, u, \omega),$$
$$u_i(t_k) = u_i^k, \quad i \in I[1, p]. \tag{5.6.5}$$

(iii) $z_i(t, s, x_i)$ *is the solution process of the auxiliary system* (5.2.2) *through* (s, x_i), $t_0 \leq s \leq t \leq t_0 + \alpha$.

(iv) $V^i \in M[J \times R^{n_i} \times \Omega, R_+]$; $\frac{\partial}{\partial t} V^i(s, z_i)$, $\frac{\partial}{\partial x} V^i(s, z_i)$ *and* $\frac{\partial^2}{\partial x^2} V(s, z_i)$ *exist and are continuous on* $J \times \mathcal{R}^{n_i}$. *Moreover,* $\frac{\partial^2}{\partial x^2} V^i(s, z_i)$ *is locally Lipschitzian in* z_i, *for each* $t_0 \leq t_k \leq s \leq t \leq t_{k+1} \leq t_0 + \alpha$, *and* $L_i^D V^i(s, t_k, y_i^k, z_i(t, s, \Delta x_i))$ *satisfies the following relation:*

$$L_i^D V^i(s, t_k, y_i^k, z_i(t, s, \Delta x_i), \omega)$$
$$\leq g^i(s, t_k, y_i^k, V^i(s, z_i(t, s, \Delta x_i), \omega), \omega), \tag{5.6.6}$$

for each $i \in I[1, p]$.

(v) *The interaction functions among the subsystems satisfy the inequality*

$$I^i(t, s, t_k, \mathbf{c}^i(s, \omega), \mathbf{d}^i(t_k, \omega))$$
$$\leq G^i(s, t_k, y^k, V(s, z(t, s, \Delta x), \omega), \omega), \tag{5.6.7}$$

where I^i *is as defined in* (5.6.3) *and*

$$V(s, \Delta x, \omega) = \left(V^1(s, \Delta x_i, \omega), V^2(s, x_2, \omega), \ldots, V^p(s, \Delta x_p, \omega)\right)^T. \tag{5.6.8}$$

(vi) $x^k(t) = x(t, t_k, x^k)$ *and* $\bar{y}^k(t) = \bar{y}(t, t_k, y^k)$ *are the solutions of* (5.6.1) *through* (t_k, x^k) *and* (t_k, y^k), *respectively, and* $V(s, z(t, s, \bar{y}^k(s) - \hat{x}(s)), \omega)$ *exists for* $s \in [t_k, t]$, $t \in J_k$ *and* $k \in I(0, N)$. *Then,*

$$V(t, \bar{y}^k(t) - x^k(t), \omega) \leq r(t, t_k, u^k) \tag{5.6.9}$$

whenever

$$V(t_k, z(t, t_k, y^k - x^k), \omega) \leq u^k, \tag{5.6.10}$$

for all $k \in I(0, N)$ *and* $t \in J_k$.

Proof. From assumptions (iii) and (iv) and by employing the standard argument, we have

$$\begin{aligned}&L_i^C V^i(s, t_k, y, x, z_i(t, s, \Delta x_i), \omega)\\&\quad \le g^i(s, t_k, y_i^k, V^i(s, z_i(t, s, \Delta x_i), \omega), \omega)\\&\quad\quad + G^i(s, t_k, y^k, V(s, z(t, s, \Delta x), \omega), \omega),\end{aligned} \tag{5.6.11}$$

for each $i \in I[1,p]$ and all $k \in I(0,N)$. Let $x^k(t)$ and $\bar{y}^k(t)$ be the solution processes of (5.4.1) through (t_k, x^k) and (t_k, y^k), respectively. Now, we set

$$m^i(s) = V^i(s, z_i(t, s, \bar{y}^k(s) - x_i^k(s)), \omega), \quad \text{for } s \in [t_k, t] \text{ and } t \in J_k,$$

and note that $m^i(t_k) = V^i(t_k, z_i(t, t_k, y_i^k - x_i^k), \omega)$, for each $i \in I[1,p]$. Now, by following the standard argument, the proof of the theorem can be completed.

Remark 5.6.2. A corollary similar to Corollary 5.2.1 can be formulated analogously.

5.7 Local Analytical Convergence Analysis for Large-Scale Algorithms

We state without proof another auxiliary comparison theorem that provides an estimate of the local analytic error for the solution process of the complex stochastic system (5.1.2) in the framework of a solution process of the large-scale hybrid subsystem (7.6.5).

Theorem 5.7.1. *Assume that*

(i) $G^i \in M[J \times R_+^p \times \Omega, R_+]$, $g^i \in M[J \times R_+ \times \Omega, R_+]$, *for each* $t \in J$, *and* $g^i(t, u_i, \omega)$ *and* $G^i(t, u, \omega)$ *are concave in* u_i *and* u, *respectively. Moreover,* $G^i(t, u, \omega)$ *is quasimonotone nondecreasing in* u, *for each* $t \in J$ *and* $i \in I[1,p]$.

(ii) $r^k(t) = r(t, t_k, u^k, \omega)$ *is the maximal solution of the following comparison system:*

$$\begin{aligned}u_i' &= g^i(t, u_i, \omega) + G^i(t, u, \omega),\\ u_i(t_k) &= u_i^k, \quad i \in I[1,p].\end{aligned} \tag{5.7.1}$$

(iii) *$z_i(t,s,x_i,\omega)$ is the solution process of the auxiliary system* (5.2.2) *through (s,x_i), $t_0 \le s \le t \le t_0+\alpha$.*

(iv) *$V^i \in M[J \times R^{n_i} \times \Omega, R_+]$; $\frac{\partial}{\partial t}V^i(s,z_i,\omega)$, $\frac{\partial}{\partial x}V^i(s,z_i,\omega)$ and $\frac{\partial^2}{\partial x^2}V(s,z_i,\omega)$ exist and are continuous on $J \times \mathcal{R}^{n_i}$. $\frac{\partial^2}{\partial x^2}V^i(s,z_i,\omega)$ is locally Lipschitzian in z_i, for each $t_0 \le t_k \le s \le t \le t_{k+1} \le t_0+\alpha$, and $L_i^D V^i(s,z_i(t,s,\Delta x_i),\omega)$ satisfies the following relation:*

$$L_i^D V^i(s,z_i(t,s,\Delta x_i),\omega) \le g^i(s,V_i(s,z_i(t,s,\Delta x_i)),\omega). \quad (5.7.2)$$

(v) *The interaction functions among the subsystems satisfy the inequality*

$$I^i(t,s,t_k,\mathbf{c}^i(t,\omega),\omega) \le G^i(s,V(s,z(t,s,\Delta x)),\omega), \quad (5.7.3)$$

where I^i is as defined in (5.6.4) *and*

$$V(s,\Delta x,\omega) = \left(V^1(s,\Delta x_i,\omega), V^2(s,\Delta x_2,\omega),\ldots,V^p(s,\Delta x_p,\omega)\right)^T. \quad (5.7.4)$$

(vi) *$x(t)$ and $x^k(t)$ are the solution processes of the continuous part of the complex system* (5.1.2) *and the stochastic large-scale hybrid system* (7.6.5) *through (t_0,x^0) and (t_k,y^k), respectively, where ${y^k}_0^N$ is a sequence in $R[(\Omega,F,P),R^n)]$ and*

$$y^k = (y_1^{kT}, y_2^{kT},\ldots,y_i^{kT},\ldots,y_p^{kT})^T,$$

where $y_i^k \in R[(\Omega,F,P),R^{n_i}]$ and $V(s,z(t,s,x^k(s)-x(s)),\omega)$ exists for $s \in [t_k,t]$, $t \in J_k$ and $k \in I(0,N)$. Then,

$$V(t,x^k(s)-x(s),\omega) \le r(t,t_k,u^k,\omega) \quad (5.7.5)$$

whenever

$$V(t_k,z(t,t_k,y^k-x(t_k)),\omega) \le u^k, \quad (5.7.6)$$

for all $k \in I(0,N)$ and $t \in J_k$.

Remark 5.7.1. A corollary similar to Corollary 5.3.1 can be formulated analogously.

Remark 5.7.2. If $f^i \equiv 0$, then Theorems 5.6.1 and 5.7.1 reduce to the usual comparison theorems for studying large-scale systems as special cases.

Remark 5.7.3. By following the arguments used in Example 5.3.1, one can illustrate the scope of Theorems 5.6.1 and 5.7.1. The details are left to the reader.

5.8 Global Analytical Convergence Analysis for Large-Scale Algorithms

Now, we are ready to discuss the error estimation results between the continuous-time and discrete large-scale algorithm (5.1.2). In addition, we formulate the convergence and relative stability results for the stochastic numerical schemes described in (5.1.1).

First, we present a very general result that establishes a theoretical/discretization error for the solution process of the complex system (5.1.1).

Theorem 5.8.1. *Assume that all the hypotheses of Theorem* 5.4.2 *are satisfied. Further, assume that*

(a) *the sequence* $\{y^k\}_0^N$ *satisfies the following condition:*

$$\|y^k - x_i^{k-1}(t_k)\| \leq \chi_i^{k-1}(N), \tag{5.8.1}$$

where $\chi_i^{k-1}(N)$ *is a discrete process defined by*

$$\chi_i^{k-1}(N) = \beta_i^{k-1}(h_N^k, N), \tag{5.8.2}$$

$\beta_i^{k-1} \in C[[0, h^0], R_+]$, $h^0 > 0$, *for each* $i \in I[1,p]$ *and all* $k \in I(1,N)$, *and* $N \in I(1,\infty)$;

(b) $w(t_k) = w(t_k, t_{k-1}, w^0)$ *is the solution process of a system of iterative process:*

$$w^k(N) = W(t_k, t_{k-1}, w^{k-1}(N)), \quad w^0(N) = w^0, \tag{5.8.3}$$

where the ith component of W *is defined by*

$$W^i(t_k, t_{k-1}, w^{k-1}(N)) = r_i(t_k, t_{k-1}, w^{k-1}(N)) + \mathcal{T}_i^k \chi_i^{k-1}(N),$$

$r^k(t)$ *is defined in Theorem* 5.7.1 *and* $\mathcal{T}_i^k = \mathcal{M}^i(t_{k+1}, t_k)$ *is a local Lipschitz constant relative to both* $V^i(s, z_i, \omega)$ *and* $z_i(t, s, x_i)$ *with respect to* x_i, z_i *and* t, *for each* $i \in I[1,p]$ *and all* $k \in I(1,N)$.

Then,

$$V(t_k, z(t_{k+1}, t_k, y^k - x(t_k), \omega)) \le w(t_k), \quad k \in I(1, N), \quad (5.8.4)$$

whenever

$$V(t_0, z(t_1, t_0, y^0 - x^0, \omega)) \le w^0. \quad (5.8.5)$$

Proof. The proof of the theorem can be formulated by following the argument used in the proof of Theorem 5.4.1. The details are left to the reader.

In the following, for the sake of completeness and to avoid monotonicity, we state without proof the remark, theorems and corollary corresponding to Remark 5.3.1, Theorems 5.4.2, 5.5.1 and 5.5.2 and Corollary 5.4.1, respectively.

Remark 5.8.1. It is important to note that the hypotheses of Theorem 5.5.1 provide a crucial iterative scheme to estimate a global error for the solution process of (5.1.2) in the framework of stochastic large-scale hybrid systems (7.6.5). Of course, this is achieved by an inequality similar to (5.4.8).

Theorem 5.8.2. *Assume that all the hypotheses of Theorem* 5.7.1 *are satisfied. Further, assume that*

(i) *the solution process* $z_i(t, s, x_i)$ *of* (5.2.2) *satisfies the following relation:*

$$\|z_i(t, s, x_i)\|^{\bar{p}} \le \lambda(t_k, t_{k-1})\gamma_i(\|x_i\|^{\bar{p}}), \quad (5.8.6)$$

where $\bar{p} \ge 1$, $\lambda \in C[R_+ \times R_+, R_+]$, *and* $\gamma_i \in C[R_+, R_+]$, *with* $\gamma_i \in \mathcal{CK}$;

(ii) *for each* $t \in J$, $V_i(t, z_i, \omega)$ *satisfies the following relations:*

$$V^i(t, z_i, \omega) \le \varphi^i(\|z_i\|^{\bar{p}}) \quad (5.8.7)$$

and

$$\theta_i(\|z_i\|^{\bar{p}}) \le V^i(t, z_i, \omega), \quad (5.8.8)$$

$\phi^i \in C[R_+, R_+]$, $\theta \in C[R_+, R_+]$, $\theta_i \in \mathcal{VK}$, *and all components of* $\phi^i \in \mathcal{CK}$;

(iii) $w(t_k) = w(t_k, t_{k-1}, w^0)$ *is the solution process of a system of iterative process:*

$$w^k(N) = W(t_k, t_{k-1}, w^{k-1}(N)), \quad w^0(N) = w^0, \tag{5.8.9}$$

where the ith component of W is defined by

$$W^i(t_k, t_{k-1}, w^{k-1}(N))$$
$$= 2^{\bar{p}} \left[(\theta_i)^{-1} (r_i(t_k, t_{k-1}, \phi(\lambda^k \gamma(w^{k-1}(N))))) + \left[\chi_i^{k-1}(N)\right]^{\bar{p}} \right],$$

$\lambda^k = \lambda(t_k, t_{k-1})$, $\phi = (\phi^1, \phi^2, \ldots, \phi^i, \ldots, \phi^p)^T$, $\gamma = (\gamma_1, \gamma_2, \ldots, \gamma_i, \ldots, \gamma_p)^T$, $r^k(t)$ *is defined in Theorem* 5.7.1 *and* χ_i^{k-1} *is defined in* (5.8.1), *for each* $i \in I[1,p]$ *and all* $k \in I(1,N)$.

Then,

$$\|y_i^k - x_i(t_k)\|^{\bar{p}} \le w_i(t_k), \quad k \in I(1,N), \tag{5.8.10}$$

whenever

$$\|y_i^0 - x_i^0\|^{\bar{p}} \le w_i^0, \quad \text{for all } i \in I[1,p]. \tag{5.8.11}$$

Corollary 5.8.1. *Assume that all the hypotheses of Theorem* 5.8.2 *are satisfied, except that the system of comparison difference equation* (5.8.9) *is replaced by*

$$w^k(N) = W(t_k, t_{k-1}, w^{k-1}(N)), w^0(N) = w^0, \tag{5.8.12}$$

where the ith component of W is defined by

$$W^i(t_k, t_{k-1}, w^{k-1}(N)) = 2^{\bar{p}}[\lambda^k \gamma((\theta_i)^{-1}(r_i(t_k, t_{k-1}, \phi(w^{k-1}(N)))))$$
$$+ \mathcal{L}^k[\chi^k(N)]^{\bar{p}}],$$

$L_*^{ik} = [L_z^i(t_k, h_N^k)]^{\bar{p}}$ *and* L_z^i *is the local Lipschitz constant of* $z_i(t,s,x_i)$ *with respect to* x_i, *for each* (t,s) *and all* $k \in I(0,N)$. *Then,*

$$\|z_i(t_k, t_{k-1}, y_i^{k-1} - x_i(t_{k-1}))\|^{\bar{p}} \le w_i(t_k), \quad k \in I(0,N), \tag{5.8.13}$$

for each $i \in I[1,p]$.

Remark 5.8.2. Based on Example 5.4.1 and others worked out in the theory of large-scale systems, another example similar to Example 5.4.1 can be constructed.

5.9 Local Convergence Analysis

Now, we present a very general result that assures Dahlquist-type convergence and the relative stability property of stochastic numerical schemes. The proofs of the results follow by imitating the well-known arguments used in Lyapunov's second method.

Theorem 5.9.1. *Under the assumptions of Theorem* 5.8.1 *with* $V^i(t, 0, \omega) \equiv 0$ (*Theorem* 5.8.2 *or Corollary* 5.8.1), *the convergence of the solution of comparison equations* (5.8.3) ((5.8.9) *or* (5.8.12)) *implies the convergence in the* p_i*th moment and the probability of stochastic numerical schemes for complex system* (5.1.2).

Theorem 5.9.2. *Under the assumptions of Theorem* 5.5.1 *with* $V^i(t, 0, \omega) \equiv 0$ (*Theorem* 5.8.2 *or Corollary* 5.8.1), *the convergence of solution of the comparison equation* (5.8.3) ((5.8.9) *or* (5.8.12)) *implies the convergence in the* p_i*th moment and the probability of stochastic numerical schemes for the complex system* (5.1.2).

5.10 Notes and Comments

In Section 5.1, a theoretical large-scale algorithm is outlined by employing the decomposition-aggregation method via the energy/Lyapunov function and the comparison methods of Ladde and Siljak [36, 37] for a large-scale complex system of differential equations with random parameters coupled with a one-step numerical complex system of difference equations in the framework of a pair of interconnected hybrid dynamic systems with the same number of components for convergence and stability analyses of the numerical schemes approximation of an initial value problem of solution processes of large-scale dynamic systems. This is motivated by the works of Ladde [23, 24], Ladde and Ladde [18, 19], and Ladde and Siljak [36, 37], as well as the classical numerical analysis work of Refs. [2, 3, 5, 9–11, 14, 16, 47]. Section 5.2 deals with assigning each component of the vector energy/Lyapunov function corresponding to each large-scale interacting pair of hybrid subsystems of overall large-scale interacting pair of hybrid systems as a subcomponent measure of the overall state, developing a very general variational comparison via vector energy/Lyapunov, and obtaining the designed conceptual state error estimate between the components of each pair of solutions of parallel operating and interacting isolated hybrid subsystem components of the overall state of the interconnected large-scale hybrid dynamic system on the currently operating interval that formed based on the knowledge of discretization process. Utilizing synchronized local error bounds on the solution process of an isolated hybrid subsystem and imitating local designed procedures, the local analytic error estimate/bounds on each solution process of large-scale isolated to global subsystems of an underlying large-scale complex system of differential equations with random parameters coupled with a one-step numerical complex system of difference equations are described in Section 5.3. The results in Sections 5.1–5.3 were adapted from Ladde and Sambandham [28, 30, 33, 34] and are extensions of the works of Ladde [23] and Ladde and Ladde [18, 19]. The results of the local analytical bounds outlined in Section 5.3 extend the global domain of operation of a one-step numerical scheme to obtain conceptual computational discretization error estimates for isolated subsystems in Section 5.4. Several feasible results, examples and remarks are presented to demonstrate the usefulness of the conceptual computational method. Sufficient conditions are given in Section 5.5, under which Dahlquist-type different types of probabilistic mode of convergence, in particular convergence in moment and in probability, are assured. The byproduct of analysis, stochastic stability, is also presented for large-scale isolated to global subsystems of an underlying large-scale complex system of differential equations with random parameters coupled with a one-step numerical complex system of difference equations. The results in Sections 5.4 and 5.5 are adapted from Ladde and Sambandham [33, 34] and are extension of the works of Ladde [23], Ladde and Siljak [36, 37], Lakshmikantham and Trigiante [39], Babuska, Prager and Vitasek [3], Henerici [14], Hayashi [13], Jankowski [15], Kloeden and Platen [16], Ohashi [42], Siljak [46], Squier [47], Wallach [49] and Woul [51]. In Sections 5.6–5.9, a local convergence analysis for large-scale algorithm is outlined. This is adapted from Ladde and Sambandham [31, 33, 34] and based on the works of Ladde and Siljak [36, 37], Nevelson and Hasminski [41] and Ladde [24] in the context of the developments in Sections 5.1–5.5.

Chapter 6

Numerical Schemes for Systems of Difference Equations of Ito Type

6.0 Introduction

The recent trend in scientific computation toward distributed processing provides a mechanism for more efficient and reliable sharing of resources to solve complex problems. In this chapter, we first formulate a very general variational comparison theorem in the context of a vector Lyapunov-like function and a system of differential inequalities for large-scale stochastic systems of Ito–Doob-type differential equations, coupled with its discretized system schemes, forming a stochastic hybrid system. We then present a procedure for developing error estimates, convergence and relative stability analysis between continuous- and discrete-time solution processes of large-scale systems of differential and difference equations generated by discretization schemes. The procedure is based on the hybrid method, composed of the variation of constants, comparison and decomposition-aggregation methods in the framework of hybrid systems. Both the system of stochastic differential equations and its theoretical iterative algorithm (numerical scheme) are decomposed into lower-order subsystems in a natural way. Each stochastic iterative subsystem component is assigned to a subprocessor for parallel distributed computation of a cost/Lyapunov-like function that is assigned to each subprocessor. The individual Lyapunov functions are stacked up to form a vector Lyapunov function, and comparison theorems are derived to establish the relative qualitative properties of the solution processes of discretized schemes with respect to their continuous-time dynamic systems.

In Section 6.1, a theoretical large-scale algorithm is formulated. In Section 6.2, a very general variational comparison theorem in the context of vector Lyapunov-like functions and a system of differential inequalities are established. Moreover, local discretization/truncation error estimates for continuous-time and numerical schemes of isolated subsystems are established. Further, we develop an auxiliary comparison result for the local analytic error estimates of the solution process isolated subsystems in the framework of hybrid systems in Section 6.3. A very general theoretical framework is developed for establishing a global theoretical/discretization error estimation of isolated subsystems in Section 6.4. In addition, several feasible results are presented. Examples are given to illustrate the usefulness of the results. In Section 6.5, the byproducts of the developed results for isolated subsystems are exhibited in a natural manner. In fact, the Dahlquist type of stochastic convergence, stability and relative stability results are presented in a systematic and unified way. The above-outlined development and analysis of numerical schemes for nonlinear isolated subsystems of differential equations of the Ito–Doob type are extended to cover an overall large-scale interconnected system of numerical schemes for stochastic differential equations of the Ito–Doob type. This is achieved in the framework summarized as follows. In the remainder of Sections 6.6–6.9, variational comparison results of Sections 6.6, 6.7 and 6.8 are extended for studying numerical schemes associated with stochastic large-scale hybrid systems in the framework of a large-scale stochastic hybrid system. The stochastic error estimates, the Dahlquist type of convergence and relative stability in moment/probability results for stochastic large-scale numerical schemes are imitated and outlined similar to their isolated subsystems in Sections 6.2–6.4. In fact, the results are parallel to the results of Sections 6.2–6.5. Examples are given to illustrate the fruitfulness of the results.

6.1 Theoretical Large-Scale Algorithms

The error estimation problem for stochastic numerical schemes with regard to a system of large-scale stochastic differential equations of

Ito type has been proposed in the framework of large-scale stochastic hybrid systems. For this purpose, it is assumed that a system of stochastic differential equations of Ito type and its numerical algorithm are of the following form:

$$\begin{aligned} dx &= \mathbf{A}(t,x)\cdot \boldsymbol{d\xi}, \quad x(t_0)=x_0,\\ \Delta y_i(t_k) &= \mathbf{B}(t_k,y)\cdot \mathbf{I}(\boldsymbol{\xi})(t_k,t_{k+1}), \quad y(t_0)=y_0,\\ y(t_0) &= x(t_0)=x_0, \quad k\in I(0,N), \end{aligned} \tag{6.1.1}$$

where $I(0,N) = \{0,1,2,\ldots,N\}$; $\mathbf{A} = (A_0^T, A_1^T, \ldots, A_j^T, \ldots, A_m^T)^T$, $A_j \in \mathcal{C}[J\times \mathcal{R}^n, \mathcal{R}^n]$, for $j \in I(0,m)$; $J = [t_0, t_0+\alpha]$ and $t_0 \in [0,\infty)$; $\mathbf{B} = (B_1^T,\ldots,B_l^T,\ldots,B_q^T)^T, B_l \in M[J\times \mathcal{R}^n, \mathcal{R}^n]$, for $l \in I(1,q)$; x_0 is an n-dimensional random vector defined on a complete probability space (Ω, F, P); $\xi(t) = (\xi_1(t), \xi_2(t), \ldots, \xi_m(t))^T$ is an m-dimensional normalized Wiener process with independent increments; $\xi(t)$ is F_t-measurable, for all $t \geq t_0$ and with F_t being an increasing family of sub-σ-algebras of F; $x_0, \xi(t)$ and ξ_k are mutually independent, for each $t \geq t_0$ and $k \geq k_0$; $\xi_0(t)= t, \boldsymbol{\xi}(t) = (\xi_0(t), \xi_1(t), \xi_2(t), \ldots, \xi_m(t))$, for $t_k \in J = [t_0, t_0+\alpha]$; $I_i(\boldsymbol{\xi})(t_k, t_{k+1}) = I_i[1]_{t_k, t_{k+1}}$ denotes the stochastic multiple integral with respect to $\boldsymbol{\xi}(t)$ over the interval $[t_k, t_{k+1}]$ with multiple-index $i \in A_\gamma, i = (j_1, j_2, \ldots, j_s, \ldots, j_l), j_s \in I(0,m)$, for $l \in I(1,\infty)$; A_γ is a given hierarchical set with $l(i) + n(i) \leq 2\gamma$; v is a multi-index of length zero; and the cardinality $A_\gamma\backslash\{v\}$ is q. By ordering the set $A_\gamma\backslash\{v\}$ in a suitable hierarchic way, let us denote $A_\gamma\backslash\{v\} = \{i_1, i_2, \ldots, i_j, \ldots, i_q\}$ and form the q-dimensional vector $I(\boldsymbol{\xi})(t_k, t_{k+1}) = (I_{i_1}(\boldsymbol{\xi})(t_k, t_{k+1}), I_{i_2}(\boldsymbol{\xi})(t_k, t_{k+1}), \ldots, I_{i_q}(\boldsymbol{\xi})(t_k, t_{k+1}))^T$, where $I_{i_1}(\boldsymbol{\xi})(t_k, t_{k+1}) = I_{(0)}$, $I_{i_2}(\boldsymbol{\xi})(t_k, t_{k+1}) = I_{(1)}, \ldots I_{i_m}(\boldsymbol{\xi})(t_k, t_{k+1}) = I_{(m)}, \ldots$; $\Delta y(t_k) = y(t_{k+1}) - y(t_k), k \in I(0,N) = [0,1,2,\ldots,N]$. It is assumed that $\mathbf{A}$ and $\mathbf{B}$ are smooth enough to ensure the existence and uniqueness of the solution processes $x(t) = x(t, t_0, x_0)$ and $y(t_k, t_0, y_0)$ of (6.1.1).

A system described by (6.1.1) may involve a large number of state variables. Using a "one shot" technique to study such a system may prove to be both analytically and computationally difficult. Therefore, a more feasible approach is developed, which consists of decomposition of a large-scale system into interconnected subsystems of lower dimension. The large system is studied in a "piece by piece" manner, and conclusions about the pieces are utilized to form conclusions about the complex system as a whole. For this purpose, the stochastic large-scale system (6.1.1) is decomposed into p-interconnected (perturbed) subsystems of the following form:

$$\begin{aligned} dx_i &= [\mathbf{a}^i(t, x_i) + \mathbf{c}^i(t, x_i)] \cdot d\boldsymbol{\xi}, \quad x_i(t_0) = x_i^0, \\ \Delta y_i(t_k) &= [\mathbf{b}^i(t_k, y) + \mathbf{d}^i(t_k, y)] \cdot I(\boldsymbol{\xi})(t_k, t_{k+1}), \quad y_i(t_0) = y_i^0, \\ y_i(t_0) &= x_i(t_0) = x_i^0, \quad k \in I(0, N) \quad \text{and } i \in I[1, p], \end{aligned} \tag{6.1.2}$$

where $\mathbf{a}^i = (a_0^{iT}, a_1^{iT}, \ldots, a_j^{iT}, \ldots, a_m^{iT})^T$ and $\mathbf{c}^i = (c_0^{iT}, c_1^{iT}, \ldots, c_j^{iT}, \ldots, c_m^{iT})^T$, $a_j^i \in C[J \times R^{n_i}, R^{n_i}]$ and $c_j^i \in C[J \times R^n, R^{n_i}]$, for $j \in I(0, m)$; $\mathbf{b}^i = (b_0^{iT}, b_1^{iT}, \ldots, b_m^{iT}, \ldots, b_l^{iT}, \ldots, b_{i_q}^{iT})^T$ and $\mathbf{d}^i = (d_0^{iT}, d_1^{iT}, \ldots, d_m^{iT}, \ldots, d_l^{iT}, \ldots, d_{i_q}^{iT})^T$, $b_l^i \in M[J \times R^{n_i}, R^{n_i}]$ and $d_l^i \in M[J \times R^n, R^{n_i}]$, for $l \in I(1, q)$; $x_0 = (x_1^{0T}, x_2^{0T}, \ldots, x_p^{0T})^T$, $x = (x_1^T, x_2^T, \ldots, x_p^T)^T$, $y = (y_1^T, y_2^T, \ldots, y_p^T)^T$ and $\sum_{i=1}^p n_i = n$, for each $i \in I[1, p]$.

The interactions (perturbations) among the p subsystems of the continuous part of system (6.1.2) are described by $\mathbf{c}^i$. The interactions (perturbations) among the p subsystems of the discrete part of system (6.1.2) are described by $\mathbf{d}^i$. The isolated subsystems corresponding to (6.1.2) are given by

$$\begin{cases} dx_i = \mathbf{a}^i(t, x_i) \cdot d\boldsymbol{\xi}, \quad x_i(t_0) = x_i^0, \\ \Delta y_i(t_k) = \mathbf{b}^i(t_k, y_i) \cdot I(\boldsymbol{\xi})(t_k, t_{k+1}), \quad y_i(t_0) = y_i^0, \\ y_i(t_0) = y_i^0, \end{cases} \tag{6.1.3}$$

respectively, for $i \in I[1, p]$ and $k \in I(0, N)$.

Remark 6.1.1. The hybrid systems defined in (7.6.1), (7.6.2) and (7.6.3), Section 7.6, corresponding to systems (6.1.1), (6.1.2) and (6.1.3) can be considered the local versions or representations of the global systems (6.1.1), (6.1.2) and (6.1.3), respectively. Moreover, we provide the solution processes of (6.1.2) and (6.1.3) with identical initial data as (t_0, x^0) and (t_0, x_i^0), respectively. The solution processes of (7.6.2) and (7.6.3) with the identical initial data (t_{k-1}, y^{k-1}) and (t_{k-1}, y_i^{k-1}) are coordinated using a hybrid of continuous- and discrete-time processes to obtain the deviation between the continuous- and discrete-time dynamic processes, which is obtained by applying a discretized process to the continuous-time dynamic process.

6.2 Local Theoretical Discretization Error Estimates for Isolated Subsystems

A comparison principle provides a very general framework for studying the qualitative properties of dynamic systems. The various forms of this kind of comparison principle have been extensively developed for continuous- and discrete-time dynamic systems. In this section, an attempt is made to formulate a variational comparison theorem for the isolated hybrid subsystems (7.6.3) of the large-scale hybrid system (7.6.2) and associated with the isolated global subsystems (6.1.3) of the large-scale subsystems (6.1.2). The results connect the three solution processes of hybrid subsystems, auxiliary systems and comparison differential equations. In addition, they provide error estimations locally.

By setting $\bar{y}_i = y_i = \mathbf{b}^i(t_k, y_i, x_i^k) \cdot \mathbf{I}(\boldsymbol{\xi})(t_k, t)$, the ith isolated hybrid subsystem (7.6.3) can be rewritten as

$$\begin{cases} dx_i = \mathbf{a}^i(t, x_i) \cdot d\boldsymbol{\xi}, \quad x_i(t_k) = x_i^k, \\ d\bar{y}_i = \mathbf{b}(t_k, y_i(k)) \cdot \mathbf{I}(\boldsymbol{\xi})(t_k, t), y_i(t_k) = y_i^k, t \in J_k, \\ y_i(t_0) = y_i^0, x_i(t_0) = x_i^0, k \in I(0, N) \quad \text{and } i \in I[1, p]. \end{cases} \tag{6.2.1}$$

In order to meet the above-stated goal, let us consider the following auxiliary system:

$$z_i' dz_i = f^i(t, z_i)dt, \quad z_i(t_0) = z_i^0, \tag{6.2.2}$$

where $f^i \in C[J \times R^{n_i}, R^{n_i}]$, and it is twice continuously differentiable with respect to z_i; the solution $z_i(t, t_0, z_i^0)$ of (6.2.2) exists for $t \geq t_0$; it is unique and continuous with respect to the initial conditions (t_0, z_{i0}) and also locally Lipschitzian in z_i^0, for $i \in I(1, p)$.

For $q_i \geq 1$ and any $V^i \in C[J \times R^{n_i}, R_+^{q_i}]$, $\frac{\partial}{\partial t}V^i(t, z_i)$, $\frac{\partial}{\partial x}V^i(t, z_i)$ and $\frac{\partial^2}{\partial x^2}V^i(t, z_i)$ exist and are continuous on $J \times R^{n_i}$, and $\frac{\partial^2}{\partial x^2}V^i(t, z_i)$ is locally Lipschitzian in z_i, for each $t_0 \leq t_k \leq s \leq t \leq t_{k+1} \leq t_0 + \alpha$ and $k \in I(0, N)$. We define the operator $D^+V^i_{(6.2.1)}$ in the context of the auxiliary system (6.2.2) as follows:

$$
\begin{aligned}
&D^{+}_{(6.2.1)}V^i(s,t_k,z_i(t,s,y_i-x_i)),\\
&\quad=\lim_{h\to 0^+}\frac{1}{h}E\Bigg[V^i(s+h,z_i(t,s+h,y_i-x_i\\
&\qquad+h\mathbf{F}^i(s,t_k,y_i,x_i)\cdot\Delta\boldsymbol{\psi}(s))),-V^i(s,z_i(t,s,y_i-x_i))|F_s\Bigg],\\
&\qquad\times\ i\in I(1,p),I[1,p],
\end{aligned}
\tag{6.2.3}
$$

where $\mathbf{F}^i(s,t_k,y_i,x_i) = (-\mathbf{a}^{iT}(s,x_i),\mathbf{b}^{iT}(t_k,y_i))^T$ and $\Delta\boldsymbol{\psi}(s) = \boldsymbol{\psi}(s+h)-\boldsymbol{\psi}(s)$.

Remark 6.2.1. Under the assumptions on f^i and V^i, we note that for each $t_0 \le t_k \le s \le t \le t_{k+1} \le t_0+\alpha$ and $k \in I(0,N)$,

$$
D^{+}_{(6.2.1)}V^i(s,t_k,z_i(t,s,\Delta x_i)) \le L_i^D V^i(s,t_k,y_i,z_i(t,s,\Delta x_i)), \tag{6.2.4}
$$

where the operator L_i^D is defined by

$$
\begin{aligned}
&L_i^D V^i(s,t_k,y_i,z_i(t,s,\Delta x_i))\\
&\quad= V_s^i(s,z_i(t,s,\Delta x_i)) + \frac{1}{2}\mathrm{tr}\left(\frac{\partial^2}{\partial z_i^2}V^i(s,z_i(t,s,\Delta x_i))\Theta_i^D(t,s,t_k,x_i,y_i)\right)\\
&\qquad+ V_{z_i}^i(s,z_i(t,s,\Delta x_i))\left[\frac{1}{2}\mathrm{tr}\left(\frac{\partial^2}{\partial x_i\partial x_i}z_i(t,s,\Delta x_i)\Lambda_i^D(s,t_k,x_i,y_i)\right)\right.\\
&\qquad\qquad\left.+\,\Phi^i(t,s,\Delta x_i)M_i^D(s,t_k,y_i,x_i)\right],
\end{aligned}
$$

where $M_i^D(s,t_k,y_i,x_i) = \lim_{h\to 0^+}\frac{1}{h}E[(\mathbf{F}^i(s,t_k,y_i,x_i)\cdot\Delta\boldsymbol{\psi}(s))(\mathbf{F}^i(s,t_k,y_i,x_i)\cdot\Delta\boldsymbol{\psi}(s))^T|F_s]$,

$$
\begin{aligned}
\Lambda_i^D(s,t_k,y_i,x_i) &= \lim_{h\to 0}\frac{1}{h}E[(\mathbf{F}^i(s,t_k,y_i,x_i)\cdot\Delta\boldsymbol{\psi}(s)\\
&\qquad\times(\mathbf{F}^i(s,t_k,y_i,x_i)\cdot\Delta\boldsymbol{\psi}(s))^T|F_s],\\
\Theta_i^D(t,s,t_k,x_i,y_i) &= \Phi^i(t,s,\Delta x_i)\Lambda^i(s,t_k,x_i,y_i)\Phi^{iT}(t,s,\Delta x_i),
\end{aligned}
$$

$\Delta x_i = y_i - x_i$ and $\Phi^i(t,s,x_i)$ is as defined before.

Remark 6.2.2. Moreover, one can define D^+V^i analogously to (6.2.3) with respect to the continuous component of (7.6.3), and its associated $L_i^D V^i$ is defined by

$$
\begin{aligned}
&L_i^D(s, z_i(t, s, \Delta x_i)) \\
&\quad = V_s^i(s, z_i(t, s, \Delta x_i)) + \frac{1}{2}\mathrm{tr}\left(\frac{\partial^2}{\partial z_i^2} V^i(s, z_i(t, s, \Delta x_i))\Delta\theta_i^D(t, s, x_i)\right) \\
&\qquad + V_{z_i}^i(s, z_i(t, s, \Delta x_i))\left[\frac{1}{2}\mathrm{tr}\left(\frac{\partial^2}{\partial x_i \partial x_i} z_i(t, s, \Delta x_i)\Delta\lambda_i^D(s, x_i)\right)\right. \\
&\qquad\qquad \left. + \Phi^i(t, s, \Delta x_i)\Delta M_i^D(s, x_i)\right], \qquad (6.2.5)
\end{aligned}
$$

where

$$
\begin{aligned}
\Delta M_i^D(s, x_i) &= \lim_{h\to 0}\frac{1}{h}E[\Delta \mathbf{a}^i(s, x_i)\cdot \Delta\boldsymbol{\xi}(s) - hf^i(s, \Delta x_i)|F_s], \\
\Delta x_i &= y_i - x_i, \\
\Delta\lambda_i^D(s, x_i) &= \lim_{h\to 0}\frac{1}{h}E[(\Delta \mathbf{a}^i(s, x_i)\cdot \Delta\boldsymbol{\xi}(s))(\Delta \mathbf{a}^i(s, x_i)\cdot \Delta\boldsymbol{\xi}(s))^T|F_s], \\
\Delta\theta_i^D(t, s, x_i) &= \Phi^i(t, s, \Delta x_i)\Delta\lambda^i(s, x_i)\Phi^{iT}(t, s, \Delta x_i), \Delta x_i = y_i - x_i, \\
\Delta\boldsymbol{\xi}(s) &= \boldsymbol{\xi}(s+h) - \boldsymbol{\xi}(s), \quad \text{and} \\
\Delta \mathbf{a}^i(s, x_i) &= (\mathbf{a}^i(s, y_i) - \mathbf{a}^i(s, x_i)).
\end{aligned}
$$

In the following, we present a basic variational comparison principle which relates the solution processes of continuous- and discrete-time subsystems by treating the ith isolated system (6.1.3), the auxiliary system (6.2.2) and the comparison differential equations as isolated systems. Moreover, the results provide the local analytic error estimates between the two isolated components, theoretical/discretization and continuous, of the ith isolated hybrid subsystem (6.2.1) treated in the framework of the ith isolated hybrid subsystem (7.6.3), Section 7.6.

Theorem 6.2.1 (Variational Comparison Theorem). *Assume that:*

(i) $g^i \in C[J \times J \times R^{n_i} \times R^{q_i}, R_+^{q_i}], g^i(t, t_k, y_i, u_i)$ *is concave and quasimonotone nondecreasing in* u_i. *For each* $(t, t_k, y_i) \in J \times J \times R^{n_i}$.

(ii) $r_i^k(t) = r_i^k(t, t_k, u_i^k)$ *is the maximal solution of the following comparison system:*

$$dz_i = g^i(t, t_k, y_i, u_i)dt, \quad u_i^k(t_k) = u_i^k, \ t \in J_k. \tag{6.2.6}$$

(iii) $z_i(t, s, x_i)$ *is the solution process of the auxiliary system* (6.2.2) *through an initial data,* (s, x_i), *for* $t_0 \le s \le t \le t_0 + \alpha$;

(iv) $V^i \in C[J \times R^{n_i}, R_+^{q_i}]$; $\frac{\partial}{\partial t}V^i(s, z_i), \frac{\partial}{\partial x}V^i(s, z_i)$, *and* $\frac{\partial^2}{\partial x^2}V(s, z_i)$ *exist and are continuous on* $J \times R^{n_i}$; $\frac{\partial^2}{\partial x^2}V^i(s, z_i)$ *is locally Lipschitzian in* z_i, *for each* $t_0 \le t_k \le s \le t \le t_{k+1} \le t_0 + \alpha$; *and* $L_i^D V^i(s, t_k, y_i, z_i(t, s, \Delta x_i))$ *defined in* (6.2.4) *satisfies the following relation* (*Remark* 6.2.1):

$$L_i^D V^i(s, t_k, y_i, z_i(t, s, \Delta x_i)) \le g^i(s, t_k, y_i, V^i(s, z_i(t, s, \Delta x_i))). \tag{6.2.7}$$

(v) $\hat{x}_i^k(t) = \hat{x}_i(t, t_k, x_i^k), \bar{y}_i^k(t) = \bar{y}_i(t, t_k, y_i^k)$ *are the solutions of* (6.2.2) *through* (t_k, x_i^k) *and* (t_k, y_i^k), *respectively, where* $\{y^k\}_0^N$ *is a sequence in* $R[(\Omega, F, P), R^n]$ *and*

$$y^k = (y_1^{kT}, y_2^{kT}, \ldots, y_i^{kT}, \ldots, y_p^{kT})^T,$$

$y_i^k \in R[(\Omega, F, P), R^{n_i}]$, *and* $E[V^i(s, z_i(t, s, \bar{y}_i^k(s) - \hat{x}_i^k(s)))]$ *exists for* $s \in [t_k, t]$, $t \in J_k$ *and* $k \in I(0, N)$. *Then,*

$$E[V^i(t, \bar{y}_i^k(t) - \hat{x}_i^k(t))|F_{t_k}] \le r_i(t, t_k, u_i^k), \quad t \in J_k \tag{6.2.8}$$

whenever

$$V^i(t_k, z_i(t, t_k, y_i^k - x_i^k)) \le u_i^k, \tag{6.2.9}$$

for $k \in I(0, N)$, $t \in J_k$ *and each* $i \in I(1, p)$.

Corollary 6.2.1. *Assume that all the hypotheses of Theorem* 6.2.1 *are satisfied, except that* (6.2.9) *and* v *are replaced by* $V^i(t_k, z_i(t, t_k, y_i^k - x_i^k)) = u_i^k$ *and* $(v)'$, *respectively.* $\hat{x}_i^k(t)$ *and* $\bar{y}_i^k(t)$ *are the solution processes of the ith isolated hybrid subsystem* (7.6.3) *described by* (6.2.1) *through* (t_k, y_i^k) *and* (t_k, y_i^k), *respectively. Then,*

$$\begin{aligned} &E[V^i(t_{k+1}, y^{k+1} - \hat{x}_i^k(t_{k+1}))|t_{k+1}F_{t_k}] \\ &\quad \le r_i(t_{k+1}, t_k, V^i(t_k, z_i(t_{k+1}, y_i^k - x_i^k))), \end{aligned} \tag{6.2.10}$$

for all $i \in I[1, p]$ *and* $k \in I(0, N)$.

Remark 6.2.3. From (6.2.8) and Corollary 6.2.1, we observe that for $t_k, t_{k+1} \in J_k$, it is obvious that

$$V^i(t_k, y_i^k - x_i^k) = E[V^i(t_k, \bar{y}_i^k(t_k))|F_{t_k}] \leq r_i(t_k, t_k, u_i^k)$$

and

$$E[V^i(t_{k+1}, y_i^{k+1} - \hat{x}_i^k(t_{k+1}^-))|F_{t_k}] \leq r_i(t_{k+1}, t_k, u_i^k).$$

This justifies Remark 6.1.1, that is, the solution processes of (7.6.2) and (7.6.3) with identical initial data as (t_{k-1}, y^{k-1}) and (t_{k-1}, y^{k-1}) are coordinated using a hybrid of continuous- and discrete-time processes to obtain the deviation the between continuous- and discrete-time dynamic processes via a discretized process applied to the continuous-time dynamic process.

6.3 Local Analytical Error Estimates for Isolated Subsystems

In this section, we state another auxiliary comparison theorem that provides an estimate of the local analytic error for the solution process of the ith isolated subsystem (6.1.3) by using the solution process of the ith isolated hybrid subsystem (7.6.3)/(6.2.1).

Theorem 6.3.1. *Assume that*

(i) $g^i \in C[J \times R_+^{q_i}, R_+^{q_i}], g^i(t, u_i)$ *is concave and quasimonotone non-decreasing in* u_i. *For each* $t \in J$, $i \in (1, p)$.

(ii) $r_i^k(t) = r_i(t, t_k, u_i^k)$ *is the maximal solution of the following comparison system:*

$$u_i{}'dz_i = g^i(t, u_i)dt, \quad u_i(t_k) = u_i^k, \tag{6.3.1}$$

$i \in I[1, p]$.

(iii) $z_i(t, s, x_i)$ *is the solution process of the auxiliary system* (6.2.2) *through* (s, x_i), $t_0 \leq s \leq t \leq t_0 + \alpha$.

(iv) $V^i \in C[J \times R^{n_i}, R_+^{q_i}]$; $\frac{\partial}{\partial t}V^i(s, z_i), \frac{\partial}{\partial x}V^i(s, z_i)$ *and* $\frac{\partial^2}{\partial x^2}V(s, z_i)$ *exist and are continuous on* $J \times R^{n_i}$; $\frac{\partial^2}{\partial x^2}V^i(s, z_i)$ *is locally Lipschitzian in* z_i, *for each* $t_0 \leq t_k \leq s \leq t \leq t_{k+1} \leq t_0 + \alpha$; *and* $L_i^D V^i(s, z_i(t, s, \Delta x_i))$ *defined in* (6.2.5) *satisfies the following relation* (*Remark* 6.2.2):

$$L_i^D V^i(s, z_i(t, s, \Delta x_i)) \leq g^i(s, z_i(t, s, x_i - \bar{x}_i)); \tag{6.3.2}$$

(v) $\hat{x}_{ix_i^0}$ *and* $\hat{x}_i^k(t)$ *are the solution processes of the continuous component of the ith isolated subsystem* (6.1.3) *and the hybrid subsystem* (7.6.3) *through* (t_0, x_i^0) *and* (t_k, y_i^k), *respectively, where* $\{y^k\}_0^N$ *is a sequence in* $R[(\Omega, F, P), R^n]$ *and*

$$y^k = (y_1^{kT}, y_2^{kT}, \ldots, y_i^{kT}, \ldots, y_p^{kT})^T,$$

$y_i^k \in R[(\Omega, F, P), R^{n_i}]$, *and* $E[V^i(s, z_i(t, s, \hat{x}_i^k(s) - \hat{x}_{ix_i^0}(s)))]$ *exists for* $s \in [t_k, t]$, $t \in J_k$ *and* $k \in I(0, N)$. *Then,*

$$E[V^i(t, \hat{x}_i^k(t) - \hat{x}_{ix_i^0}(t))|F_{t_k}] \leq r_i(t, t_k, u_i^k), \tag{6.3.3}$$

whenever

$$V^i(t_k, z_i(t, t_k, y_i^k - \hat{x}_{ix_i^0}(t_k))) \leq u_i^k, \quad t \in J_k, \tag{6.3.4}$$

for all $i \in I[1, p]$ *and* $k \in I(0, N)$.

Proof. Let $\hat{x}_{ix_i^0}(t)$ and $\hat{x}_i^k(t)$ be the solution processes of (6.1.3) and (7.6.3) through (t_0, x_i^0) and (t_{k-1}, y_i^{k-1}), respectively. We define

$$m^i(s) = V^i(s, z_i(t, s, \hat{x}_i^k(s) - \hat{x}_{ix_i^0}(s))), \quad \text{for } t_k \leq s \leq t \leq t_{k+1}. \tag{6.3.5}$$

By using the fundamental properties of the solution process of (6.2.2), we note that

$$m^i(t) = V^i(t, z_i(t, t, \hat{x}_i^k(t) - \hat{x}_{ix_i^0}(t)))$$

and

$$m^i(t_k) = V^i(t_k, z_i(t, t_k, y_i^k - \hat{x}_{ix_i^0}(t_k))). \tag{6.3.6}$$

The rest of the proof of the theorem follows by imitating the standard argument.

Corollary 6.3.1. *Assume that all the hypotheses of Theorem* 6.3.1 *are satisfied, except that* (6.3.6) *is replaced by*

$$V^i(t_k, z_i(t_{k+1}, t_k, y_i^k - \hat{x}_{ix_i^0}(t_{k+1}))) = u_i^k, \tag{6.3.7}$$

for all $i \in I[1, p]$ *and* $k \in I(0, N)$. *Then,*

$$\begin{aligned} &E[V^i(t_{k+1}, \hat{x}_i^k(t_{k+1}) - \hat{x}_{ix_i^0}(t_{k+1}))|F_{t_k}] \\ &\quad \leq r_i(t_{k+1}, t_k, V^i(t_k, z_i(t_{k+1}, t_k, y_i^k - \hat{x}_{ix_i^0}(t_k))), \end{aligned} \tag{6.3.8}$$

where $r_i(t_{k+1}, t_k, u_i^k)$ *is the solution process of* (6.3.1) *through* (t_k, u_i^k), *for all* $i \in I[1, p]$ *and* $k \in I(0, N)$.

Remark 6.3.1. If $f^i \equiv 0$, then Theorems 6.2.1 and 6.3.1 reduce to the usual comparison theorem as special cases.

Example 6.3.1. (i) For $i \in I[1,p]$, the comparison functions $g^i(t, u_i) \equiv 0$ are admissible in Theorem 6.3.1. In this case, (6.3.8) can be reduced to

$$\begin{aligned} E[V^i(t_{k+1}, \hat{x}_i^k(t_{k+1}) - \hat{x}_{ix_i^0}(t_{k+1}))|F_{t_k}] \\ \leq V^i(t_k, z_i(t_{k+1}, t_k, y_i^k - \hat{x}_{ix_i^0}(t_k))). \end{aligned} \tag{6.3.9}$$

(ii) Let $G^i(t)$ be $q_i \times q_i$ matrices and continuous, whose off-diagonal entries are nonnegative, for $i \in I[1,p]$. The comparison functions $g^i(t, u_i) = G^i(t)u_i$ are also admissible in Theorem 6.3.1. Here, $r_i(t, t_0, u_i^0) = \Phi_c^i(t, s)u_i^0$, where $\Phi_c^i(t, s)$ is the fundamental matrix solution $du_i = G^i(t)u_i dt, \Phi_c^i(s, s) = I_{q_i \times q_i}$ (an identity matrix). Hence, (6.3.8) reduces to

$$\begin{aligned} E[V^i(t_{k+1}, \hat{x}_i^k(t_{k+1}) - \hat{x}_{ix_i^0}(t_{k+1}))|F_{t_k}] \\ \leq \Phi_c^i(t_{k+1}, t_k)V^i(t_k, z_i(t_{k+1}, t_k, y_i^k - \hat{x}_{ix_i^0}(t_k))). \end{aligned} \tag{6.3.10}$$

Problem 6.3.1. For $i \in I[1,p]$, the functions $V^i(s, z_i) = ||z_i||$ and $V^i(s, z_i) = z_i^T P_i z_i$ are admissible in Theorem 6.2.1 and (6.3.1) with $q_i = 1$. Here, P_i is a nonnegative definite $n_i \times n_i$ matrix, for each $i \in I[1,p]$.

Remark 6.3.2. From various combinations of functions in Theorems 6.2.1 and 6.3.1, Corollaries 6.2.1 and 6.3.1 and Remarks 6.2.1, 6.2.2 and 6.3.1, and Examples 6.3.1 and 6.3.2, one can obtain various possible special estimates of type (6.2.8), (6.2.10), (6.3.8), (6.3.9) and (6.3.10). These particular estimates provide a basis for comparison with the existing results in the literature. This will be illustrated subsequently. The details are left to the reader.

We state without proof an auxiliary result with regard to the comparison equations. This result will be used subsequently. It is a very simple observation about the nondecreasing property of the maximal solution of the comparison system that is based on a well-known result in the theory of systems of differential inequalities.

Lemma 6.3.1. *Let hypothesis (i) in Theorem* 6.2.1 (*Theorem* 6.3.1) *be satisfied. Let* $r_{iu}^k(t) = r_i(t, t_k, u_i^k)$ *and* $r_{iv}^k(t) = r_i(t, t_k, v_i^k)$ *be the maximal solutions of comparison differential equations* (6.2.6) *and* (6.3.1) *through* (t_k, u_i^k) *and* (t_k, v_i^k), *respectively. Then,*

$$r_{iu}^k(t) \leq r_{iv}^k(t), \quad \text{for} \quad t \geq t_k, \tag{6.3.11}$$

whenever

$$u_i^k \leq v^k. \tag{6.3.12}$$

6.4 Global Theoretical Discretization Error Estimates for Isolated Subsystems

In this section, we propose to study the error estimates between the continuous- and discrete-time subsystems of isolated hybrid subsystems. The obtained results are applied to study the error estimates in numerical analysis of stochastic for initial value problems of Ito type. For each ith isolated hybrid subsystem (7.6.3)/(6.2.1), we associate a vector Lyapunov-like function.

We are ready to present a very general result that establishes a global theoretical/discretization error estimate for the solution process of the ith subsystem (6.1.3).

Theorem 6.4.1. *Assume that all the hypotheses of Theorem* 6.3.1 *are satisfied. Further, assume that*

(a) *the sequence* $\{y^k\}_0^N$ *satisfies the following condition:*

$$E[\|y_i^k - \hat{x}_i^{k-1}(t_k)\| | F_{t_{k-1}}] \leq \chi_i^{k-1}(N), \tag{6.4.1}$$

where $\chi_i^{k-1}(N)$ *is an* $F_{t_{k-1}}$*-measurable discrete process defined by*

$$\chi_i^{k-1}(N) = \beta_i^{k-1}(h_N^k, N), \tag{6.4.2}$$

$\beta_i^{k-1} \in C[[0, h^0], R_+]$, $h^0 > 0$, *for each* $i \in I[1, p]$ *and for all* $k \in I(1, N)$, *and* $N \in I(1, \infty)$;

(b) $w_i(t_k) = w_i(t_k, t_{k-1}, w_i^0)$ *is the solution process of an iterative process*:

$$w_i^k(N) = W^i(t_k, t_{k-1}, w_i^{k-1}(N)), \quad w_i^0(N) = w_i^0, \tag{6.4.3}$$

where $W^i(t_k, t_{k-1}, w_i^{k-1}(N)) = r_i(t_k, t_{k-1}, w_i^{k-1}(N)) + \mathcal{J}_i^k \chi_i^{k-1}(N)$. $r_i(t_k)$ *is defined in Theorem* 6.3.1, $\mathcal{J}_i^k = \mathcal{M}^i(t_{k+1}, t_k)$ *is a* q_i*-dimensional local Lipschitz constant vector relative to both* $V^i(s, z_i)$ *and* $z_i(t, s, x_i)$ *with respect to* x_i, z_i *and* t, *for each* $i \in I[1, p]$ *and all* $k \in I(1, N)$. *Then*,

$$E[V^i(t_k, z_i(t_{k+1}, t_k, y_i^k - \hat{x}_{ix_i^0}(t_k)))] \leq w_i(t_k), \quad k \in I(1, N), \tag{6.4.4}$$

whenever

$$V^i(t_0, z_i(t_1, t_0, y_i^0 - x_i^0)), \leq w^0, \tag{6.4.5}$$

for each $i \in I(1, p)$.

Proof. The immediate goal is to establish a system of difference inequality with respect to

$$E[V^i(t_k, z_i(t_{k+1}, t_k, y_i^k - \hat{x}_{ix_i^0}(t_k)))|F_{t_{k-1}}].$$

For this purpose, we set $V^i(t_{k-1}, z_i(t_k, t_{k-1}, y_i^{k-1} - \hat{x}_{ix_i^0}(t_{k-1}))) = u_i^{k-1}$ and apply Corollary 6.3.1 to obtain an estimate (6.3.8), which can be written as

$$\begin{aligned} &E[V^i(t_k, \hat{x}_i^{k-1}(t_k) - \hat{x}_{ix_i^0}(t_k))|F_{t_{k-1}}] \\ &\quad \leq r_i(t_k, t_{k-1}, V^i(t_{k-1}, z_i(t_k, t_{k-1}, y_i^{k-1} - \hat{x}_{ix_i^0}(t_{k-1})))). \end{aligned} \tag{6.4.6}$$

From (6.3.1), by noting the fact about the uniqueness of the solution of (6.2.2),

$$V^i(t_k, \hat{x}_i^{k-1}(t_k) - \hat{x}_{ix_i^0}(t_k)) = V^i(t_k, z_i(t_k, t_k, \hat{x}_i^{k-1}(t_k) - \hat{x}_{ix_i^0}(t_k))),$$

and using the Lipschitzian property of $V^i(s, z_i)$ and $z_i(t, s, x_i)$ in z_i, coupled with algebraic manipulations, we arrive at

$$\begin{aligned} &E[V^i(t_k, z_i(t_{k+1}, t_k, y_i^k - \hat{x}_{ix_i^0}(t_k)))|F_{t_{k-1}}] \\ &\quad \leq E[V^i(t_k, \hat{x}_i^{k-1}(t_k) - \hat{x}_{ix_i^0}(t_k))|F_{t_{k-1}}] + \mathcal{M}^i(t_{k+1}, t_k)\chi_i^{k-1}(N). \end{aligned} \tag{6.4.7}$$

From (6.4.6) and (6.4.7), we obtain the desired system of difference inequality:

$$\begin{aligned} &E[V^i(t_k, z_i(t_{k+1}, t_k, y_i^k - \hat{x}_{ix_i^0}(t_k)))|F_{t_{k-1}}] \\ &\leq r_i(t_k, t_{k-1}, V^i(t_{k-1}, z_i(t_k, t_{k-1}, y_i^{k-1} - \hat{x}_{ix_i^0}(t_{k-1})))) \\ &\quad + \mathcal{J}_i^k \chi_i^{k-1}(N). \end{aligned} \tag{6.4.8}$$

Now, we set

$$m_i(k) = E[V^i(t_k, z_i(t_{k+1}, t_k, y_i^k - \hat{x}_{ix_i^0}(t_k)))|F_{t_{k-1}}]$$

and note that $m_i(0) = V^i(t_0, z_i(t_1, t_0, y_i^0 - x_i^0))$ and

$$m_i(k-1) = V^i(t_{k-1}, z_i(t_k, t_{k-1}, y_i^{k-1} - \hat{x}_{ix_i^0}(t_{k-1}))).$$

From these considerations and the definition of W^i in (6.4.3), inequalities (6.4.8) and (6.4.5) reduce to

$$m_i(k) \leq W^i(t_k, t_{k-1}, m_i(k-1)), k \in I(1, N) \tag{6.4.9}$$

and

$$m_i(0) \leq w_i^0, \tag{6.4.10}$$

respectively, for each $i \in I(1,p)$. Now, by the applications of Lemma 6.3.1, the system of difference inequalities and the comparison theorem, we obtain the desired inequality:

$$\begin{aligned} &E[V^i(t_k, z_i(t_{k+1}, t_k, y_i^k - \hat{x}_{ix_i^0}(t_k)))] \\ &\leq w_i(t_k), k \in I(0, N) \quad \text{and } i \in I[1, p]. \end{aligned}$$

This completes the proof of the theorem.

Remark 6.4.1. It is important to note that the hypotheses of Theorem 6.4.1 provide a crucial iterative scheme to estimate a global error for the solution process of (6.1.3) in the framework of the hybrid subsystems (7.6.3). Of course, this is achieved by inequality (6.4.8).

One can pose a question: under less demanding and more feasible conditions on functions, is there a better way to obtain a system of difference inequalities similar to (6.4.8)? In the following, we make a few attempts to provide solutions to this problem.

Theorem 6.4.2. *Assume that all the hypotheses of Theorem* 6.3.1 *are satisfied. Further, assume that*

(i) *the solution process* $z_i(t,s,x_i)$ *of* (6.2.2) *satisfies the following relation:*

$$\|z_i(t,s,x_i)\|^{p_i} \leq \lambda_i(t_k,t_{k-1})\gamma_i(\|x_i\|^{p_i}), \tag{6.4.11}$$

where $p_i \geq 1$, $\lambda_i \in C[R_+ \times R_+, R_+]$ *and* $\gamma_i \in C[R_+, R_+]$, *with* $\gamma_i \in \mathcal{CK}$;

(ii) *for each* $t \in J, V^i(t,z_i)$ *satisfies the following relations:*

$$V^i(t,z_i) \leq \varphi^i(\|z_i\|^{p_i}) \tag{6.4.12}$$

and

$$\vartheta_i(\|z_i\|^{p_i}) \leq \sum_{j=1}^{q_i} V_j^i(t,z_i), \tag{6.4.13}$$

$\varphi^i \in C[R_+, R_+^{q_i}]$ *and* $\vartheta \in C[R_+,R_+], \vartheta_i \in \mathcal{VK}$ *and all components* $\varphi^i \in \mathcal{CK}$;

(iii) $w_i(t_k) = w_i(t_k,t_{k-1},w_i^0)$ *is the solution process of a scalar iterative process:*

$$w_i^k(N) = W^i(t_k,t_{k-1},w_i^{k-1}(N)), \quad w_i^0(N) = w_i^0, \tag{6.4.14}$$

where

$$\begin{aligned} &W^i(t_k,t_{k-1},w_i^{k-1}(N)) \\ &\quad = 2^{p_i}\left[(\vartheta_i)^{-1}\left(\sum_{j=1}^{q_i} r_{ij}\left(t_k,t_{k-1},\varphi^i\left(\lambda_i^k\gamma_i(w_i^{k-1}(N))\right)\right)\right)\right. \\ &\qquad \left. + \left[\chi_i^{k-1}(N)\right]^{p_i}\right], \quad \lambda_i^k = \lambda_i(t_k,t_{k-1}), \quad \text{for each } i \in I[1,p], \\ &\qquad \text{and all } k \in I(1,N). \end{aligned}$$

Then,

$$E[\|y_i^k - \hat{x}_{ix_i^0}(t_k)\|^{p_i}] \leq w_i(t_k), \quad k \in (1,N), \tag{6.4.15}$$

whenever

$$\|y_i^0 - x_i^0\|^{p_i} \leq w_i^0, \tag{6.4.16}$$

for each $i \in I[1,p]$.

Proof. Again, the immediate goal is to establish a scalar difference inequality with respect to $E[\|y_i^k - \hat{x}_{ix_i^0}(t_k)\|^{p_i}|F_{t_{k-1}}]$. As before, by following the proof of Theorem 6.3.1, we arrive at (6.4.6). Now, from (6.4.11), (6.4.12) and Lemma 6.3.1, we get

$$\begin{aligned} & r_i(t_k, t_{k-1}V^i(t_{k-1}, z_i(t_k, t_{k-1}, y_i^{k-1} - \hat{x}_{ix_i^0}(t_{k-1})))) \\ & \quad \leq r_i(t_k, t_{k-1}, \varphi^i(\lambda_i^k \gamma_i(\|y_i^{k-1} - \hat{x}_{ix_i^0}(t_{k-1})\|^{p_i}))). \end{aligned} \tag{6.4.17}$$

From (6.4.13) and the property of ϑ_i, we have

$$\begin{aligned} & E[\|\bar{x}_i^{k-1}(t_k) - \bar{x}_{ix_i^0}(t_k)\|^{p_i}|F_{t_{k-1}}] \\ & \quad \leq (\vartheta_i)^{-1} \left(\sum_{j=1}^{q_i} E[V_j^i(t_k, \hat{x}_i^{k-1}(t_k) - \hat{x}_{ix_i^0}(t_k))|F_{t_{k-1}}] \right). \end{aligned} \tag{6.4.18}$$

We note that

$$\begin{aligned} & \|y_i^k - \hat{x}_{ix_i^0}(t_k)\|^{p_i} \\ & \quad \leq 2^{p_i} \left[\|y_i^k - \hat{x}_i^{k-1}(t_k)\|^{p_i} + \|\hat{x}_i^{k-1}(t_k) - \hat{x}_{ix_i^0}(t_k)\|^{p_i} \right]. \end{aligned} \tag{6.4.19}$$

From (6.4.1), (6.4.6), (6.4.17), (6.4.18) and (6.4.19), one can obtain the following scalar difference inequality:

$$\begin{aligned} & E[\|y_i^k - \hat{x}_{ix_i^0}(t_k)\|^{p_i}|F_{t_{k-1}}] \\ & \quad \leq 2^{p_i} \left[(\vartheta)^{-1} \left(\sum_{j=1}^{q_i} r_{ij} \left(t_k, t_{k-1}, \varphi^i \left(\lambda_i^k \gamma_i(\|y_i^{k-1} - \hat{x}_{ix_i^0}(t_{k-1})\|^{p_i}) \right) \right) \right) \right. \\ & \qquad \left. + \left[\chi_i^{k-1}(N) \right]^{p_i} \right]. \end{aligned} \tag{6.4.20}$$

This, together with the definition of W^i in (6.4.14), leads to the following scalar difference inequality:

$$m_i(k) \leq W^i(t_k, t_{k-1}, m_i(k-1)),$$

with

$$m_i(0) \leq w_i^0.$$

Now, by imitating the proof of Theorem 6.4.1, the proof of the theorem follows immediately.

Corollary 6.4.1. *Assume that all the hypotheses of Theorem* 6.4.2 *are satisfied, except that the scalar comparison difference equation* (6.4.14) *is replaced by*

$$w_i^k(N) = W^i(t_k, t_{k-1}, w_i^{k-1}(N)), \quad w_i^0(N) = w_i^0, \tag{6.4.21}$$

where

$$\begin{aligned} &W^i(t_k, t_{k-1}, w_i^{k-1}(N)) \\ &= 2^{p_i}\left[\lambda_i^k \gamma_i \left((\vartheta_i)^{-1}\left(\sum_{j=1}^{q_i} r_{ij}(t_k, t_{k-1}, \varphi^i(w_i^{k-1}(N)))\right)\right)\right. \\ &\qquad \left. + \mathcal{L}_i^{k-1}\left[\chi_i^{k-1}(N)\right]^{p_i}\right], \end{aligned}$$

$\mathcal{L}_i^k = [L_z^i(t_k, h_N^k)]^{p)i}$, L_z^i *is a local Lipschitz constant of* $z_i(t, s, x_i)$ *with respect to* x_i, *for each* (t, s) *and all* $i \in I[1, p]$ *and* $k \in I(1, N)$.

Then,

$$E[\|z_i(t_k, t_{k-1}, y_i^{k-1} - \hat{x}_{ix_i^0}(t_{k-1}))\|^{p_i}] \leq w_i(t_k), \quad k \in I(0, N). \tag{6.4.22}$$

Proof. Again, the goal is to establish a scalar difference inequality with respect to $E[\|z_i(t_{k+1}, t_k, y_i^k - \hat{x}_{ix_i^0}(t_k))\|^{p_i}|F_{t_{k-1}}]$. As before, by following the proof of Theorem 6.4.1, we arrive at (6.4.6). From (6.4.12) and Lemma 6.3.1, we have

$$\begin{aligned} &r_i\left(t_k, t_{k-1}, V^i\left(t_{k-1}, z_i(t_k, t_{k-1}, y_i^{k-1} - \hat{x}_{ix_i^0}(t_{k-1}))\right)\right) \\ &\quad \leq r_i\left(t_k, t_{k-1}, \varphi^i\left((\|z_i(t_k, t_{k-1}, y_i^{k-1} - \hat{x}_{ix_i^0}(t_{k-1}))\|^{p_i}\right)\right). \end{aligned} \tag{6.4.23}$$

By imitating a similar argument as in Theorem 6.4.2, using the Lipschitzian property of the solution process of (6.2.2), assumption

(i) and (6.4.1), we have

$$\|z_i(t_{k+1}, t_k, y_i^k - \hat{x}_{ix_i^0}(t_k))\|^{p_i}$$
$$\leq 2^{p_i}\left[\mathcal{L}_i^k(N)[\chi_i^{k-1}(N)]^{p_i} + \lambda_i^k \gamma_i \left(\|\hat{x}_i^{k-1}(t_k) - \hat{x}_{ix_i^0}(t_k)\|^{p_i}\right)\right]. \tag{6.4.24}$$

From (6.4.6), (6.4.18), (6.4.23), (6.4.24) and the definition of W^i in (6.4.21), we arrive at the following scalar difference inequality:

$$E[\|z_i(t_{k+1}, t_k, y_i^k - \hat{x}_{ix_i^0}(t_k))\|^{p_i}|F_{t_{k-1}}]$$
$$\leq W(t_k, t_{k-1}, \|z_i(t_k, t_{k-1}, y_i^{k-1} - \hat{x}_{ix_i^0}(t_{k-1}))\|).$$

This establishes the desired scalar difference inequality. Now, the rest of the proof can be completed by imitating the argument used in Theorem 6.4.1. The details are left to the reader.

The following example illustrates the scope of Theorems 6.4.1 and 6.4.2 and Corollary 6.4.1.

Example 6.4.1. (i) Assume that the hypotheses of Theorem 6.4.2 are satisfied in the context of the following considerations: suppose that for each $i \in I[1,p]$, $f^i(t, z_i) = A^i(t)z_i$ in (6.2.2), where A^i is a continuous $n_i \times n_i$ matrix function. $\Phi^i(t,s)$ is a fundamental matrix solution of

$$dz_i = A^i(t)z_i dt, \Phi^i(s,s) = I_{n_i \times n_i} \quad \text{(an identity matrix)}.$$

(ii) Further, assume that $g^i(t, u_i) \equiv 0$. In this case, $r_i(t, t_0, u_i^0) = u_i^0, z_i(t, s, x_i) = \Phi^i(t,s)x_i$. Then, (6.3.9) reduces to

$$E[V^i(t_k, \hat{x}_i^{k-1}(t_k) - \hat{x}_{ix_i^0}(t_k))F_{t_{k-1}}]$$
$$\leq V^i(t_{k-1}, \Phi^i(t_k, t_{k-1})(y_i^{k-1} - \hat{x}_{ix_i^0}(t_{k-1}))), \tag{6.4.25}$$

for each $i \in I[1,p]$ and all $k \in I(1,N)$. We note that $r_i(t, t_0, u_i^0)$ is nondecreasing in u_i^0. Furthermore, $\gamma_i = s, \lambda_i^k = \|\Phi^i(t_k, t_{k-1})\|^{p_i}$.

(iii) We assume that condition (ii) of Theorem 6.4.2 is satisfied. Then, the comparison functions $W^i(t_k, t_{k-1}, w_i^{k-1}(N))$ in (6.4.3), (6.4.14) and (6.4.21) reduce to

$$w_i^k(N) = \mathcal{J}_i^k \chi_i^{k-1}(N) + w_i^{k-1}(N), \tag{6.4.26}$$

$$w_i^k(N) = 2^{p_i} \left[(\vartheta_i)^{-1} \left(\sum_{j=1}^{q_i} \varphi_j^i(\lambda_i^k w_i^{k-1}(N)) \right) + [\chi_i^{k-1}(N)]^{p_i} \right] \tag{6.4.27}$$

and

$$w_i^k(N) = 2^{p_i} \left[\lambda_i^k (\vartheta_i)^{-1} \left(\sum_{j=1}^{q_i} \varphi_j^i(w_i^{k-1}(N)) \right) + \mathcal{L}_i^k [\chi_i^{k-1}(N)]^{p_i} \right], \tag{6.4.28}$$

respectively, for all $k \in I(1, N)$ and for each $i \in I[1, p]$.

Example 6.4.2. Assume the conditions as described in Example 6.4.1(a) and consider $g^i(t, u_i) = G^i(t)u_i$, where $G^i(t)$ is a $q_i \times q_i$ as defined in Example 6.3.1(ii). Again, in this case, $r_i(t, t_0, u_i^0) = \Phi_c^i(t, s)u_i^0$ is nondecreasing in u_i^0. We assume that condition (ii) of Theorem 6.4.2 is satisfied. Then, (6.3.10) reduces to

$$\begin{aligned} &E[V^i(t_k, \hat{x}_i^{k-1}(t_k) - \hat{x}_{ix_i^0}(t_k))F_{t_{k-1}}] \\ &\quad \leq \Phi_c^i(t_k, t_{k-1}) V^i(t_{k-1}, \Phi^i(t_k, t_{k-1})(y_i^{k-1} - \hat{x}_{ix_i^0}(t_{k-1}))), \end{aligned} \tag{6.4.29}$$

and in this case, the comparison functions $W^i(t_k, t_{k-1}, w_i^{k-1}(N))$ in (6.4.3), (6.4.14) and (6.4.21) become

$$w_i^k(N) = \Phi_c^i(t_k, t_{k-1}) w_i^{k-1}(N) + \mathcal{J}_i^k \chi_i^{k-1}(N), \tag{6.4.30}$$

$$\begin{aligned} w_i^k(N) = 2^{p_i} \Bigg[&(\vartheta_i)^{-1} \left(\sum_{j=1}^{q_i} \sum_{j=1}^{q_i} \Phi_{cjl}^i(t_k, t_{k-1}) \varphi_l^i \left(\lambda_i^k w_i^{k-1}(N) \right) \right) \\ &+ [\chi_i^k(N)]^{p_i} \Bigg] \end{aligned} \tag{6.4.31}$$

and

$$w_i^k(N) = 2^{p_i}\left[\lambda_i^k(\vartheta_i)^{-1}\left(\sum_{j=1}^{q_i}\sum_{j=1}^{q_i}\Phi_{cjl}^i(t_k,t_{k-1})\varphi_l^i\left(w_i^{k-1}(N)\right)\right)\right.$$
$$\left.+\mathcal{L}_i^k[\chi_i^k(N)]^{p_i0}\right], \tag{6.4.32}$$

respectively, for all $i \in I[1,p]$ and $k \in I(1,N)$.

Example 6.4.3. We pick $V^i(s,z_i) = \|z_i\|^2$ and assume the conditions as described in Example 6.3.1(i). In this case, V^i and $z_i(t,s,x_i) = \Phi^i(t,s)x_i$ are continuously differentiable with respect to (s,x_i). Also, $L_i^D V^i(s,z_i(t,s,\Delta x_i))$, which is defined in (6.4.17), reduces to

$$2(\Phi^i(t,s)\Delta x_i)^T\Phi^i(t,s)\Delta M^i(s,x_i) + \mathrm{tr}(\Phi^i(t,s)\Delta\lambda^i(s,x_i)\Phi^i(t,s)), \tag{6.4.33}$$

where

$$\Delta M^i(s,x_i) = [\Delta a_0^i(s,x_i) - A^i(s)\Delta x_i],$$
$$\Delta\lambda^i(s,x_i) = \Delta\sigma^i(s,x_i)\Delta\sigma^{iT}(s,x_i)$$

and $\sigma^i = [a_1^i, a_2^1, \ldots, a_m^i]$ is an $n \times m$ matrix whose elements are as defined in (6.1.2). It is assumed that

$$2(\Phi^i(t,s)\Delta x_i)^T\Phi^i(t,s)\Delta M^i(s,x_i) + \mathrm{tr}(\Phi^i(t,s)\Delta\lambda^i(s,x_i)\Phi^i(t,s))$$
$$\leq u^i(s)\|\Phi^i(t,s)x_i - \bar{x}_i\|^2, \tag{6.4.34}$$

where μ^i is an integrable function on the given interval J_k. Here,

$$g^i(s,u_i) = \mu^i(s)u_i \quad \text{with}$$

$$\Phi_c^i(t_k,t_{k-1}) = \exp\left[\int_{t_{k-1}}^{t_k}\mu^i(s)ds\right] \quad \text{and} \quad q_i = 1.$$

Here, $p_i = 2$,

$$\varphi^i(s) = s = \vartheta_i(s), \quad r_i(t, t_0, u_i^0) = \Phi_c^i(t_k, t_{k-1})u_i^0$$

is nondecreasing in u_i^0. In this case, (6.3.10) becomes

$$\begin{aligned} &E[\|\hat{x}_i^{k-1}(t_k) - \hat{x}_{ix_i^0}(t_k)\|^2 F_{t_{k-1}}] \\ &\quad \leq \Phi_c^i(t_k, t_{k-1})\|\Phi^i(t_k, t_{k-1})(y_i^{k-1} - \hat{x}_{ix_i^0}(t_{k-1}))\|^2. \end{aligned} \tag{6.4.35}$$

The comparison functions in (6.4.3), (6.4.14) and (6.4.21) are

$$\exp\left[\int_{t_{k-1}}^{t_k} \mu^i(s)ds\right] w_i^{k-1}(N) + \mathcal{J}_i^k \chi_i^{k-1}(N), \tag{6.4.36}$$

$$4\left[\exp\left[\int_{t_{k-1}}^{t_k} \mu^i(s)ds\right] \lambda_i^k w_i^{k-1}(N) + \left[\chi_i^{k-1}(N)\right]^2\right] \tag{6.4.37}$$

and

$$4\left[\lambda_i^k \exp\left[\int_{t_{k-1}}^{t_k} \mu^i(s)ds\right] w_i^{k-1}(N) + \mathcal{L}_i^k \left[\chi_i^{k-1}(N)\right]^2\right], \tag{6.4.38}$$

respectively. The solutions of (6.4.26), (6.4.30), (6.4.36), (6.4.37) and (6.4.38) have the following form:

$$w_i(t_k) = \prod_{j=1}^{k} H^i(t_j, t_{j-1})w_i^0 + \sum_{l=1}^{k} \prod_{j=l+1}^{k} H^i(t_j, t_{j-1})v_i^l(N), \tag{6.4.39}$$

where H^i and ν_i^l stand for the coefficients of functions with respect to homogeneous system/scalar difference equations and non-homogeneous/perturbation functions in their respective comparison equations. Thus, (6.4.39) provides the estimates corresponding to (6.4.4), (6.4.15) and (6.4.22), respectively.

Remark 6.4.2. In addition to the conditions in Example 6.4.3, if $f^i(t, z_i) \equiv 0$, then (6.4.35), (6.4.36), (6.4.37) and (6.4.38) lead to

$$E[\|\hat{x}_i^{k-1}(t_k) - \hat{x}_{ix_i^0}(t_k)\|^2 F_{t_{k-1}}] \leq \Phi_c^i(t_k, t_{k-1})\|y_i^{k-1} - \hat{x}_{ix_i^0}(t_{k-1})\|^2, \tag{6.4.40}$$

$$\exp\left[\int_{t_{k-1}}^{t_k} \mu^i(s)ds\right] w_i^{k-1}(N) + \chi_i^{k-1}(N), \tag{6.4.41}$$

$$4\left[\exp\left[\int_{t_{k-1}}^{t_k} \mu^i(s)ds\right] w_i^{k-1}(N) + [\chi_i^{k-1}(N)]^2\right] \tag{6.4.42}$$

and

$$4\left[\exp\left[\int_{t_{k-1}}^{t_k} \mu^i(s)ds\right] w_i^{k-1}(N) + [\chi_i^{k-1}(N)]^2\right], \tag{6.4.43}$$

respectively. The solutions corresponding to these comparison functions have the same form as in (6.4.39).

Remark 6.4.3. The feasibility of hypothesis (a) of Theorem 6.4.1 can be justified by the following well-known examples:

(a) If $b^i(t_{k-1}, y_i^{k-1}, h_N^{k-1}) = \sum_{j=1}^{m} \left(\frac{h_N^i}{j!}\hat{x}_i^{(j)k-1}(t_{k-1})\right), k \in I(0, N)$ with $m > 0$, then

$$\chi_i^k(N) = 0\left(h_N^{m+1}\right) h_N^k.$$

(b) If $b^i(t_{k-1}, y_i^{k-1}, h_N^{k-1}) = g_i(y_i^{k-1}) + \left(h_N^{k-1}\right)^{\alpha} G_i(t_{k-1}, y_i^{k-1}, h_N^{k-1})$ $+ 0\left(\left(h_N^{m+1}\right)^{\beta}\right)$, where $\alpha \in R$ and $\beta > 0, g_i : R^{n_i} \to R^{n_i}, G_i : J \times R^{n_i} \times [0, h_0] \to R^{n_i}$ and satisfy the relation

$$\|b^i(t_{k-1}, y_i^{k-1}, h_N^{k-1}) - \hat{x}_i^{k-1}(t_k)\| \leq \left(h_N^{k-1}\right)^{\alpha} \gamma_i^k(h_N^k, N),$$

for some nonnegative function γ_i^k, for each $i \in I[1, p], k \in I(1, N)$, then we have

$$\chi_i^k(N) = \left(h_N^{k-1}\right)^{\alpha} \gamma_i^k(h_N^k, N).$$

In addition to Remark 6.4.3, by employing comparison Theorem 6.3.1, we also provide the sufficient conditions for the feasibility of hypothesis (a) of Theorem 6.4.1 in the context of the isolated hybrid subsystem (7.6.3)/(6.2.1).

Lemma 6.4.1. *Let the hypotheses of Theorem* 6.3.1 *and assumption* (ii) *of Theorem* 6.4.2 *be satisfied. Then,*

$$E[\|y_i^k - \hat{x}_i^{k-1}(t_k)\|F_{t_{k-1}}] \leq \chi_i^{k-1}(N), \tag{6.4.44}$$

where $y^k = (y_1^{kT}, y_2^{kT}, \ldots, y_i^{kT}, \ldots, y_p^{kT})^T, y_i^k \in R^{n_i}$ *are as defined before and* $\chi_i^{k-1}(N)$ *is defined by*

$$\begin{aligned}\chi_i^{k-1}(N) &= \beta_i^{k-1}(h_N^k, N)\\ &= (\vartheta)^{-1}\left(\sum_{j=1}^{q_i} r_{ij}(t_k, t_{k-1}, \varphi^i(\|z_i(t_k, t_{k-1}, 0)\|))\right).\end{aligned} \tag{6.4.45}$$

Proof. Let $\hat{x}_i^{k-1}(t)$ and $\hat{y}_i^{k-1}(t)$ be the solution processes of the ith isolated hybrid subsystem (7.6.3) described by (6.2.1) through (t_{k-1}, y_i^{k-1}) and (t_{k-1}, y_i^{k-1}), respectively, where $\{y^k\}_0^N$ is a sequence in R^n and $y^k = (y_1^{kT}, y_2^{kT}, \ldots, y_i^{kT}, \ldots, y_p^{kT})^T$, where $y_i^k \in R^{n_i}$. From (6.3.8) and (6.3.10), one can obtain

$$\begin{aligned}&E[V^i(t_k, \bar{y}_i^{k-1}(t_k) - \hat{x}_i^{k-1}(t_k))F_{t_{k-1}}]\\ &\quad \leq r_i(t_k, t_{k-1}, V^i(t_{k-1}, z_i(t_k, t_{k-1}))),\end{aligned}$$

for all $i \in I[1,p]$ and $k \in I(1,N)$. With this, together with the definition of $\bar{y}_i = y + (t - t_{k-1})b^i(t_{k-1}, y_i^{k-1}, h^{k-1})$, we arrive at

$$\begin{aligned}&E[V^i(t_k, y_i^{k1} - \hat{x}_i^{k-1}(t_k))F_{t_{k-1}}]\\ &\quad \leq r_i(t_k, t_{k-1}, V^i(t_{k-1}, z_i(t_k, t_{k-1}, 0))).\end{aligned} \tag{6.4.46}$$

From (6.4.12), (6.4.13) and (6.4.46), we obtain the desired inequality

$$E[\|y_i^k - \hat{x}_i^{k-1}(t_k)\|F_{t_{k-1}}] \leq \chi_i^{k-1}(N),$$

where $\chi_i^{k-1}(N)$ is as defined in (6.4.45). This completes the proof of the lemma.

Remark 6.4.4. We remark that the full force of Theorem 6.3.1 is not completely utilized. Its particular case (Theorem 6.3.1) has been used in Lemma 6.3.1 to establish the feasibility of assumption (a) of Theorem 6.4.1.

6.5 Convergence Analysis

Now, we present a very general result that provides sufficient conditions under which Dahlquist-type convergence in moment and probability is assured. Moreover, we also analyze the stability and convergence of the results.

Theorem 6.5.1. *Under the assumptions of Theorem* 6.4.1 (*Theorem* 6.4.2 *or Corollary* 6.4.1), *the convergence of solution of the comparison difference equations* (6.4.3) (6.4.14) *or* (6.4.21) *implies the convergence of one-step numerical schemes for the ith isolated subsystem* (6.1.3).

Proof. The proof of the theorem follows from relations (6.4.4) and (6.4.5) ((6.4.15) and (6.4.16) or (6.4.22) and (6.4.16)), which play an important role in the study of difference or differential equations. The convergence/stability of the solution comparison (6.4.3) ((6.4.14) or (6.4.21)) implies stochastic mean/mean-square convergence/stability of the solution process of (6.1.3). We follow the standard argument used in the qualitative study of the systems of stochastic differential/difference equations of Ito–Doob type.

We present a few examples illustrating the role and scope of Theorem 6.5.1 in qualitative analysis of large-scale stochastic discretization process of continuous-time processes in the framework of stochastic large-scale hybrid dynamic systems in the areas of convergence and stability analysis of numerical schemes as well as applied statistical data analysis.

Example 6.5.1. The convergence of solutions of the comparison system of difference equations in (6.4.26), (6.4.30), (6.4.36) and (6.4.41) in Examples 6.4.1, 6.4.2 and 6.4.3, corresponding to the comparison system (6.3.3) in Theorem 6.3.1, guarantees the convergence in moment of the stochastic numerical scheme in (6.2.3). The solutions of (6.4.26), (6.4.30), (6.4.36) and (6.4.41) have the form of (6.4.39), that is,

$$w_i(t_k) = \prod_{j=1}^{k} H^i(t_j, t_{j-1}) w_i^0 + \sum_{l=1}^{k} \prod_{j=l+1}^{k} H^i(t_j, t_{j-1}) \nu_i^l(N). \quad (6.5.1)$$

Therefore, the convergence of solutions of systems (6.4.26), (6.4.30), (6.4.36) and (6.4.41) is assured by the convergence of

$$\sum_{l=1}^{k} \prod_{j=l+1}^{k} H^i(t_j, t_{j-1}) \nu_i^l(N). \quad (6.5.2)$$

Of course, this expressions depends on the maximal solutions of comparison functions and the solution of auxiliary systems.

Example 6.5.2. Similarly, the convergence of solutions of the comparison system of difference equations in (6.4.27), (6.4.31), (6.4.37) and (6.4.42) ((6.4.28), (6.4.32), (6.4.38) and (6.4.43)) in Examples 6.4.1, 6.4.2 and 6.4.3, corresponding to the comparison systems (6.4.14) (6.4.21) in Theorem 6.4.2 (Corollary 6.4.1), guarantees the convergence in moment of the stochastic numerical scheme in (6.1.3). The solutions of (6.4.27), (6.4.31), (6.4.37) and (6.4.42) ((6.4.28), (6.4.32), (6.4.38) and (6.4.43)) have the form of (6.4.46). Therefore, the convergence of solutions of systems (6.4.26), (6.4.30), (6.4.36) and (6.4.41) is assured by the convergence of

$$\sum_{l=1}^{k} \prod_{j=l+1}^{k} H^i(t_j, t_{j-1}) \nu_i^l(N). \quad (6.5.3)$$

A sufficient condition can be given for the convergence of (6.5.3) by using Lemma 6.4.1. Finally, we present a very general result that provides sufficient conditions under which the relative stability property of the numerical one-step scheme (6.1.3) is guaranteed. In addition, an example is given to illustrate the scope of the result.

Theorem 6.5.2. *Under the assumptions of Theorem* 6.4.1 (*Theorem* 6.4.2 *or Corollary* 6.4.1), *the convergence of the solution of comparison equations* (6.4.3) ((6.4.14) *or* (6.4.21)) *implies the relative stability of one-step methods of the numerical scheme for the ith isolated subsystem* (6.1.3).

Example 6.5.3. Examples 6.5.1(I) and 6.5.2 provide the basis for the relative stability property of the one-step numerical scheme (6.1.3). From Example 6.5.2 and Lemma 6.4.1, the comparison solution process (6.4.39) converges as

$$w_i^0 \to 0 \text{ and } \sum_{l=1}^{k} \prod_{j=l+1}^{k} H^i(t_j, t_{j-1}) \nu_i^l(N) \to 0 \text{ as } N \to \infty. \tag{6.5.4}$$

6.6 Local Convergence Analysis for Large-Scale Algorithms

In this section, we develop a few auxiliary comparison results analogous to the results in Section 6.2. For this purpose and for the sake of simplicity, we associate a scalar Lyapunov-like function ($q_i = 1$) to each isolated hybrid subsystem (7.6.3), Section 7.6.

By following the argument used to derive (6.2.1), one can derive the form of large-scale hybrid system (7.6.2) as

$$\begin{cases} dx_i = [\mathbf{a}^i(t, x_i) + \mathbf{c}^i(t, x)] \cdot d\boldsymbol{\xi}, x_i(t_k) = x_i^k, t \in J_k, \\ d\bar{y}_i(t_k) = [\mathbf{b}^i(t_k, y_i) + \mathbf{d}^i(t_k, y)] \cdot \mathbf{I}(\boldsymbol{\xi})(t_k, t_{k+1}), y_i(t_k) = y_i^k, \\ y_i(t_0) = y_i^0, x_i(t_0) = x_i^0, k \in I(0, N) \quad \text{and} \quad i \in I(0, p). \end{cases} \tag{6.6.1}$$

We use the same auxiliary system of differential equations (6.2.2). As stated before, we associate a scalar Lyapunov-like function to each isolated stochastic hybrid subsystem (7.6.3). Similar to the definition $D^+_{(6.6.1)} V^i(s, t_k, z_i(t, s, \Delta x_i))$, one can define $D^+_{(6.6.1)} V^i(s, t_k, z_i(t, s, \Delta x_i))$. With respect to each subsystem of large-scale systems (6.6.1), operators similar to the operators L_i^D's defined in (6.2.4) can also be defined analogously and are denoted by L_i^C's, and

$$D^+_{(6.6.1)} V^i(s, t_k, x, z_i(t, s, \Delta x_i)) \leq L_i^C V^i(s, t_k, x, y, z_i(t, s, \Delta x_i)), \tag{6.6.2}$$

where L_i^C is defined and can be represented as

$$
\begin{aligned}
&L_i^C V^i(s, t_k, x, y, z_i(t, s, \Delta x_i)) \\
&\quad = L_i^D V^i(s, t_k, y_i, z_i(t, s, \Delta x_i)) + I^i(t, s, t_k, c^i(s, x), d^i(t_k, y)),
\end{aligned} \tag{6.6.3}
$$

$$
\begin{aligned}
&I^i(t, s, t_k, c^i(s, x), d^i(t_k, y)) \\
&\quad = \frac{1}{2}\mathrm{tr}\left(\frac{\partial^2}{\partial z_i^2} V^i(s, z_i(t, s, \Delta x)\right) \Theta_i^C(t, s, t_k, x, y)) \\
&\qquad + V_{z_i}^i(s, z_i(t, s, \Delta x_i)) \left[\frac{1}{2}\mathrm{tr}\left(\frac{\partial^2}{\partial x_i \partial x_i} z_i(t, s, \Delta x_i) \Lambda_i^C(s, t_k, x, y)\right)\right. \\
&\qquad \left. + \Phi^i(t, s, \Delta x_i) M_i^C(s, t_k, x, y)\right],
\end{aligned}
$$

$$
\begin{aligned}
M_i^C(s, t_k, x, y) &= \lim_{h \to 0^+} \frac{1}{h} E[C^i(s, t_k, x, y) \cdot \Delta\psi(s) - hf^i(s, \Delta x_i)|F_s], \\
\Lambda_i^C(s, t_k, x, y) &= \lim_{h \to 0^+} \frac{1}{h} E[([F^i(s, t_k, x_i, y_i) + C^i(s, t_k, x, y)] \cdot \Delta\psi(s)) \\
&\quad \times ([F^i(s, t_k, x_i, y_i) + C^i(s, t_k, x, y)] \cdot \Delta\psi(s))^T |F_s] \\
&\quad - \Lambda_i^D(s, t_k, x_i, y_i), \\
\Theta_i^C(t, s, t_k, x, y) &= \Phi^i(t, s, \Delta x_i) \Lambda_i^C(s, t_k, x, y) \Phi^{iT}(t, s, \Delta x_i).
\end{aligned}
$$

$$
\Delta x_i = y_i - x_i \text{ and } C^i(s, t_k, x, y) = (-c^i(s, x), d^i(t_k, y))^T.
$$

Remark 6.6.1. Moreover, one can define $D^+ V^i$ analogously to (6.6.2) with respect to the continuous component of (6.6.1)/(6.1.2), and its corresponding $L_i^C V^i$ is defined by

$$
\begin{aligned}
&L_i^C V^i(s, x, z_i(t, s, \Delta x_i)) \\
&\quad = L_i^D V^i(s, z_i(t, s, \Delta x_i)) + I^i(t, s, t_k, c^i(t, x)),
\end{aligned} \tag{6.6.4}
$$

where

$$
\begin{aligned}
I^i(t, s, t_k, c^i(t, x)) &= \frac{1}{2}\mathrm{tr}\left(\frac{\partial^2}{\partial z_i^2} V^i(s, z_i(t, s, \Delta x_i)) \Delta\theta_i^C(t, s, x)\right), \\
&\quad + V_{z_i}^i(s, z_i(t, s, \Delta x_i))
\end{aligned}
$$

$$\times \left[\frac{1}{2}\text{tr}\left(\frac{\partial^2}{\partial x_i \partial x_i} z_i(t,s,\Delta x_i)\Delta\lambda_i^C(s,x)\right),\right.$$

$$\left. + \ \Phi^i(t,s,\Delta x_i)\Delta M_i^C(s,x)\right],$$

$$\Delta M_i^C(s,x) = \lim_{h\to 0^+}\frac{1}{h}E[\Delta c^i(t,x)\cdot\Delta\xi(s) - hf^i(s,\Delta x_i)|F_s],$$

$$\Delta\lambda_i^C(s,x) = \lim_{h\to 0^+} E[([\Delta a^i(s,x_i) + \Delta c^i(t,x)]\cdot\Delta\xi(s))$$

$$\times ([\Delta a^i(s,x_i) + \Delta c^i(t,x)]\cdot\Delta\xi(s))^T|F_s]$$

$$- \ \Delta\lambda_i^D(s,x_i),$$

$$\Delta\theta_i^C(t,s,x_i) = \Phi^i(t,s,\Delta x_i)\Delta\lambda_i^C(s,x_i)\Phi^{iT}(t,s,\Delta x_i),$$

$$\Delta\xi(s) = \xi(s+h) - \xi(s), \Delta c^i(t,x) = [c^i(t,y) - c^i(t,x)],$$

$$\Delta a^i(s,x_i) = (a^i(s,y_i) - a^i(s,x_i)), \quad \text{and} \quad \Delta x_i = y_i - x_i.$$

Remark 6.6.2. A remark similar to Remark 6.2.1 can be formulated analogously.

In the following, we present comparison theorems for stochastic large-scale hybrid system (6.6.1). They are stated with brief outlines of their proofs.

The following comparison theorem provides an estimate of the local theoretical/discretization error for the solution process of the large-scale stochastic system (6.1.2) in the framework of a solution process of the corresponding large-scale hybrid system (6.6.1). We simply briefly sketch the proof without further details.

Theorem 6.6.1. *Assume that*

(i) $G^i \in C[J \times J \times R^n \times R_+^p, R_+]$, $g^i \in C[J \times J \times R^{n_i} \times R_+, R_+]$, *for each* $(t, t_k, y_i^k) \in J \times J \times R^{n_i}$, *are concave in* u_i *and* u, *respectively; moreover,* $G^i(t, t_k, y^k, u)$ *is quasimonotone nondecreasing in* u, *for each* $(t, t_k, y^k) \in J \times J \times R^n$ *and* $i \in I[1, p]$.

(ii) $r^k(t) = r(t, t_0, u^k)$ *is the maximal solution of the following comparison system:*

$$du_i = (g^i(t, t_k, y_i^k, u_i) + G^i(t, t_k, y^k, u))dt,$$
$$u_i(t_k) = u_i^k, \quad i \in I[1, p]. \tag{6.6.5}$$

(iii) $z_i(t, s, x_i)$ *is the solution process of the auxiliary system* (6.2.2) *through* $(s, x_i), t_0 \le s \le t \le t_0 + \alpha$.

(iv) $V^i \in C[J \times R^{n_i}, R_+]$; $\frac{\partial}{\partial t}V^i(s, z_i), \frac{\partial}{\partial x}V^i(s, z_i)$ *and* $\frac{\partial^2}{\partial x^2}V^i(s, z_i)$ *exist and are continuous on* $J \times \mathcal{R}^{n_i}$; $\frac{\partial^2}{\partial x^2}V^i(s, z_i)$ *is locally Lipschitzian in* z_i, *for each* $t_0 \le t_k \le s \le t \le t_{k+1} \le t_0 + \alpha$; *and* $L_i^D V^i(s, t_k, y_i^k, z_i(t, s, \Delta x_i))$ *satisfies the following relation:*

$$L_i^D V^i(s, t_k, y_i^k, z_i(t, s, \Delta x_i))$$
$$\le g^i(s, t_k, y_i^k, V^i(s, z_i(t, s, \Delta x_i))), \tag{6.6.6}$$

for each $i \in I[1, p]$.

(v) *For each* $i \in I[1, p]$, *the interaction rate functions among the subsystems satisfy the inequality*

$$I^i(t, s, t_k, c^i(s, x), d^i(t_k, y)) \le G^i(s, t_k, y^k, V(s, z(t, s, \Delta x))), \tag{6.6.7}$$

where I^i *is as defined in* (6.6.3) *and*

$$V(s, \Delta x) = ((V^1(s, \Delta x_1), V^2(s, \Delta x_2), \ldots, V^p(s, \Delta x_p)))^T. \tag{6.6.8}$$

(vi) $x^k(t) = x(t, t_k, x^k)$ *and* $\bar{y}^k(t) = \bar{y}(t, t_k, y^k)$ *are the solutions of* (6.6.1) *through* (t_k, x^k) *and* (t_k, y^k), *respectively; furthermore,*

$$E[V(s, z(t, s, \bar{y}^k(s) - \hat{x}^k(s)))],$$

exists for $s \in [t_k, t]$, $t \in J_k$, *and* $k \in I(0, N)$.

Then,

$$E[V(t, \bar{y}^k(t) - x^k(t)) | F_{t_k}] \le r(t, t_k, u^k), \tag{6.6.9}$$

whenever

$$V(t_k, z(t, t_k, y^k - x^k)) \le u^k, \tag{6.6.10}$$

for all $k \in I(0, N)$ *and* $t \in J_k$.

Proof. From assumptions (iii) and (iv) and employing the standard argument, we have

$$L_i^C V^i(s, t_k, y, x, z_i(t, s, \Delta x_i))$$
$$\leq g^i(s, t_k, y_i^k, V^i(s, z_i(t, s, \Delta x_i))) + G^i(s, t_k, y^k, V(s, z(t, s, \Delta x))), \tag{6.6.11}$$

for each $i \in I[1, p]$ and all $k \in I(0, N)$. Let $x^k(t)$ and $\bar{y}^k(t)$ be the solution processes of (6.6.1) through (t_k, x^k) and (t_k, y^k), respectively. Now, we set

$$m^i(s) = V^i(s, z_i(t, s, \bar{y}_i^k(s) - x_i^k(s))), \quad \text{for } s \in [t_k, t] \text{ and } t \in J_k$$

and note that $m^i(t_k) = V^i(t_k, z_i(t, t_k, y_i^k - x_i^k))$, for each $i \in I(1, p)$. Now, by following the standard argument, the proof of the theorem can be completed.

Corollary 6.6.1. *Assume that all the hypotheses of Theorem* 6.6.1 *are satisfied, except that* (6.6.9) *and* (*v*) *are replaced by* $V^i(t_k, z_i(t, t_k, y_i^k - x_i^k)) = u_i^k$ *and, for each* $I[1, p], x_i^k(t)$ *and* $\bar{y}_i^k(t)$ *are the solution processes of the ith hybrid subsystems described by* (6.6.1) *through* (t_k, y_i^k) *and* (t_k, y_i^k), *respectively.*

Then,

$$E[V^i(t_{k+1}, y_i^{k+1} - \hat{x}_i^k(t_{k+1}))|F_{t_k}]$$
$$\leq r_i(t_{k+1}, t_k, V^i(t_k, z_i(t_{k+1}, t_k, y_i^k - x_i^k))), \tag{6.6.12}$$

for all $i \in I[1, p]$ *and* $k \in I(0, N)$.

Remark 6.6.3. If $f^i \equiv 0$, then Theorem 6.6.1 reduces to the usual comparison theorem for studying large-scale systems as special cases.

6.7 Local Analytical Convergence Analysis for Large-Scale Algorithms

We state without proof another auxiliary comparison theorem that provides an estimate of the local analytic error for the solution process of the complex stochastic system (6.1.2) in the framework of the solution process of large-scale hybrid subsystem (7.6.2).

Theorem 6.7.1. *Assume that*

(i) $G^i \in C[J \times R_+^p, R_+], g^i \in C[J \times R_+, R_+]$, *for each* $t \in J, g^i(t, u_i)$ *and* $G^i(t, u)$ *are concave in* u_i *and* u, *respectively; moreover,* G^i *is quasimonotone nondecreasing in* u, *for each* $t \in J$ *and* $i \in I[1, p]$.

(ii) $r^k(t) = r(t, t_k, u^k)$ *is the maximal solution of the following comparison system:*

$$du_i = (g^i(t, u_i) + G^i(t, u))dt, u_i(t_k) = u_i^k, \quad \text{for each } i \in I[1, p]. \tag{6.7.1}$$

(iii) $z_i(t, s, x_i)$ *is the solution process of the auxiliary system* (6.2.2) *through* $(s, x_i), t_0 \le s \le t \le t_0 + \alpha$.

(iv) $V^i \in C[J \times R^{n_i}, R_+]$; $\frac{\partial}{\partial t}V^i(s, z_i)$, $\frac{\partial}{\partial x}V^i(s, z_i)$ *and* $\frac{\partial^2}{\partial x^2}V^i(s, z_i)$, *are locally Lipschitzian in* z_i, *for each* $t_0 \le t_k \le s \le t \le t_{k+1} \le t_0 + \alpha$; *and* $L_i^D V^i(s, z_i(t, s, \Delta x_i))$ *satisfies the following relation:*

$$L_i^D V^i(s, z_i(t, s, \Delta x_i)) \le g^i(s, V^i(s, z_i(t, s, \Delta x_i))). \tag{6.7.2}$$

(v) *The interaction functions among the subsystems satisfy the inequality*

$$I^i(t, s, t_k, c^i(t, x)) \le G^i(s, V(s, z(t, s, \Delta x))), \tag{6.7.3}$$

where I^i *is as defined in* (6.6.4) *and*

$$V(s, \Delta x) = (V^1(s, \Delta x_1), V^2(s, \Delta x_2), \ldots, V^p(s, \Delta x_p))^T. \tag{6.7.4}$$

(vi) $x(t)$ *and* $x^k(t)$ *are the solution processes of the continuous components of the complex system* (6.1.2) *and the stochastic large-scale hybrid system* (7.6.2) *through* (t_0, x^0) *and* (t_k, y^k), *respectively, where* $\{y^k\}_0^N$ *is a sequence in* $R[(\Omega, F, P), R^n]$ *and*

$$y^k = (y_1^{kT}, y_2^{kT}, \ldots, y_i^{kT}, \ldots, y_p^{kT})^T,$$

$y_i^k \in R[(\Omega, F, P), R^{n_i}]$, *and* $E[V(s, z(t, s, x^k(s) - x(s)))]$ *exists for* $s \in [t_k, t]$, $t \in J_k$ *and* $k \in I(0, N)$.

Then,

$$E[V(t, x^k(s) - x(s))|F_{t_k}] \leq r(t, t_k, u^k), \quad t \in J_k, \tag{6.7.5}$$

whenever

$$V(t_k, z(t, t_k, y^k - x(t_k))) \leq u_i^k, \tag{6.7.6}$$

for all $k \in I(0, N)$.

Corollary 6.7.1. *Assume that all the hypotheses of Theorem* 6.7.1 *are satisfied, except that* (6.7.6) *is replaced by*

$$V^i(t_k, z_i(t_{k+1}, t_k, y_i^k - x_i(t_{k+1}))) = u_i^k, \tag{6.7.7}$$

for all $i \in I[1, p]$ *and* $k \in I(0, N)$.
Then,

$$\begin{aligned} &E[V^i(t_{k+1}, \hat{x}_i^k(t_{k_1}))|F_{t_l}] \\ &\quad \leq r_i(t_{k+1}, t_k, V^i(t_k, z_i(t_{k+1}, t_k, y_i^k - \hat{x}_{ix_i^0}(t_k)))), \end{aligned} \tag{6.7.8}$$

where $r_i(t_{k+1}, t_k, u_i^k)$ *is the solution process of* (6.7.1) *through* (t_k, u_i^k), *for all* $i \in I[1, p]$ *and* $k \in I(0, N)$.

Remark 6.7.1. If $f^i \equiv 0$, then Theorem 6.7.1 reduces to the usual comparison theorem for studying large-scale systems as special cases.

Example 6.7.1. (i) For $i \in I[1, p]$, the comparison rate functions $g^i(t, u_i) \equiv 0, G^i(t, u) = G^i u$ and the off-diagonal element are nonnegative and admissible in Theorem 6.2.1. In this case, (6.4.27) can be reduced to

$$\begin{aligned} &E[V(t_{k+1}, \hat{x}_i^k(t_{k+1}) - \hat{x}_{ix_i^0}(t_{k+1}))|F_{t_k}] \\ &\quad \leq \Phi(z_i(t_k, z_i(t_{k+1}, t_k, y_i^k - \hat{x}_i(t_k)))u_i^k), \end{aligned} \tag{6.7.9}$$

whenever u_i^k is as given in (6.7.7).

6.8 Global Analytical Convergence Analysis for Large-Scale Algorithms

In this section, we propose to study the error estimates between the continuous- and discrete-time ith subsystems of large-scale subsystems (6.1.2). The obtained results are applied to study the error estimate in the numerical analysis of stochastic for initial value problems of Ito type. For each ith hybrid subsystem (6.1.2), we associate a vector Lyapunov-like function.

We are ready to present a very general result that establishes a global theoretical/discretization error estimate for the solution process of the large-scale system (6.1.1).

Theorem 6.8.1. *Assume that all the hypotheses of Theorem* 6.6.1 *are satisfied. Further, assume that*

(a) *the sequence* $\{y^k\}_0^N$ *satisfies the following condition:*

$$E[\|y_i^k - x_i^{k-1}(t_k)\||F_{t_{k-1}}] \leq \chi_i^{k-1}(N), \qquad (6.8.1)$$

where $\chi_i^{k-1}(N)$ *is an* $F_{t_{k-1}}$*-measurable discrete process defined by*

$$\chi_i^{k-1}(N) = \beta_i^{k-1}(h_N^k, N), \qquad (6.8.2)$$

$\beta_i^{k-1} \in C[[0, h^0], R_+]$, $h^0 > 0$, *for each* $i \in I[1, p]$ *and all* $k \in I(1, N)$, *and* $N \in I(1, \infty)$;

(b) $w(t_k) = w(t_k, t_{k-1}, w^0)$ *is the solution process of a system of iterative process:*

$$w^k(N) = W(t_k, t_{k-1}, w^{k-1}(N)), \quad w^0(N) = w^0, \qquad (6.8.3)$$

where the i*th component of* W *is defined by*

$$W^i(t_k, t_{k-1}, w^{k-1}(N)) = r_i(t_k, t_{k-1}, w^{k-1}(N)) + \mathcal{J}_i^k \chi_i^{k-1}(N), \qquad (6.8.4)$$

$r^k(t)$ *is defined in Theorem* 6.6.1 *an* $\mathcal{J}_i^k = \mathcal{M}^i(t_{k+1}, t_k)$ *is a local Lipschitz constant relative to both* $V^i(s, z_i)$ *and* $z_i(t, s, x_i)$ *with respect to* x_i, z_i *and* t, *for each* $i \in I[1, p]$ *and all* $k \in I(1, N)$.

Then,

$$E[V(t_k, z(t_{k+1}, y^k - x(t_k)))] \leq w(t_k), k \in I(1, N), \qquad (6.8.5)$$

whenever

$$V\left(t_0, z(t_1, t_0, y^0 - x^0)\right) \leq w^0. \qquad (6.8.6)$$

Proof. The proof of the theorem can be formulated by following the argument used in the proof of Theorem 6.4.1. The details are left to the reader.

Remark 6.8.1. It is important to note that the hypotheses of Theorem 6.8.1 provide a crucial iterative scheme to estimate a global error for the solution process of (6.1.1) in the framework of the hybrid subsystems (7.6.2). Of course, this is achieved by an inequality similar to inequality (6.4.8). Further, Remark 6.4.2 can be reformulated analogously.

In the following, for that sake of completeness and to avoid monotonicity, we state a result similar to Theorem 6.4.2 without proof to estimate a global error for the ith solution process of subsystem (6.1.2) in the framework of the ith stochastic large-scale hybrid subsystem (7.6.2).

Theorem 6.8.2. *Assume that all the hypotheses of Theorem* 6.3.1 *are satisfied. Further, assume that*

(i) *the solution process* $z_i(t, s, x_i)$ *of* (6.2.2) *satisfies the following relation:*

$$\|z_i(t, s, x_i)\|^{\bar{p}} \leq \lambda(t_k, t_{k-1})\gamma_i(\|x_i\|^{\bar{p}}), \tag{6.8.7}$$

where $\bar{p} \geq 1, \lambda \in C[R_+ \times R_+, R_+]$ *and* $\gamma_i \in C[R_+, R_+]$, *and* $\gamma_i \in \mathcal{CK}$;

(ii) *for each* $t \in J, V^i(t, z_i)$ *satisfies the following relations:*

$$V^i(t, z_i) \leq \varphi^i(\|z_i\|^{\bar{p}}), \tag{6.8.8}$$

and

$$\vartheta_i(\|z_i\|^{\bar{p}}) \leq V^i(t, z_i), \tag{6.8.9}$$

$\varphi^i \in C[R_+, R_+]$ *and* $\vartheta \in C[R_+, R_+], \vartheta_i \in \mathcal{VK}$ *and all components of* $\varphi^i \in \mathcal{CK}$;

(iii) $w(t_k) = w(t_k, t_{k-1}, w^0)$ *is the solution process of a system of iterative process:*

$$w^k(N) = W(t_k, t_{k-1}, w^{k-1}(N)), \quad w^0(N) = w^0, \tag{6.8.10}$$

where the ith component of W *is defined by*

$$W^i(t_k, t_{k-1}, w^{k-1}(N))$$

$$= 2^{\bar{p}} \left[(\vartheta)^{-1} \left(r_i(t_k, t_{k-1}, \varphi(\lambda^k \gamma(w^{k-1}(N)))) \right) + \left[\chi_i^{k-1}(N) \right]^{\bar{p}} \right],$$

$$\lambda^k = \lambda(t_k, t_{k-1}), \quad \varphi = (\varphi^1, \varphi^2, \ldots, \varphi^i, \ldots, \varphi^p)^T,$$

$$\gamma = (\gamma_1, \gamma_2, \ldots, \gamma_i, \ldots, \gamma_p)^T.$$

$r^k(t)$ *is defined in Theorem* 6.3.1, *and* χ_i^{k-1} *is defined in* (6.6.1), *for each* $i \in I[1,p]$ *and all* $k \in I(1, N)$.

Then,

$$E[\|y_i^k - x_i(t_k)\|^{\bar{p}}] \leq w_i(t_k), \quad k \in I(1, N), \tag{6.8.11}$$

whenever

$$\|y_i^0 - x_i^0\|^{\bar{p}} \leq w_i^0, \quad \text{for all } i \in I[1,p]. \tag{6.8.12}$$

Corollary 6.8.1. *Assume that all the hypotheses of Theorem* 6.8.2 *are satisfied, except that the system of comparison difference equation* (6.8.9) *is replaced by*

$$w^k(N) = W(t_k, t_{k-1}, w^{k-1}(N)), \quad w^0(N) = w^0, \tag{6.8.13}$$

where the ith component of W *is defined by*

$$W^i(t_k, t_{k-1}, w^{k-1}(N))$$

$$= 2^{\bar{p}} \left[\lambda^k \gamma \left((\vartheta_i)^{-1} (r_i(t_k, t_{k-1}, \varphi(w^{k-1}(N)))) \right) + \mathcal{L}^k \left[\chi^k(N) \right]^{\bar{p}} \right], \tag{6.8.14}$$

$\mathcal{L}_*^{ik} = \left[L_z^i(t_k, h_N^k) \right]^{\bar{p}}$ *and* L_z^i *is a local Lipschitz constant of* $z_i(t, s, x_i)$ *with respect to* x_i, *for each* (t, s) *and all* $k \in I(0, N)$.

Then,

$$\|z(t_k, t_{k-1}, y_i^{k-1} - x_i(t_{k-1}))\|^{\bar{p}} \leq w_i(t_k), \quad k \in I(0, N), \tag{6.8.15}$$

for each $i \in I[1,p]$.

Remark 6.8.2. An example similar to Example 6.4.1 can be constructed based on Example 6.4.1 and those worked out in the theory of large-scale systems.

6.9 Local Convergence Analysis

Now, we are ready to discuss the error estimation results between the continuous-time and discrete large-scale systems (6.1.2). In addition, we formulate the convergence, relative stability and stability results for the stochastic numerical schemes described in (6.2.1).

We present a very general result that assures Dahlquist-type convergence and the relative stability property of the stochastic numerical schemes. The proofs of the results follow by imitating the well-known arguments used in Lyapunov's Second Method.

Theorem 6.9.1. *Under the assumptions of Theorem* 6.8.1/ *Theorem* 6.8.2 *or Corollary* 6.8.1, *the convergence of solution of the comparison equations* (6.8.3) ((6.8.9) *or* (6.8.12)) *implies the convergence of the stochastic numerical schemes for the complex system* (6.1.2).

Theorem 6.9.2. *Under the assumptions of Theorem* 6.8.1/ *Theorem* 6.8.2 *or Corollary* 6.8.1, *the convergence of solution of the comparison equations* (6.8.3) ((6.8.14) *or* (6.4.21)) *implies the relative stability of the stochastic of numerical scheme for the complex system* (6.1.2).

6.10 Notes and Comments

In Section 6.1, a theoretical large-scale algorithm is outlined by employing the decomposition-aggregation method via the energy/Lyapunov function and the comparison methods of Ladde and Siljak [36, 37], Ladde and Sambandham [31, 33, 34] and Ladde and Ladde [18, 19] for a large-scale complex stochastic system of differential equations of Ito–Doob type coupled with a one-step Euler-type stochastic numerical complex system of difference equations in the framework of a pair of interconnected stochastic hybrid dynamic systems with the same number of components for convergence and stability analyses of the numerical scheme approximation of an initial value problem of the solution process of large-scale stochastic dynamic systems of Ito–Doob type. This is motivated by the works of Ladde [23, 24], Ladde and Ladde [18, 19], Ladde and Siljak [36, 37], and Lakshmikantham and Trigiante [39] and the classical numerical analysis works of Babuska, Prager and Vitasek [3], Henerici [14]. Hayashi [13], Jankowski [15], Ohashi [42], Siljak [46], Squier [47], Wallach [49] and Woul [51]. Sections 6.1–6.9 are adapted from Ladde [23, 24]. This extends the work of Lakshmikantham and Trigiante [39] and the classical numerical analysis works of Babuska, Prager and Vitasek [3], Henerici [14]. Hayashi [13], Jankowski [15], Ohashi [42], Siljak [46], Squier [47], Wallach [49] and Woul [51].

Chapter 7

Discrete-Time Probabilistic, Stochastic Dynamic Modeling and Statistical Data Analyses

7.0 Introduction

In Section 7.1, by applying the Liouville-type theorem in the theory of dynamic systems, the Liouville-type theorem for finding the joint probability distribution of the solution processes of the system of difference equations with random parameters is outlined. In Section 7.2, we briefly outline, via a few illustrations, a conceptual framework for developing probability distributions associated with solutions of either difference or differential equations under given arbitrary initial data. Currently, a theoretical foundation is under investigation. Section 7.3 deals with the properties of a short-run market and is analyzed considering the development of the short-run market under the influence of finite-state Markovian structural perturbations. The stochastic convergence and stability properties are analyzed. Sufficient conditions are given in terms of an intensity-associated Markov process and coefficient rate matrices. The scope and significance of mathematical results are exhibited by presenting several examples in each section. In Section 7.4, results parallel to those in Section 7.3 are developed under the influence of a discrete-time Gaussian noise process. Section 7.5 deals with the problem of dynamic modeling of a local sample mean and covariance statistic process. In Section 7.6, the continuous-time stochastic approximation is reformulated in the framework of a hybrid dynamic process.

7.1 Development of Probability Distribution Model via Discrete-Time Iterative Process with Deterministic Parameters — I

In this section, we present a method of determining the probability distribution of the solution process of (1.1.1). Of course, this method is not applicable to a very general system of difference equations with random parameters (1.1.1), but it will be limited to a particular class of systems of difference equations with random parameters of the following type:

$$x_k = h(k-1, x_{k-1}, A(\omega)), \quad x_{k_0} = x_0, \quad k \geq k_0 + 1, \tag{7.1.1}$$

where $x_0 \in \mathbb{R}^n$ and $A \in \mathbb{R}^m$ are random vectors defined on a complete probability space, $(\Omega, \mathcal{F}, P)$. It is assumed that the joint probability density function (JPDF) of x_0 and A, $p_{x_0,A}(x_0, a, k_0)$, is given. It is also assumed that $h(k, x, a)$ is continuously differentiable with respect to $(x, a)^T$. We denote by $x_k = x(k, k_0, x_0, A)$ a solution process of (7.1.1).

The main objective of this section is to determine the marginal probability density function (MPDF) $p_X(k, x)$ of the solution process of (7.1.1). To do this, we utilize the following system obtained from (7.1.1):

$$z_k = H(k-1, z_{k-1}), \quad z_{k_0} = z_0, \quad k \geq k_0 + 1, \tag{7.1.2}$$

where $z_k \in \mathbb{R}^{n+m}$, $z_k = [x_k^T, A^T]^T$, $H = [h^T, A^T]^T$ and $z_0 = [x_0^T, A^T]^T$.

Further, we establish several results necessary for the development of the PDF algorithm.

The differentiability of the solution process, z_k, of (7.1.2) with respect to the initial data z_0 is outlined in Theorem A.1.1. This theorem stochasticizes and also provides a rigorous and systematic proof.

The development of the algorithm in our subsequent work requires the knowledge of the structure of the fundamental matrix solution of

(A.1.2) in the context of the original system (7.1.1). For this purpose, we decompose the initial data $y_{0_i} = e_i$ in the following manner: for $1 \leq i \leq n$, we decompose y_0 in (A.1.2) and denote it as

$$y_0 = e_i = \begin{bmatrix} \hat{e}_i \\ 0 \end{bmatrix},$$

where $0 \in \mathbb{R}^m$ and $\hat{e}_i$ is an n-dimensional vector with the ith component 1 and 0 elsewhere.

For $n+1 \leq i \leq m+n$, we decompose y_0 in a similar manner and denote it as

$$y_0 = e_i = \begin{bmatrix} 0 \\ \hat{e}_i \end{bmatrix}, \quad 1 \leq i \leq m,$$

where $0 \in \mathbb{R}^n$ and $\hat{e}_i$ is an m-dimensional vector with ith component 1 and 0 elsewhere.

The following result gives specific information regarding each vector $\frac{\partial z}{\partial z_{0_i}}$, $1 \leq i \leq m+n$, and hence we obtain the structure of the fundamental matrix solution of (A.1.2) in the context of (7.1.1).

Lemma 7.1.1. *Assume that the hypotheses of Theorem* A.1.1 *hold. Then,*

(i) *for*

$$y_0 = \begin{bmatrix} \hat{e}_i \\ 0 \end{bmatrix} \in \mathbb{R}^{n+m}, \quad 1 \leq i \leq n,$$

the solution of (A.1.2) *is given by*

$$y_H^i(k) = \begin{bmatrix} u_H^i(k) \\ 0 \end{bmatrix},$$

where $u_H^i(k)$ *is the solution of the system of homogeneous difference equations*

$$u_k = h_x(k-1, x_{k-1}, A)u_{k-1}, \quad u_{k_0} = u_0 = \hat{e}_i; \tag{7.1.3}$$

(ii) *for*

$$y_0 = \begin{bmatrix} 0 \\ \hat{e}_j \end{bmatrix} \in \mathbb{R}^{n+m}, \quad 1 \leq j \leq m,$$

the solution of (A.1.2) *is given by*

$$y^j_{NH}(k)\begin{bmatrix} w^j_{NH}(k) \\ \hat{e}_j \end{bmatrix},$$

where $w^j_{NH}(k)$ *is the solution of the system of nonhomogeneous difference equations*

$$\begin{aligned} v_k &= h_x(k-1, x_{k-1}, A)v_{k-1} \\ &\quad + \frac{\partial h}{\partial A_j}(k-1, x_{k-1}, A), \quad v_{k_0} = v_0 = 0. \end{aligned} \tag{7.1.4}$$

Moreover, the fundamental matrix solution, Φ_H, *of* (A.1.2) *is composed of column vectors* $y^i_H(k)$, $1 \le i \le n$, *and* $y^j_{NH}(k)$, $1 \le j \le m$*:*

$$\begin{aligned} &\Phi_H(k, k_0, z_0) \\ &\quad = [y^1_H(k), y^2_H(k), \ldots, y^n_H(k), y^1_{NH}(k), y^2_{NH}(k), \ldots, y^m_H(k)]^T \\ &\quad = \begin{bmatrix} \frac{\partial x}{\partial x_0}(k, k_0, x_0, A) & \frac{\partial x}{\partial A}(k, k_0, x_0, A) \\ \mathbf{0} & \mathbf{I} \end{bmatrix}, \end{aligned} \tag{7.1.5}$$

where $\frac{\partial x}{\partial x_0}(k, k_0, x_0, A) = \Phi_h(k, k_0, x_0)$ *is the fundamental matrix solution of the discrete linear system* (7.1.3), I *is an* $m \times m$ *identity matrix;* $\mathbf{0}$ *is an* $m \times n$ *zero matrix; and* $\frac{\partial x}{\partial A}(k, k_0, x_0, A)$ *is an* $n \times m$ *matrix whose column vectors are the solutions of* (7.1.4).

Proof. For (i), consider the system (A.1.2) with $1 \le i \le n$ and denote

$$y_0 = \begin{bmatrix} \hat{e}_i \\ 0 \end{bmatrix} \quad \text{and} \quad y = \begin{bmatrix} u \\ v, \end{bmatrix},$$

where $u \in \mathbb{R}^n$ and $v \in \mathbb{R}^m$. Using this notation, (A.1.2) can be written as the following system:

$$\begin{aligned} \begin{bmatrix} u_k \\ v_k \end{bmatrix} &= \begin{bmatrix} h_x(k-1, x_{k-1}, A) & h_A(k-1, x_{k-1}, A) \\ 0 & I \end{bmatrix} \begin{bmatrix} u_{k-1} \\ v_{k-1} \end{bmatrix}, \\ \begin{bmatrix} u_{k_0} \\ v_{k_0} \end{bmatrix} &= \begin{bmatrix} \hat{e}_i \\ 0 \end{bmatrix}, \end{aligned} \tag{7.1.6}$$

or, equivalently,

$$u_k = h_x(k-1, x_{k-1}, A)u_{k-1} + h_A(k-1, x_{k-1}, A)v_{k-1}, \quad u_{k_0} = \hat{e}_i,$$
$$v_k = v_{k-1}, \quad v_{k_0} = 0. \tag{7.1.7}$$

From $v_{k_0} = 0$, we have $v_k = 0$, and hence (7.1.7) reduces to (7.1.3). Therefore, by the uniqueness of the solution of (A.1.2), $1 \le i \le n$, we have

$$y_H^i(k) = \begin{bmatrix} u_k \\ v_k \end{bmatrix} = \begin{bmatrix} u_H^i(k) \\ 0 \end{bmatrix},$$

where $u_H^i(k)$ is the solution of (7.1.3).

For (ii), consider system (A.1.2) with $n+1 \le i \le n+m$, and denote y_0 and y as

$$y_0 = \begin{bmatrix} 0 \\ \hat{e}_j, \end{bmatrix} \quad \text{with } 1 \le j \le m, \text{ and } y = \begin{bmatrix} w \\ v \end{bmatrix},$$

where $w \in \mathbb{R}^n$ and $v \in \mathbb{R}^m$. In this setting, system (A.1.2) can be rewritten as

$$w_k = h_x(k-1, x_{k-1}, A)w_{k-1} + h_A(k-1, x_{k-1}, A)v_{k-1}, \quad w_{k_0} = 0,$$
$$v_k = v_{k-1}, v_{k_0} = \hat{e}_j. \tag{7.1.8}$$

By solving the second equation in (7.1.8), we obtain

$$v_k = \hat{e}_j, \quad k \ge k_0.$$

From this, the first equation in (7.1.8) reduces to

$$w_k = h_x(k-1, x_{k-1}, A)w_{k-1} + h_A(k-1, x_{k-1}, A)\hat{e}_j, \quad w_{k_0} = 0, \tag{7.1.9}$$

and hence

$$w_k = h_x(k-1, x_{k-1}, A)w_{k-1} + \frac{\partial h}{\partial A_j}(k-1, x_{k-1}, A), \quad w_{k_0} = 0.$$

From this, we can conclude that the solution of (A.1.2), $n+1 \le i \le n+m$, is

$$y_{NH}^j(k) = \begin{bmatrix} w_{NH}^j(k) \\ \hat{e}_j \end{bmatrix},$$

where $w_{NH}^j(k)$ is the solution of (7.1.4).

For $1 \leq i \leq n$ and $1 \leq j \leq m$, $y^i(H(k))$ and $y^j_{NH}(k)$ are linearly independent solutions of (A.1.2). Therefore, the fundamental matrix solution of (A.1.2) has the following structure:

$$\Phi_H(k, k_0, z_0)$$
$$= \begin{bmatrix} u^1_H(k) & u^2_H(k) & \cdots & u^n_H(k) & w^1_{NH}(k) & w^2_{NH}(k) & \cdots & w^m_{NH}(k) \\ 0 & 0 & \cdots & 0 & 1 & 0 & \cdots & 0 \\ 0 & 0 & \cdots & 0 & 0 & 1 & \cdots & 0 \\ \vdots & \vdots & & \vdots & \vdots & \vdots & & \vdots \\ 0 & 0 & \cdots & 0 & 0 & 0 & \cdots & 1 \end{bmatrix},$$

which can be represented as in (7.1.5).

We are ready to develop an algorithm for the determination of the MPDF $p_X(k, x)$ for a solution of (7.1.1) given the JPDF $p_{X_0,A}(x_0, a, k_0)$. For this purpose, we utilize Theorem A.1.2.

From (A.1.20), it is clear that the determination of $p_X(k, x)$ reduces to the task of constructing a mapping like G with the appropriate properties that maps the initial condition and/or parameter probability space to the solution probability space. Note that the image and domain of G must be of the same dimension. Also, note that, in our description of system (7.1.1), we have not required that all the components of x_0 be random variables. To maintain the proper dimension for the domain of G, we augment the vector that is obtained by replacing the deterministic components of x_0 with the components of the random parameter vector $A(\omega)$.

In preparation for the construction of the mapping G, we develop the following notation.

Let l denote the total number of random components of x_0. Denote by C^l_n the combination of l elements from the set

$$I(1, n) = \{1, 2, 3, \ldots, n\}$$

that corresponds to the random components of x_0 and define it

$$C^l_n = \{c_1, c_2, \ldots, c_l\}. \tag{7.1.10}$$

Using the components of z_0 and (7.1.10), we define vectors $\bar{x}_0$, $\bar{z}_0$ and B as

$$\begin{cases} \bar{x}_0 = (x_{0c_1}, x_{0c_2}, \ldots, x_{0c_l})^T, \\ \bar{z}_0 = (x_{0c_1}, x_{0c_2}, \ldots, x_{0c_l}, A_1, A_2, \ldots, A_{n-l})^T, \\ B = (A_{n-l+1}, A_{n-l+2}, \ldots, A_m)^T. \end{cases} \tag{7.1.11}$$

Note that we have assumed $l + m \geq n$; therefore, $\bar{z}_0$ is well-defined. In the case when $l = n$, the introduction of B vector is not necessary.

Now, we present a result that ensures the existence of a transformation with the desired properties described in Theorem A.1.2.

Lemma 7.1.2. *Let the assumptions of Theorem* A.1.1 *be satisfied. Further, assume that for $k \geq k_0$, the vectors*

$$u_H^i(k) = \prod_{s=k_0+1}^{k} h_x(s, x_{s-1}, A)\hat{e}_{c_i}, \quad 1 \leq i \leq l,$$

and

$$w_{NH}^j(k) = \sum_{s=k_0+1}^{k} \left(\prod_{l=k_0+1}^{k} h_x^{-1}(l-1, x_{l-1}, A) \right) \times \frac{\partial h}{\partial A_j}(s-1, x_{s-1}, A), \quad 1 \leq j \leq n-l,$$

are linearly independent. Then, there exists a transformation G defined on $\Re^{l+m}$ into itself such that

(i) *G is one-to-one and onto;*
(ii) *G^{-1} exists and is continuously differentiable.*

Proof. Let $x_k = x(k, k_0, x_0, A)$ be a solution process of (7.1.1), and let l denote the number of random components of x_0. Now, by following the notations and definitions in (7.1.10) and (7.1.11), we define a transformation $G : \Re^{l+m} \to \Re^{l+m}$:

$$G(\bar{z}_0, B) = (x^T(k, k_0, x_0, A), B^T)^T. \tag{7.1.12}$$

G is a well-defined mapping. From the application of Theorem A.1.1, one can easily see that

$$\frac{\partial G(\bar{z}_0, B)}{\partial(\bar{z}_0, B)} = \frac{\partial(x^T(k), B^T)^T}{\partial(\bar{z}_0, B)}$$

exists and is continuous.

By the assumption of the theorem, one can conclude that, for $k \geq k_0 + 1$, the column vectors of the lower triangular block matrix

$$\frac{\partial(x(k), B)}{\partial(\bar{z}_0, B)} = \begin{bmatrix} \dfrac{\partial x(k, k_0, x_0, A)}{\partial \bar{z}_0} & \dfrac{\partial x(k, k_0, x_0, A)}{\partial B} \\ 0_{(m+l-n)\times n} & I_{(m+l-n)\times(m+l-n)} \end{bmatrix} \tag{7.1.13}$$

are linearly independent, that is, $\begin{bmatrix} \frac{\partial x(k,k_0,x_0,A)}{\partial x_{0c_i}} \\ 0 \end{bmatrix} \in \Re^{l+m}$, for $1 \leq i \leq l$,

$$\begin{bmatrix} \dfrac{\partial x(k, k_0, x_0, A)}{\partial A_j} \\ \hat{e}_j \end{bmatrix} \in \Re^{l+m}, \quad \text{for } 1 \leq j \leq n - l, \quad \text{and}$$

$$\begin{bmatrix} \dfrac{\partial x(k, k_0, x_0, A)}{\partial A_j} \\ \hat{e}_j \end{bmatrix} \in \Re^{l+m},$$

for $n - l + 1 \leq j \leq m$, are linearly independent, where $\{\hat{e}_1, \hat{e}_2, \ldots, \hat{e}_j, \ldots, \hat{e}_{m+l-n}\}$ is a standard basis for $\mathbb{R}^{m+l-n}$. From the structure of the matrix defined in (7.1.13), one can see that it is non-singular. Therefore, G is a one-to-one and onto map. From Theorem A.1.2, G^{-1} exists and is continuously differentiable. This completes the proof.

Remark 7.1.1. From Lemma 7.1.2, we observe that

$$\left|\frac{\partial G(\bar{z}_0, B)}{\partial(\bar{z}_0, B)}\right| = \left|\frac{\partial x(k, k_0, x_0, A)}{\partial \bar{z}_0}\right|. \tag{7.1.14}$$

Also, define

$$L(x, B, k, k_0) = \left|\frac{\partial G^{-1}(x, B)}{\partial(x, B)}\right|. \tag{7.1.15}$$

The following result provides an important recurrence relationship for the expression in (7.1.14).

Lemma 7.1.3. *Assume that the hypotheses of Lemma 7.1.2 hold. Let $\bar{z}_0$ be a vector determined by a combination C_n^l defined in (7.1.10). Then, there exists a unique $n \times n$ matrix $\Phi_{c_l}^k$ such that*

$$|\Phi_{c_l}^k| = \left|\frac{\partial G(\bar{z}_0, B)}{\partial(\bar{z}_0, B)}\right|,$$

and $|\Phi_{c_l}^k|$ is the solution process of

$$r_k = |h_x(k-1, x_{k-1}, A)|r_{k-1} + q(k), \quad r_{k_0} = 0, \tag{7.1.16}$$

where

$$q(k) = \sum_{j=1}^{n-l} q_j(k)$$

and

$$\begin{aligned} q_j(k) = \Big| & u_H^{c_1}(k)\ u_H^{c_2}(k)\ \cdots\ u_H^{c_l}(k)\ w_{NH}^1(k)\ w_{NH}^2(k)\ \cdots\ w_{NH}^{j-1}(k) \\ & h_x w_{NH}^{j+1}(k-1)\ \cdots\ h_x w_{NH}^{n-l}(k-1)\Big|, \end{aligned}$$

where $u_H^{c_i}$ is the solution of (7.1.3), $1 \le i \le l$, and w_{NH}^j is the solution of (7.1.4), $1 \le j \le n-l$.

Proof. We note that the matrix $\Phi_{c_l}^k$ can also be constructed in the following manner. From the fundamental matrix solution of (A.1.2),

$$\begin{aligned} &\Phi_H(k, k_0, z_0) \\ &= \begin{bmatrix} u_H^1(k) & u_H^2(k) & \cdots & u_H^n(k) & w_{NH}^1(k) & w_{NH}^2(k) & \cdots & w_{NH}^m(k) \\ \mathbf{0} & \mathbf{0} & \cdots & \mathbf{0} & \hat{e}_1 & \hat{e}_2 & \cdots & \hat{e}_m \end{bmatrix}, \end{aligned}$$

we delete the columns that correspond to the components of x_0 that are deterministic and we delete the rows $n+1 < n+2, \ldots, n+(n-l)$.

The resulting matrix, defined by $\Phi^k_{c_l}$, has the following structure:

$$\Phi^k_{c_l} = \begin{bmatrix} u_H^{c_1}(k) & u_H^{c_2}(k) & \cdots & u_H^{c_1}(k) & w_{NH}^1(k) & \cdots & w_{NH}^{n-l}(k) & \cdots & w_{NH}^m(k) \\ \mathbf{0} & \mathbf{0} & \cdots & \mathbf{0} & \bar{e}_1 & \cdots & \bar{e}_{n-l} & \cdots & \bar{e}_m \end{bmatrix}$$

$$= \begin{bmatrix} \dfrac{\partial x(k)}{\partial \bar{z}_0} & \dfrac{\partial x(k)}{\partial B} \\ 0_{(m+l-n)\times n} & I_{(m+l-n)\times(m+l-n)} \end{bmatrix}$$

$$= \left[\frac{\partial G(\bar{z}_0, B)}{\partial(\bar{z}_0, B)}\right], \tag{7.1.17}$$

where $\bar{e}_i$ is an $m+l-n$-dimensional vector with the ith component 1 and 0 elsewhere, for $1 \leq i \leq m+l-n$.

From (7.1.14) and the evaluation of the determinant of (7.1.17), we obtain

$$|\Phi^k_{c_l}| = |u_H^{c_1}(k)\ u_H^{c_2}(k)\ \cdots\ u_H^{c_l}(k)\ w_{NH}^1(k)\ \cdots\ w_{NH}^{n-l}(k)|$$

$$= \left|\frac{\partial G(\bar{z}_0, B)}{\partial(\bar{z}_0, B)}\right|. \tag{7.1.18}$$

We note that $|\Phi^{k_0}_{c_l}| \neq 0$.

Using the assumption, $u_H^{c_i}(k)$, $l \leq i \leq l$, are solutions of (7.1.3), and $w_{NH}^j(k)$, $1 \leq j \leq n-l$, are solutions of (7.1.4). Hence, (7.1.18) can be written as

$$|\Phi^k_{c_l}| = \Bigg| h_x u_H^{c_1}(k-1)\ h_x u_H^{c_2}(k-1)\ \cdots$$

$$h_x u_H^{c_1}(k-1)\ \left(h_x w_{NH}^1(k-1) + \frac{\partial h}{\partial A_l}\right)\ \cdots$$

$$\left(h_x w_{NH}^{n-l}(k-1) + \frac{\partial h}{\partial A_{n-l}}\right)\Bigg|,$$

where $h_x = h_x(k-1, x_{k-1}, A)$. Applying the basic properties of determinants and using the definition of $\Phi^{k-1}_{c_l}$, one can obtain

$$|\Phi^k_{c_l}| = |h_x||\Phi^{k-1}_{c_l}| + q(k),$$

where

$$q(k) = \sum_{j=1}^{n-l} q_j(k)$$

and

$$q_j(k) = \Big| u_H^{c_1}(k) \ \cdots \ u_H^{c_l}(k) \ h_x w_{NH}^1(k-1) \ \cdots \\ h_x w_{NH}^j(k-1) h_x w_{NH}^{j+1}(k-1) \cdots \ h_x w_{NH}^{n-l}(k-1) \Big|.$$

This completes the proof.

In the special case when $n = \ell$, Lemma 7.1.3 gives us the following result.

Corollary 7.1.1. *Let the hypotheses of Lemma* 7.1.2 *be satisfied, with* $l = n$. *Then,*

$$|\Phi_{c_n}^k| = \left| \frac{\partial x_k}{\partial x_0} \right|,$$

and $\Phi_{c_n}^k$ *is a solution of*

$$r_k = |h_x(k-1, x_{k-1}, A)| r_{k-1}, \quad r_{k_0} = 1. \tag{7.1.19}$$

Proof. The proof parallels the proof of Lemma 7.1.3. In this case,

$$\begin{aligned} |\Phi_{c_n}^k| &= |u_H^1(k) \ u_H^2(k) \ \cdots \ u_H^n(k)| \\ &= |h_x u_{k-1}^1 \ h_x u_{k-1}^2 \ \cdots \ h_x u_{k-1}^n| \\ &= |h_x||\Phi_{c_n}^{k-1}|, \\ |\Phi_{c_n}^{k_0}| &= 1, \end{aligned}$$

where

$$h_x = h_x(k-1, x_{k-1}, A).$$

Another interesting case of Lemma 7.1.3 is satisfied with $l = n - 1$.

Corollary 7.1.2. *Let the hypotheses of Lemma* 7.1.3 *be satisfied with* $l = n - 1$. *Then,* $|\Phi_{c_{n-1}}^k| = \left|\frac{\partial x_k}{\partial \bar{z}_0}\right|$ *is a solution of*

$$\begin{cases} r_k = |h_x(k-1, x_{k-1}, A)| r_{k-1} + |u_H^{c_1}(k) \ u_H^{c_2}(k) \ \cdots \ u_H^{c_{n-1}}(k) \ \frac{\partial h}{\partial A_1}|, \\ r_{k_0} = 0. \end{cases} \tag{7.1.20}$$

Proof. In this special case of Lemma 7.1.3,

$$\begin{aligned}
|\Phi^k_{c_{n-1}}| &= \left|u_H^{c_1}(k)\ u_H^{c_2}(k)\ \cdots\ u_H^{c_{n-1}}(k)\ w_{NH}^1(k)\right| \\
&= \left|h_x u^{c_1}_{k-1}\ h_x u^{c_2}_{k-1}\ \cdots\ h_x u^{c_{n-1}}_{k-1}\ \left(h_x u^1_{k-1} + \frac{\partial h}{\partial A_1}\right)\right| \\
&= \left|h_x u^{c_1}_{k-1}\ h_x u^{c_2}_{k_1}\ \cdots\ h_x u^{c_{n-1}}_{k-1}\ h_x w^1_{k-1}\right| \\
&\quad + \left|h_x u^{c_1}_{k-1}\ h_x u^{c_2}_{k-1}\ \cdots\ h_x u^{c_{n-1}}_{k-1}\ \frac{\partial h}{\partial A_1}\right| \\
&= |h_x|\left|\Phi^{k-1}_{c_{n-1}}\right| + \left|u_H^1(k)\ u_H^2(k)\ \cdots\ u_H^{n-1}(k)\ \frac{\partial h}{\partial A_1}\right|.
\end{aligned}$$

Now, we are ready to state and prove a result that provides a method for determining the MPDF of the solution process of (7.1.1).

Theorem 7.1.1. *Let the hypotheses of Lemma 7.1.3 be satisfied. Assume that the JPDF $p_{\bar{Z}_0,B}(x_0,a,k_0)$ is given. Then, JPDF $p_{X,B}(x,b,k)$ is given by*

$$\begin{aligned}
p_{X,B}(x,b,k) &= p_{\bar{Z}_0,B}(G^{-1}(x(k,k_0,x,b),b,k)) \\
&\quad \times L(x,b,k,k_0), \quad k \geq k_0+1, \qquad (7.1.21)
\end{aligned}$$

and the marginal JPDF of $x(k,k_0,x_0,A)$ is given by

$$p_X(x,k) = \int p_{X,B}(x,b,k)db, \qquad (7.1.22)$$

where $L(x,b,k,k_0)$ is defined in (7.1.15).

Proof. From Remark 7.1.1 and Lemma 7.1.3, we have

$$\left|\frac{\partial G(\bar{z}_0,B)}{\partial(\bar{z}_0,B)}\right| = \sum_{s=k_0+1}^{k}\left(\prod_{j=s}^{k}|h_x(j,x_{j-1},A)|\right)q(s), \qquad (7.1.23)$$

where q is defined in (7.1.16). Set

$$J(k,k_0,\bar{z}_0,B) = \left|\frac{\partial G(\bar{z}_0,B)}{\partial(\bar{z}_0,B)}\right|, \quad k \geq k_0+1. \qquad (7.1.24)$$

It is known that

$$L(x,B,k,k_0) = \left(\frac{1}{J(k,k_0,\bar{z}_0,B)}\right)\Bigg|_{\bar{z}_0=x^{-1}(k,k_0,x,B)}$$
$$= \left(\frac{1}{J(k,k_0,x^{-1}(k,k_0,x,B),B)}\right). \quad (7.1.25)$$

From (7.1.24), (7.1.25) and (A.1.20), the JPDF of X and B is given by

$$p_{X,B}(k,k_0,x,b) = p_{\bar{z}_0,B}(G^{-1}(x(k,k,x,b),b,k))L(x,b,k,k_0).$$

This establishes the first part of the conclusion of the theorem. The second part follows from the definition of the marginal JPDF. This completes the proof of the theorem.

Example 7.1.1. Consider the following linear system:

$$x_k = \alpha(k-1,A(\omega))x_{k-1}, \quad x_{k_0} = x_0, \quad (7.1.26)$$

where $x \in \Re^n$, α is an $n \times n$ matrix function and $A \in \Re^m$. From (7.1.26), we obtain the following $(n+m)$-dimensional system:

$$z_k = H(k, z_{k-1}), \quad z_{k_0} = z_0, \quad (7.1.27)$$

where $z_k = [x_k^T, A^T(\omega)]^T$, $z_0 = [x_0^T, A^T(\omega)]^T$ and

$$H(k-1,z_{k-1},A(\omega)) = [(\alpha(k-1,A(\omega))x_{k-1})^T, A^T(\omega)]^T. \quad (7.1.28)$$

It is clear that

$$h(k-1,x_{k-1},A(\omega)) \equiv \alpha(k-1,A(\omega))x_{k-1}$$

is continuously differentiable with respect to z, and an application of Lemma 7.1.1 gives us that $\frac{\partial z}{\partial z_{0i}}$ is the solution of

$$y_k = H_z(k-1,x_{k-1},A(\omega))y_{k-1}$$
$$= \begin{bmatrix} \alpha(k-1,A(\omega)) & \frac{\partial(\alpha(k-1,A(\omega))x_{k-1})}{\partial A} \\ 0_{m\times n} & I \end{bmatrix} y_{k-1},$$
$$y_{k_0} = y_{0_i} = e_i, \quad (7.1.29)$$

where $e_i \in \Re^{m+n}$ and H is defined in (7.1.28). Lemma 7.1.3 gives us a more specific structure for the solution of (7.1.29) with respect to the value of i. For $1 \leq i \leq n$, the solution of (7.1.29) is defined as

$$y_H^i(k) = \left[u_H^{iT}(k)\ 0^T\right]^T,$$

where $u_H^i(k)$ is the solution of

$$v_k = \alpha(k-1, A(\omega))v_{k-1}, \quad v_{k_0} = e_i, \tag{7.1.30}$$

and $e_i \in \Re^n$, that is,

$$u_H^i = \prod_{l=k_0+1}^{k} \alpha(l-1, A(\omega))e_i. \tag{7.1.31}$$

On the other hand, for $n+1 \leq i \leq n+m$, the solution of (7.1.29) has the following structure:

$$y_{NH}^j(k) = w_{NH}^{jT}(k)\ \hat{e}_j^T, \quad j = i-n,$$

where $w_{NH}^j(k)$ is the solution of

$$v_k = \alpha(k-1, A(\omega))v_{k-1} + \frac{\partial(\alpha(k-1, A(\omega)))x_{k-1}}{\partial A_j}, \quad v_{k_0} = 0. \tag{7.1.32}$$

Using the variation of parameters method, the solution of (7.1.32) is

$$w_{NH}^j(k) = \sum_{s=k_0+1}^{k} Y_k Y_s^{-1} \left(\frac{\partial\alpha(s-1, A(\omega))}{\partial A_j} \prod_{l=k_0+1}^{s} \alpha(l-1, A(\omega))x_0\right), \tag{7.1.33}$$

where

$$Y_k = \prod_{l=k_0+1}^{k} \alpha(l-1, A(\omega))$$

and

$$Y_s^{-1} = \prod_{l=0}^{s-k_0} \alpha^{-1}(s-l, A(\omega)).$$

Recall that $\Phi_H(k, k_0, z_0)$ is defined in terms of the solutions of (7.1.30) and (7.1.32), or, equivalently, in terms of (7.1.31) and (7.1.3), as follows:

$$\Phi_H(k,k_0,z_0)$$
$$= \begin{bmatrix} u_H^1(k) & \cdots & u_H^n(k) & w_{NH}^1(k) & w_{NH}^2(k) & \cdots & w_{NH}^m(k) \\ 0 & \cdots & 0 & 1 & 0 & \cdots & 0 \\ 0 & \cdots & 0 & 0 & 1 & \cdots & 0 \\ \vdots & & \vdots & \vdots & \vdots & & \vdots \\ 0 & \cdots & 0 & 0 & 0 & \cdots & 1 \end{bmatrix}. \tag{7.1.34}$$

Let the set $C_n^l = (c_1, \ldots, c_n)$ be the random components of x_0. The existence of the mapping G defined in Lemma 7.1.2 is ensured if the column vectors $c_1, \ldots, c_n$ of the matrix α evaluated at k_0+1 and the vectors

$$\frac{\partial h(k_0, x_0, A(\omega))}{\partial A_j} \equiv \frac{\partial(\alpha(k_0, A(\omega))x_{k_0})}{\partial A_j}$$
$$= \frac{\partial(\alpha(k_0, A(\omega)))}{\partial A_j} x_{k_0}, \quad 1 \le j \le n-l,$$

are linearly independent.

Under these assumptions, the application of Lemma 7.1.3 yields that

$$|\Phi_{c_l}^k| = \left|\frac{\partial G(\bar{z}_0, B)}{\partial(\bar{z}_0, B)}\right|$$

is the solution of

$$r_k = |\alpha(k-1, A(\omega))| r_{k-1} + q(k), \quad r_{k_0} = |\Phi_C^{k_0}|, \tag{7.1.35}$$

where

$$q(k) = \sum_{p=1}^{n-l} q_p(k),$$

$$q_p(k) = \Big| u_H^{c_1}(k) \cdots u_H^{c_1}(k)\ \alpha(k-1, A(\omega)) w_{NH}^1(k-1)\ \cdots$$

$$\alpha(k-1, A(\omega))w_{NH}^{p-1}(k-1)\frac{\partial\alpha(k-1, A(\omega))}{\partial A_p}\,x_{k-1}$$

$$\left(\alpha(k-1, A(\omega))w_{NH}^{p+1}(k-1) + \left(\frac{\partial\alpha}{\partial A_p}\right)x_{k-1}\right)\cdots$$

$$\left(\alpha(k-1), A(\omega))w_{NH}^{n-l}(k-1) + \left(\frac{\partial\alpha}{\partial A_{n-l}}\right)x_{k-1}\right)\Bigg|,$$

where $u_H^{c_i}(k)$, $1 \leq i \leq l$, is defined in (7.1.31), and $w_{NH}^j(k)$, $1 \leq j \leq m$, is defined in (7.1.33).

An application of the method of variation of parameters to (7.1.35) yields the following solution:

$$|\Phi_{c_l}^k| = \sum_{s=k_0+1}^{k} Y_k Y_s^{-1} q(s) + Y_k Y_{k_0}^{-1}|\Phi_{c_l}^{k_0}|, \qquad (7.1.36)$$

where

$$Y_k = \prod_{l=k_0+1}^{k} |\alpha(l-1, A(\omega))|$$

and

$$Y_s^{-1} = \prod_{l=0}^{s-k_0} |\alpha^{-1}(s-l, A(\omega))|.$$

Utilizing (7.1.24), (7.1.25) and (7.1.36), we obtain the following expression for the MPDF of the solution of (7.1.26):

$$p_X(x) = \int_{B_r}\cdots\int_{B_1} p_{X,B}(k, k_0, x, b)\,db_1\cdots db_r, \qquad (7.1.37)$$

where

$$B = (A_{l+1}, A_{l+2}, \ldots, A_m), \quad r = m - l,$$

and

$$p_{X,B}(k, k_0, x, b) = p_{\bar{Z}_0,B}(k, k_0, h^{-1}(k, k_0, x, b), a(b)) \cdot \frac{1}{|\Phi_{c_l}^k|}\,.$$

In the special case considered in Corollary 7.1.1, when $l = n$, the calculation of $|\Phi^k_{c_l}|$ is straightforward. In this case, the initial condition of (7.1.35) is

$$r_{k_0} = |\Phi^{k_0}_C| = 1,$$

and $q(k)$ in (7.1.37) is identically zero. Therefore, the solution of (7.1.35), (7.1.36), reduces to

$$|\Phi^k_{c_n}| = \prod_{l=k_0+1}^{k} |\alpha(l-1, A(\omega))||\alpha^{-1}(k_0, A(\omega))|,$$

and the MPDF (7.1.37) reduces to

$$p_X(x) = \int_A p_{X_0,A}(k, k_0, G^{-1}(k, k_0, x, a), a) \times \frac{1}{\prod_{l=k_0+1}^{k} |\alpha(l-1, a)||\alpha^{-1}(k_0, a)|} da.$$

In the case when x_0 has only one deterministic component, that is, when $l = n-1$, $r_{k_0} = |\Phi^{k_0}_{c_{n-1}}| = 0$ and $h(k)$ in (7.1.36) has only one term,

$$q_1(k) = \left| \prod_{l=k_0+1}^{k} \alpha(l-1, A(\omega)) e_{c_1} \cdots \prod_{l=k_0+1}^{k} \alpha(l-1, A(\omega)) e_{c_{n-1}} \times \frac{\partial \alpha(k-1, A(\omega))}{\partial A_1} \prod_{s=k_0+1}^{k} \alpha(s-1, A(\omega)) x_0 \right|. \quad (7.1.38)$$

This case was considered in Corollary 7.1.2. Under these assumptions, the solution of (7.1.35) is

$$|\Phi^k_{c_{n-1}}| = \sum_{s=k_0+1}^{k} \prod_{l=k_0+1}^{k} |\alpha(l-1, A(\omega))| \prod_{l=0}^{s-k_0} |\alpha(s-l, A(\omega))| q_1(l), \quad (7.1.39)$$

where $q_1(k)$ is defined in (7.1.38). Therefore, by Theorem 7.1.1, the marginal probability function, $P_X(x)$, of the solution of (7.1.39) in this case is

$$p_X(x) = \int_{B_r} \cdots \int_{B_1} p_{X,B}(k, k_0, (x, b))db_1 \cdots db_r,$$

where

$$p_{X,B}(k, k_0, x, b) = p_{\bar{Z}_0,B}(k, k_0, G^{-1}(k, k_0, x, b), a(b)) \frac{1}{|\Phi^k_{c_{n-1}}|}$$

and $|\Phi^k_C|$ is defined in (7.1.39).

Remark 7.1.2. If $\alpha(k-1, A(\omega)) \equiv \alpha(A(\omega))$, the solution of (7.1.35) reduces to

$$|\Phi^k_{c_l}| = \sum_{s=k_0+1}^{k} |\alpha(A(\omega))|^{k-s} q(s). \tag{7.1.40}$$

In the cases considered in Corollaries 7.1.1 and 7.1.2, (7.1.40) reduces to

$$|\Phi^k_{c_n}| = |\alpha(A(\omega))|^{k-k_0}$$

and

$$|\Phi^k_{c_n}| = \sum_{s=k_0+1}^{k} |\alpha(A(\omega))|^{k-s} q_1(s),$$

respectively, where $q_1(k)$ is defined as

$$q_1(k) = \Big| [\alpha A(\omega)]^{k-k_0} e_{c_1} \cdots [\alpha(A(\omega))]^{k-k_0} e_{c_{n-1}} \times \frac{\partial \alpha(A(\omega))}{\partial A_1} [\alpha(A(\omega))]^{k-k_0} x_0 \Big|.$$

In the following example, we convert an nth order equation to a linear system of equations and apply the analysis developed in Example 7.1.1.

Example 7.1.2. Consider the following nth-order difference equation:

$$x_k + \alpha_1(k-1, A_1(\omega))x_{k-1} + \alpha_2(k-1, A_2(\omega))x_{k-2} + \cdots + a_n(k-1, A_n(\omega))x_{k-n} = 0,$$

or, equivalently,

$$\begin{aligned} x_k = &-\alpha_1(k-1, A_1(\omega))x_{k-1} - \alpha_2(k-1, A_2(\omega))x_{k-2} - \cdots \\ &- \alpha_n(k-1, A_n(\omega))x_{k-n}. \end{aligned} \tag{7.1.41}$$

By defining

$$\eta_{k-1} = [x_{k-1}, x_{k-2}, \ldots, x_{k-n}]^T$$

and

$$\eta_k = [x_k, x_{k-1}, \ldots, x_{k-(n-1)}]^T,$$

one can construct a system of equations that is equivalent to (7.1.41) in terms of η_{k-1} and η_k as follows:

$$\begin{aligned} \eta_k &= \gamma(k-1, A(\omega))\eta_{k-1}, \\ \eta_{k_0} &= [x_{k_0+n-1}, x_{k_0+n-2}, \ldots, x_{k_0}]^T, \end{aligned} \tag{7.1.42}$$

where

$$\gamma(k-1, A(\omega))$$

$$= \begin{bmatrix} -\alpha_1(k-1, A_1) & -\alpha_2(k-1, A_2) & \cdots & -\alpha_{n-1}(k-1, A_{n-1}) & -\alpha_n(k-1, A_n) \\ 1 & 0 & \cdots & 0 & 0 \\ 0 & 1 & \cdots & 0 & 0 \\ 0 & 0 & \cdots & 0 & 0 \\ \vdots & \vdots & & \vdots & \vdots \\ 0 & 0 & \cdots & 1 & 0 \end{bmatrix}$$

and

$$A(\omega) = [A_1(\omega), A_2(\omega), \ldots, A_n(\omega)]^T.$$

Our previous analysis of a linear system in Example 7.1.1 can be applied to system (7.1.42). For this system, (7.1.27) is defined as

$$\begin{aligned} z_k = [&-\alpha_1(k-1, A_1(\omega))\eta_{k-1}^1, \ldots, -\alpha_n(k-1, A_n(\omega))\eta_{k-1}^n, \\ &\eta_{k-1}^1, \eta_{k-1}^2, \ldots, \eta_{k-1}^{n-1}, A_1(\omega), A_2(\omega), \ldots, A_n(\omega)]^T, \end{aligned}$$

where

$$z_k = [\eta_k^T, A^T(\omega)]^T$$

and

$$z_{k_0} = z_0 = [\eta_{k_0}^T, A^T(\omega)]^T.$$

From our previous analysis, we know that $\frac{\partial z}{\partial z_{0_i}}$ is the solution of the system

$$y_k = \left[\begin{array}{ccccc}
-\alpha_1 & \cdots & -\alpha_{n-1} & -\alpha_n & \frac{\partial}{\partial A_1}\alpha(k-1, A_1(\omega))\eta_{k-1}^1 \\
1 & \cdots & 0 & 0 & 0 \\
0 & \cdots & 0 & 0 & 0 \\
0 & \cdots & 0 & 0 & 0 \\
\vdots & & \vdots & \vdots & \vdots \\
0 & \cdots & 1 & 0 & 0 \\
0 & \cdots & \cdots & 0 & 1 \\
\vdots & & & \vdots & 0 \\
\vdots & & & \vdots & \vdots \\
0 & \cdots & \cdots & 0 & 0
\end{array}\right.$$

$$\left.\begin{array}{cc}
\cdots & \frac{\partial}{\partial A_1}\alpha(k-1, A_1(\omega))\eta_{k-1}^n \\
\cdots & 0 \\
\cdots & 0 \\
\cdots & 0 \\
 & \vdots \\
\cdots & 0 \\
\cdots & 0 \\
\cdots & 0 \\
 & \vdots \\
\cdots & 1
\end{array}\right] y_{k-1},$$

$$y_{k_0} = y_0 = e_i, \quad e_i \in \Re^{2n}, \tag{7.1.43}$$

where

$$-\alpha_i \equiv -\alpha_i(k-1, A_i(\omega)).$$

Furthermore, for $1 \le i \le n$,

$$y_H^i(k) = [u_H^{iT}, 0^T]^T$$

is the solution of (7.1.43), where

$$u_H^i(k) = \prod_{l=k_0+1}^{k} \gamma(l-1, A(\omega))e_i, \quad e_i \in \Re^n, \tag{7.1.44}$$

and for $n+1 \le i \le 2n$,

$$y_{NH}^j(k) = [w_{NH}^{jT}, \hat{e}_j^T]^T, \quad j = i-n,$$

is the solution of (7.1.43), where

$$w_{NH}^j(k) = \sum_{s=k_0+1}^{k} Y_k Y_s^{-1} \left(\frac{\partial \gamma(s-1, A)}{\partial A_j} \prod_{l=k_0+1}^{s} \gamma(l-1, A(\omega))\eta_0 \right), \tag{7.1.45}$$

where

$$Y_k = \prod_{l=k_0+1}^{k} \gamma(l-1, A(\omega))$$

and

$$Y_s^{-1} = \prod_{l=0}^{s-k_0} \gamma(s-l, A(\omega)).$$

As in the previous example, we require the linear independence of the column vectors $c_1, c_2, \ldots, c_l$ of γ and the vectors

$$\frac{\partial(\gamma(k_0, A(\omega))\eta_0)}{\partial A_j}, \quad 1 \le j \le n-l,$$

where $c = \{c_1, c_2, \ldots, c_l\}$ denote the random components of η_0. This ensures the existence of the mapping G defined in Lemma 7.1.2.

Subsequent application of Lemma 7.1.3 gives us that

$$|\Phi_{c_l}^k| = \left|\frac{\partial G(\bar{z}_0, B)}{\partial(\bar{z}_0, B)}\right|$$

is the solution of

$$r_k = |\gamma(k-1, A(\omega))| r_{k-1} + q(k), \quad r_{k_0} = |\Phi_c^{k_0}|, \tag{7.1.46}$$

where

$$q(k) = \sum_{p=1}^{n-l} q_p(k),$$

$$q_p(k) = \left| u_H^{c_1}(k) \;\cdots\; u_H^{c_l}(k)\; \gamma(k, A(\omega)) w_{NH}^1(k) \;\cdots\; \gamma(k, A(\omega)) w_{NH}^{p-1}(k) \right.$$
$$\left. \frac{\partial \gamma(k-1, A)}{\partial A_p} \prod_{l=k_0+1}^{k} \gamma(l-1, A(\omega))\eta_0 \; w_{NH}^{p+1}(k) \;\cdots\; w_{NH}^{n-l}(k) \right|,$$

and $u_H^i(k)$ and $w_{NH}^j(k)$ are defined in (7.1.44) and (7.1.45), respectively. Therefore, using the variation of parameters method, we obtain

$$|\Phi_{c_l}^k| = \sum_{s=k_0+1}^{k} Y_k Y_s^{-1} q(s) + Y_k Y_{k_0}^{-1} |\Phi_C^{k_0}|, \tag{7.1.47}$$

where

$$Y_k = \prod_{l=k_0+1}^{k} |\gamma(l-1, A(\omega))|,$$

$$Y_s^{-1} = \prod_{l=0}^{s-k_0} |\gamma^{-1}(l-1, A(\omega))|$$

and $q(s)$ is defined in (7.1.46).

Utilizing (7.1.47) and applying Theorem 7.1.1, we arrive at the MPDF $p_\eta(\eta)$ for the solution of (7.1.41):

$$p_\eta(\eta) = \int_{B_r} \cdots \int_{B_l} p_{\eta,B}(k, k_0, \eta, b) db_1, \ldots, db_r,$$

where

$$p_{\eta,B}(k, k_0, \eta, b) = p_{\bar{z}_0,B}(k, k_0, h^{-1}(k, k_0, \eta, b), a(b)) \cdot \frac{1}{|\Phi^k_{c_l}|}.$$

The MPDF for the solution of $x(k)$ of the original equation (7.1.41) can be determined from the following relation:

$$f_X(x) = \int \cdots \int f_\eta(\eta) d\eta_{n-1} \cdots d\eta_2.$$

Remark 7.1.3. For $n = 2$ and $\gamma(k, A(\omega)) \equiv \gamma(A(\omega))$, consider the second-order difference equation

$$x_k = -\alpha_1(A_1(\omega))x_{k-1} - x_{k-2}, \tag{7.1.48}$$

or, equivalently,

$$\begin{aligned} \eta_k &= \Gamma(A(\omega))\eta_{k-1}, \\ \eta_{k_0} &= \eta_0 = [x_{k_0+1}, x_{k_0}]^T, \end{aligned} \tag{7.1.49}$$

where

$$\Gamma(A(\omega)) = \begin{bmatrix} -\alpha_1(A_1(\omega)) & 1 \\ 1 & 0 \end{bmatrix}.$$

Without loss of generality, assume that $\eta_0^2 = x_{k_0+1}$ is degenerate. In this case, (7.1.27) is a three-dimensional system defined as

$$\begin{aligned} z_k &= [-\alpha_1(A_1(\omega))\eta^1_{k-1} - \eta^2_{k-1}, \eta^1_{k-1}, A_1(\omega)]^T, \\ z_{k_0} &= z_0 = [\eta_0^T, A^T(\omega)]^T, \end{aligned}$$

where

$$A(\omega) = A_1(\omega).$$

In this case, (7.1.43) has the following structure:

$$\begin{aligned} y_k &= \begin{bmatrix} -\alpha_1(A_1(\omega) & 1 & -\frac{\partial}{\partial A_1}(\alpha_1(A_1(\omega)))\eta^1_{k-1} \\ 1 & 0 & 0 \\ 0 & 0 & 1 \end{bmatrix} y_{k-1}, \\ y_{k_0} &= y_0 = e_i, \quad e_i \in \Re^e. \end{aligned} \tag{7.1.50}$$

Furthermore, for

$$y_0 = [\hat{e}_i^T, 0]^T, \quad 1 \le i \le 2,$$

$$y_H^i(k) = [u_H^{iT}, 0]^T$$

is the solution of (7.1.50), where

$$u_H^i(k) = [\Gamma(A(\omega))]^{k-k_0} e_i, \quad e_i \in \Re^2, \tag{7.1.51}$$

and for

$$y_0 = [0, 0, 1]^T,$$
$$y_{NH}^1(k) = [w_{NH}^{1T}(k), 1]^T,$$

where

$$w_{NH}^1(k) = \sum_{s=k_0+1}^{k} [\Gamma(A(\omega))]^{k-s-1} \left[\frac{\partial}{\partial A_1}([\Gamma(A(\omega))]^{s-k_0})\eta_0\right]. \tag{7.1.52}$$

To satisfy the assumptions of Lemma 7.1.2, we require the linear independence of the vectors

$$\prod_{s=k_0+1}^{k} h_X(s, \eta_{s-1}, A)\hat{e}_1 \equiv [\Gamma(A(\omega))]^{k-k_0-1} \begin{bmatrix} -\alpha(A_1(\omega)) \\ 1 \end{bmatrix} \tag{7.1.53}$$

and

$$\begin{aligned} &\sum_{l=k_0+1}^{k} \left(\prod_{s=l+1}^{k} h_X(s, \eta_{s-1}, A)\right) \frac{\partial h(l-1, \eta_{l-1}, A)}{\partial A_1} \\ &\equiv [\Gamma(A(\omega))]^{k-k_0-1} \left[\begin{bmatrix} \frac{\partial(-\alpha_1(A_1(\omega)))}{\partial A_1}\eta_{k_0}^1 \\ 0 \end{bmatrix}\right. \\ &\quad \left. + \sum_{l=k_0+2}^{k} [\Gamma^{-1}(A(\omega))]^{l-k_0-1} \begin{bmatrix} \frac{\partial(-\alpha_1(A_1(\omega)))}{\partial A_1} & 0 \\ 0 & 0 \end{bmatrix} \begin{bmatrix} \eta_{l-1}^1 \\ \eta_{l-1}^2 \end{bmatrix}\right]. \end{aligned} \tag{7.1.54}$$

Since

$$\eta_{l-1} = [\Gamma(A(\omega))]^{l-k_0-1}\eta_{k_0}$$

and Γ is a symmetric matrix, (7.1.54) reduces to

$$[\Gamma(A(\omega))]^{k-k_0-l}\begin{bmatrix}\frac{\partial(-\alpha_1(A_1(\omega)))}{\partial A_1}\\ 0\end{bmatrix} + \sum_{l=k_0+2}^{k}[\Gamma^{-1}(A(\omega))]^{k-k_0-1}$$

$$\times [\Gamma(A(\omega))]^{k-k_0-1}\begin{bmatrix}\frac{\partial(-\alpha_1(A_1(\omega)))}{\partial A_1} & 0\\ 0 & 0\end{bmatrix}\begin{bmatrix}\eta^1_{l-1}\\ \eta^2_{l-1}\end{bmatrix}$$

$$= (k-k_0)\begin{bmatrix}\frac{\partial(-\alpha_1(A_1(\omega)))}{\partial A_1}\\ 0\end{bmatrix}. \tag{7.1.55}$$

Therefore, (7.1.53) and (7.1.55) will be linearly independent if the vectors $[-\alpha_1(A_1(\omega)), 1]^T$ and $\left[\frac{\partial(-\alpha_1(A_1(\omega)))}{\partial A_1}\eta^1_{k_0}, 0\right]^T$ form a basis for $\Re^2$, or, equivalently, if

$$\frac{\partial(-\alpha_1(A_1(\omega)))}{\partial A_1}\eta^1_{k_0} \neq 0.$$

An application of Lemma 7.1.3, in particular Corollary 7.1.2, gives us that $|\Phi^k_{c_l}|$, defined as

$$|\Phi^k_C| = \left|\frac{\partial G(\bar{z}_0, B)}{\partial(\bar{z}_0, B)}\right|,$$

is the solution of

$$r_k = |\Gamma(A(\omega))|r_{k-1} + \left|u^1_H(k)\begin{bmatrix}\frac{\partial(-\alpha_1(A_1(\omega)))}{\partial A_1}\eta^1_{k_0}\\ 0\end{bmatrix}\right|. \tag{7.1.56}$$

An application of the variation of parameters formula yields

$$|\Phi^k_{c_l}| = \sum_{s=k_0+1}^{k}|\Gamma(A(\omega))|^{k-k_0-1}q_1(k), \tag{7.1.57}$$

where

$$q_1(k) = \left| [\Gamma(A(\omega))]^{k-k_0-1}\hat{e}_1 \begin{bmatrix} \dfrac{\partial(-\alpha_1(A_1(\omega)))}{\partial A_1}\eta^1_{k_0} \\ 0 \end{bmatrix} \right|.$$

From Theorem 7.1.1, the marginal JPDF of the solution of (7.1.49) is defined as

$$p_H(\eta) = p_{H_0}(k, k_0, G^{-1}(k, k_0, \eta))\frac{1}{|\Phi^k_{c_l}|}, \tag{7.1.58}$$

where $|\Phi^k_{c_l}|$ is defined in (7.1.57), and hence the marginal JPDF of solution (7.1.48) is defined as

$$p_X(x) = p_{H_1}(\eta_1) = \int p_H(\eta)d\eta_2.$$

7.2 Extension of Classical Probability Distribution Model with Random Parameters — II

In this section, we briefly present a few illustrations outlining a conceptual framework for developing probability distributions associated with either difference or differential equations. The theoretical foundation is currently under investigation. This will be reported later. In the following examples, we discuss a procedure to find a particular solution of a linear homogeneous deterministic differential equation. This idea leads to the formulation of an initial value problem in the study of differential equations and its applications for the derivation of a distribution function. Illustrations are used to exhibit the underlying idea for generating original distributions. To motivate interest, we determine existing simple, well-known distributions via our approach.

In the following, we exhibit example for deterministic initial-value problem for linear differential equations.

Example 7.2.1. Initial-value problem for time-varying linear differential equations: Let us formulate an initial-value problem for a first-order linear homogeneous differential of type as

$$dx = f(t)xdt, x(t_0) = x_0, \tag{7.2.1}$$

where the rate function f is assumed to be continuous on its domain and $J = [a, a + \alpha], t, x, t_0, x_0$ are real numbers belonging to R. The problem of finding a solution to (7.2.1) is referred to as an **initial value problem** (IVP). Its solution is represented by $x(t) = x(t, t_0, x_0)$, for $t \geq t_0$ and $t, t_0 \in J$. In fact, in closed form, the solution $x(t) = x(t, t_0, x_0)$ of the IVP (7.2.1) is represented by $x(t) = x(t, t_0, x_0) = \exp[\int_{t_0}^{t} [f(s)]ds]x_0 = \Phi(t, t_0)x_0$. This solution is termed as a **particular solution** of (7.2.1). $\Phi(t, t_0)$ defined in (7.2.1) is called a **normalized fundamental solution** of (7.2.1) because $\Phi(t, t_0)$ has an algebraic inverse ($\Phi(t, t_0) \neq 0$ on J) and $\Phi(t_0, t_0) = 1$. We further assume that

$$\lim_{t\to\infty} \sup \left[\frac{1}{t - t_0} \int_{t_0}^{t} \frac{\partial}{\partial x} f(s, x)\right] = -\gamma < 0, \qquad (7.2.2)$$

for some γ being a real positive number.

The solution of the IVP (7.2.1) possesses the following conceptual computational properties:

P_1: The IVP (7.2.1) consists of a unique initial value problem denoted as $x(t, t_0, x_0) \equiv x(t)$.

P_2: The solution process is continuously differentiable with respect to initial data, that is,

$$\begin{aligned}
\frac{\partial}{\partial x_0} x(t) &\equiv \frac{\partial}{\partial x_0} x(t, t_0, x_0) \\
&= \frac{\partial}{\partial x_0}\left[\exp\left[\int_{t_0}^{t} [f(s)]ds\right] x_0\right] = \frac{\partial}{\partial x_0}[\Phi(t, t_0)x_0] \\
&= \exp\left[\int_{t_0}^{t} [f(s)]ds\right] = \Phi(t, t_0)
\end{aligned}$$

and

$$\begin{aligned}
\frac{\partial}{\partial t_0} x(t) &\equiv \frac{\partial}{\partial t_0} x(t, t_0, x_0) \\
&= \frac{\partial}{\partial t_0}\left[\exp\left[\int_{t_0}^{t} [f(s)]ds\right] x_0\right] = \frac{\partial}{\partial t_0}[\Phi(t, t_0)x_0] \\
&= -\left[\exp\left[\int_{t_0}^{t} [f(s)]ds\right] f(t_0)x_0\right] = -\frac{\partial}{\partial t_0}[\Phi(t, t_0)f(t_0)x_0]
\end{aligned}$$

If and only if

$$dx = -xf(s)ds, x(t_0) = x_0, \quad \text{for } s \leq t_0 \leq t, \tag{7.2.3}$$

where $s, t_0, t \in J; x(t, t_0, x_0)$ is the forward solution of the IVP (7.2.1), and

$$\begin{aligned}\exp\left[-\int_{t_0}^{t}[f(u)]du\right]\frac{\partial}{\partial t_0}x(t) &\equiv \exp\left[-\int_{t_0}^{t}[f(u)du\right]\frac{\partial}{\partial t_0}x(t,t_0,x_0)\\ &= \frac{d}{dt_0}x(t,t_0,x_0),\end{aligned}$$

it is known that

$$\frac{\partial}{\partial t_0}\Phi(t,t_0), \Phi(t,t_0), \Phi(t_0,t_0) = 1 \quad \text{and} \quad \frac{\partial}{\partial t_0}\Phi(t,t_0) = \Phi(t,t_0).$$

Under an asymptotic stability condition of the trivial solution of (7.2.1), it is obvious that the state transition operator $\Phi(t_0, t_0)$ satisfies the probability distribution properties function.

In the following, we exhibit examples of initial-value problems for differential equations with random parameters and with known distributions.

Based on a numerical technique developed by Bellomo and Pistone and Harlow and Delph, we obtain the JPDF for the dependent variables or the MPDF for the individual dependent variables.

Example 7.2.2. Consider the exponential growth model of the form

$$\frac{dN}{dt} = rN, N(0) = N_0.$$

Here, r and N_0 can be treated as random parameters.

Assume that the initial condition N_0 is a deterministic constant and r is a random variable and independent of time t. Let $r \sim \text{Normal}(\bar{r}, \sigma^2)$.

Our aim is to study the mean and variance of $\ln \frac{N}{N_0}$, that is, $E\left[\ln \frac{N}{N_0}\right]$ and $V\left[\ln \frac{N}{N_0}\right]$.

The density function of the population of size N at time t is given as follows:

$$\begin{aligned} f_N(N) &= f_r\left(\frac{1}{t}\ln\frac{N}{N_0}\right)\frac{1}{N_t}, \quad -\infty < N < \infty, \\ &= \frac{1}{\sqrt{2\pi}\sigma}\frac{1}{N_t}\exp\left[-\frac{1}{2\sigma^2}\left\{\frac{1}{t}\ln\frac{N}{N_0} - \bar{r}\right\}\right], \ -\infty < \frac{1}{t}\ln\frac{N}{N_0} < \infty. \end{aligned}$$

Therefore, $E\left[\ln\frac{N}{N_0}\right] = \text{tr}$ and $V\left[\ln\frac{N}{N_0}\right] = t^2\sigma^2$. We assume that the initial condition N_0 is a deterministic constant and r follows Unif(1, 2) and independent of time; that is, $r \sim$ Unif(1, 2).

Then,

$$f_N(N) = \frac{1}{tN}; \quad 1 < \frac{1}{t}\ln\frac{N}{N_0} < 2.$$

Therefore,

$$E\left[\ln\frac{N}{N_0}\right] = \frac{3}{2}t \quad \text{and} \quad V\left[\ln\frac{N}{N_0}\right] = \frac{1}{12}t^2.$$

Example 7.2.3. Consider the logistic growth model,

$$\frac{dN}{dt} = rN\left(1 - \frac{N}{K}\right), \quad N(0) = N_0 > 0.$$

Here, r, K and N_0 can be treated as random parameters.

Assume that r and K are random variables and N_0 is a deterministic constant.

The joint density function of (N, r) is given by

$$\begin{aligned} f_{N,r}(t, N, r) &= f_{r,K}(r, K \text{ in terms of } r, N, N_0) \\ &\quad \times \frac{N_0^2(e^{rt} - 1)e^{rt}}{(N_0e^{rt} - N)^2}. \end{aligned}$$

Assume that r and $K \sim$ i.i.d. with Unif(1, 2). Then, the joint density of (N, r) is

$$f_{N,r}(t, N, r) = \frac{N_0^2(e^{rt} - 1)e^{rt}}{(N_0e^{rt} - N)^2}, \quad 1 < r < 2, \quad 1 < \frac{NN_0(1 - e^{rt})}{(N - N_0e^{rt})}.$$

The marginal density $f_N(t, N)$ can be obtained by integrating $f_{N,r}(t, N, r)$ with r. Similar results can also be obtained for the competing species model.

7.3 Formulation of Prototype Model of Short-Run Market Equilibrium under Markovian Structural Perturbations and Qualitative Analysis — I

This section is devoted to a short-run dynamic market in the area of economics under random perturbations. In this work, internal structural perturbations are introduced via random parametric excitations. The stochastic stability analysis of a short-run market equilibrium under random structural perturbations is outlined. Furthermore, the roles and scopes of the smooth auxiliary system, the variational methods of constants and the comparison method in the 21st century are illustrated in a unified and systematic way.

The stability analysis of the short-run market equilibrium process with random perturbations first involves using the following process:

$$p_k = \frac{\beta}{\alpha} p_k^*, \tag{7.3.1}$$

where p_k = price, p_k^* = anticipated or expected price and μ_k = is a random disturbance. Here, β and α are the rate coefficients for the supply and demand functions, respectively. We assume that the random disturbances are serially uncorrelated but do not necessarily have equal variance. The variables p_{0k} and p_k^* are deviations from the equilibrium price. A variety of extrapolative expectations are considered for p_k^*, all of which are shown to be special cases of the following distributed-lag model:

$$p_k = a_1 p_{k-1} + a_2 p_{k-2} + \cdots + a_n p_{k-n}, \tag{7.3.2}$$

for which $n \geq 2$. A short-run market equilibrium process under internal random structural perturbations or random parametric variation is formulated. Assume that the coefficients of the supply and demand functions and thus the coefficients of (7.3.2) are represented by a finite-state Markov process with s possible states. For example, assume that the supply sources are finite in number and that their availability is modeled by a Markov process, $\eta(k)$, with the transition probability matrix $\Pi = (\pi_{ij})_{s\times s}$. As a prototype model, consider a model with two suppliers, A and B. Table 7.1 could represent the values of the Markov state η.

Table 7.1: Display of Two Suppliers Markov State.

Supplier A	Supplier B	η
Available	Available	4
Available	Not available	3
Not available	Available	2
Not available	Not available	1

Under the generalized scenario of an s-dimensional Markov process, the n-dimensional lag equation (7.3.2) becomes

$$p_k = a_1(\eta(k))p_{k-1} + a_2 p_{k-2} + \cdots + a_n p_{k-n}, \tag{7.3.3}$$

which can be rewritten as a system of difference equations,

$$\Delta y(k+1) = A(\eta(k))y(k), \tag{7.3.4}$$

where

$$y(k) = [p_{k-n}, p_{k-n+1}, \ldots, p_{k-1}]^T \tag{7.3.5}$$

and

$$A(\eta(k))_{n\times n} = \begin{bmatrix} -1 & 1 & 0 & \ldots & 0 & 0 \\ 0 & -1 & 1 & \ldots & 0 & 0 \\ 0 & 0 & -1 & \ldots & 0 & 0 \\ \vdots & \vdots & \vdots & \ddots & \vdots & \vdots \\ 0 & 0 & 0 & \ldots & -1 & 1 \\ a_n & a_{n-1} & a_{n-2} & \ldots & a_2 & a_1(\eta(k)) \end{bmatrix}, \tag{7.3.6}$$

where the discrete-time Markov process $\eta(k)$, in particular the Markov chain, is defined on the complete probability space $(\Omega, \mathfrak{F}, P)$ with a finite number of states s.

We consider an auxiliary discrete-time process associated with (7.3.4) as

$$\Delta m(k) = -Im(k), m(k_0) = m_0, \tag{7.3.7}$$

where, without loss of generality, $m_0 > 0$. Further, note that for the smooth auxiliary IVP associated with (7.3.4), we have a closed-form solution represented as

$$m(k, k_0, m_0) = \Phi(k, k_0)m_0 = (1)^k I m_0, \quad \text{for } k \geq k_0. \tag{7.3.8}$$

Moreover, from Lemmas 1.1.1, 1.1.2 and 1.1.3, we have

$$\frac{\partial}{\partial m_0} m(k, k_0, m_0) = (1)^k I = 1I \quad \text{and}$$
$$\frac{\partial^2}{\partial m_0 \partial m_0} m(k, k_0, m_0) = 0, \quad \text{for } k \geq k_0. \tag{7.3.9}$$

Remark 7.3.1. The smooth auxiliary system (7.3.7) reduces to the classical case of investigating a very trivial system. This exhibits the roles and scopes of the auxiliary system of constants and the variational comparison method in a unified and systematic way.

Procedure for developing the variational comparison approach: We consider the following type of energy or Lyapunov function in the context of the auxiliary system (7.3.7) for the short-run market equilibrium dynamic system (7.3.4) as follows.

The L-operator for (7.3.4) in the context of the auxiliary system discrete-time process (7.3.7) is given by, for $p = k - 1$ and $p + 1 = k, m(k, k, m_0) = m_0$ and be a solution process of (7.3.4) and (7.3.7), respectively. From Remark 7.3.1, following the proof of Theorem 1.2.3, in particular, Corollary 1.2.5,

$$\begin{aligned}
&V(k-1, m(k, k-1, y), \omega) \\
&\quad = \alpha(\eta(k)) m(k, k-1, y)^T m(k, k-1, y) \\
&\quad = V(k-1, x, \eta(k)) = \alpha(\eta(k)) y^T y,
\end{aligned} \tag{7.3.10}$$

where $\alpha(\eta(k)) > 0$, for $k \geq k_0$, with a diagonal matrix D, which has positive entries along the diagonal and zero as the off-diagonal entries, and applying the Corollaries 1.1.2 and 1.1.3 in the context of (7.3.4), we have

$$\begin{aligned}
V(p+1, m(k, p+1, y(p+1))) = V(p+1, m(k, p+1, y(p) \\
+ A(\eta(p))),
\end{aligned}$$

and also noting

$$V(p, m(k, p, y(p))) = V(p, m(k, p+1, y(p) + A(\eta(p))y(p)))),$$

we have

$$\begin{aligned}
&E[V(p+1, m(k, p+1, y(p+1))) - V(p, m(k, p, y)|\mathcal{F}_p)] \\
&\quad = \Delta_{(7.3.4)} V_i(p, m(k, p, y)) \\
&\quad = \sum_{j=1}^{s} \pi_{ij} [\mathcal{L}_a V_j(p, m(k, p, y)) + \mathcal{L}_e V_j(p, m(k, p, y)) \\
&\qquad + \mathcal{L}_0 V_j(p, m(k, p, y)) - V_i(p, m(k, p, y))] \\
&\quad = \sum_{j=1}^{s} \pi_{ij} [m^T(k, p, y) A^T(i) D(j) m(k, p, y) \\
&\qquad + m^T(k, p, y) D(j) A(i) m(k, p, y) \\
&\qquad + m^T(k, p, y) A^T(i) D(j) A(i) m(k, p, y) \\
&\qquad - m^T(k, p, y) D(i) m(k, p, y)] \\
&\quad = \sum_{j=1}^{s} \pi_{ij} [y^T A^T(i) D(j) y + y^T D(j) A(i) y \\
&\qquad + y^T A^T(i) D(j) A(i) y - y^T D(i) y] \\
&\quad = \sum_{j=1}^{s} \pi_{ij} [y^T (A^T(i) D(j) + D(j) A(i) \\
&\qquad + A^T(i) D(j) A(i)) y - y^T D(i) y].
\end{aligned} \tag{7.3.11}$$

Hence, setting $\Gamma(j) = (\sqrt{D(j)})^{-1}[A^T(i)D(j) + D(j)A(i) + A^T(i)D(j)A(i)]\sqrt{D(j)}$, we have

$$\Delta_{(7.3.4)} V_i(p, y) = \sum_{j=1}^{s} \pi_{ij} [(\sqrt{D(j)}y)^T \Gamma(j) \sqrt{D(j)} y - y^T D(i) y],$$

$$\Delta_{(7.3.4)} V_i(p, m(k, p, y)) \leq \sum_{j=1}^{s} \pi_{ij} \lambda_M(\Gamma(i)) V_j(p, y) - \sum_{j=1}^{s} \pi_{ij} V_i(p, y),$$

which can be rewritten as

$$\Delta_{(7.3.4)}V(p,y) \leq GV(p,y), \tag{7.3.12}$$

where $\lambda_M(\Gamma(j))$ is the largest eigenvalue of $\Gamma(j)$ and $i, j \in \{1, 2, \ldots, s\} = I(1, s)$; G is a matrix with its elements, g_{ij}, defined by

$$g_{ij} = \begin{cases} -(1 - \pi_{i1}\lambda_M\Gamma(i)), & \text{for } j = i, \\ \pi_{ij}\lambda_M(i), & \text{for } j \neq i. \end{cases} \tag{7.3.13}$$

Under this setup, the comparison equation becomes

$$\Delta u(p+1) = Gu(p), u(k_0) = u_0. \tag{7.3.14}$$

In the following, we provide sufficient conditions for the short-run market equilibrium dynamic system (7.3.4) under Markovian random fluctuations in the coefficient rate matrix of (7.3.6) via the method of energy or Lyapunov function in the context of the variational comparison approach.

7.3.1 *Mean-square convergence and stability analysis*

In this section, the procedure of applying the convergence analysis of (7.3.4) is outlined. The scope and significance of the variational comparison method are illustrated, for which we utilize the different forms of Lyapunov-like functions and derive different sets of sufficient conditions for the mean-square convergence of the short-run market equilibrium dynamic system (7.3.4) under Markovian random structural perturbations.

Mean-Square Convergence — 1: First, let us consider a Lyapunov-like function as in (7.3.10), wherein α is a positive scalar function, This comparison difference equation (7.3.14) is identical to the comparison equation in Theorem 1.2.3, Corollaries 1.2.5 and 1.2.6. The mean-square convergence conditions in the context of (7.3.4) reduce to:

Stability condition for comparison iterates process (7.3.14):

$$C_1\colon\ -1+\pi_{ii}\lambda_M\Gamma(i) > -1, \tag{7.3.15}$$

$$C_2\colon\ 1 > \pi_{ii}\lambda_M\Gamma(i) \tag{7.3.16}$$

and

$$-\pi_{ii}\lambda_M\Gamma(i)+1 > d_i^{-1}\sum_{i=1,i\neq j}^{s}\pi_{ij}\lambda_M\Gamma(i),\ \text{for all } j\in I(1,s). \tag{7.3.17}$$

Mean-Square Convergence — 2: Another set of convergence and stability conditions for (7.3.4) is obtained in the context of Remark 7.3.1 and the following Lyapunov function:

$$V(k-1,m(k,k-1,y),\omega) = V(k-1,y,\eta(k)) = y^T D(\eta(k))y, \tag{7.3.18}$$

where D is an $s\times s$ diagonal matrix random function of a finite state s Markov chain, with its state transition constant intensity matrix being $\Pi=(\pi_{ij})_{s\times s}$ and

$$D(\eta(k)) = \operatorname{diag}\{d_1(\eta(k)),\ldots,d_n(\eta(k))\}, \tag{7.3.19}$$

where $d_i(\eta(k)) > 0$, for $1\leq i\leq n$. The difference Δ-operator of (7.3.4) in the context of the energy or Lyapunov function (7.3.18) is expressed by following a similar argument as above for (7.3.11). Then, we have

$$\Delta_{(7.3.4)}V_i(p,m(k,p,y)),\quad \text{for } i,j\in\{1,2,\ldots,s\}$$

$$\sum_{j=1}^{s}\pi_{ij}[y^T(A^T(i)D(j)+D(j)A(i)+A^T(i)D(j)A(i))-y^TD(i)y]. \tag{7.3.20}$$

where $D(i)$ and $D(j)$ are $s\times s$ values of the diagonal matrix random functions that are defined at the ith and jth states of the finite states $i,j\in\{1,2,\ldots,s\}$ of the discrete-time Markov chain η state.

By imitating the above-described steps from (7.3.11) to (7.3.14), we arrive at as follows:

$$\begin{aligned}\Gamma(j) = (\sqrt{D(j)})^{-1}[&A^T(i)D(j)+D(j)A(i)\\ &+A^T(i)D(j)A(i)](\sqrt{D(j)})^{-1}\end{aligned} \tag{7.3.21}$$

and noting that

$$\left(\sqrt{D(i)}\right)^{-1} = \text{diag}\left\{\sqrt{d_1(i)}, \sqrt{d_2(i)}, \ldots, \sqrt{d_n(i)}\right\}$$

and

$$\left(\sqrt{D(j)}\right)^{-1} = \text{diag}\left\{\sqrt{d_1(j)}, \sqrt{d_2(j)}, \ldots, \sqrt{d_n(j)}\right\}$$

are the $s \times s$ values of the diagonal matrix random functions that are defined at the ith and jth states of the finite states $i, j \in \{1, 2, \ldots, s\}$ of the discrete-time Markov chain η state, we have

$$\Delta_{(7.3.4)} V_i(p, y) = \sum_{j=1}^{s} \pi_{ij} \left[(\sqrt{D(j)}y)^T \Gamma(j) \sqrt{D(j)}y - y^T D(i) y\right],$$

and hence, we have a system of difference inequality as before in (7.3.12), which can be rewritten as

$$\Delta_{(7.3.4)} V(p, y) \leq GV(p, y), \tag{7.3.22}$$

where $\lambda_M(\Gamma(j))$ is the largest eigenvalue of $\Gamma(j)$ and $i, j \in \{1, 2, \ldots, s\} = I(1, s)$; G is a matrix with its elements, g_{ij}, as defined in (7.3.13) with exactly similar comparison equation (7.3.14) with parametric coefficient rate functions. As a result of this, the mean square convergence and stability conditions C_1 and C_2 of the comparison equation assures the convergence and stability properties.

7.3.2 *Influence of stochastic and discrete-time processes*

In the following, we want to analyze the influence of a discrete-time dynamic process. For this purpose, we characterize the dynamic component associated with a purely discrete-time process as follows.

We redefine the aggregate discrete-time rate matrix in (7.3.21) into two components, namely (a) a purely discrete-time matrix and (b) a discrete-time process-generated matrix. (7.3.18) can be written as $\Gamma(j) = \Psi(j) + \Lambda(j)$,

$$\Psi(j) = \left(\sqrt{D(j)}\right)^{-1} [A^T(i)D(j) + D(j)A(i)] \left(\sqrt{D(j)}\right)^{-1},$$

$$\Lambda(j) = \left(\sqrt{D(j)}\right)^{-1} A^T(i)D(j)A(i) \left(\sqrt{D(j)}\right)^{-1},$$

and let $\lambda_M(\Psi(j))$ and $\lambda_M(\Lambda(j))$ be the largest eigenvalues of $\Psi(j)$ and $\Lambda(j)$, respectively, $j \in I(1, s)$.

Under this type of decomposition of the difference operator of comparison rate matrix, the system of difference inequality and the corresponding comparison systems of difference equation yields as follows:

$$\Delta_{(7.3.4)} V(p, y) \leq GV(p, y); \tag{7.3.23}$$

$$g_{ij} = \begin{cases} -(1 - \pi_{ii}[\lambda_M(\Psi(i)) + \lambda_M(\Lambda(i))]), & \text{for } j = i; \\ \pi_{ij}[\lambda_M(\Psi(j)) + \lambda_M(\Lambda(j))], & \text{for } j \neq i; \end{cases} \tag{7.3.24}$$

$$\Delta u(p + 1) = Gu(p), u(k_0) = u_0, \tag{7.3.25}$$

where $u, u_0 \in R^s$.

The solution process of the comparison equation (7.3.25) satisfies the convergence and stability conditions as listed below.

$$C_1^*: -1 + \pi_{ii}[\lambda_M(\Psi(i)) + \lambda_M(\Lambda(i))] > -1; \tag{7.3.26}$$

$$C_2^*: 1 > \pi_{ii}[\lambda_M(\Psi(i)) + \lambda_M(\Lambda(i))] \tag{7.3.27}$$

and

$$-\pi_{ii}\lambda_M(\Psi(i)) + \lambda_M(\Lambda(i)) + 1 > \sum_{i=1, i\neq j}^{s} \pi_{ij}[\lambda_M(\Psi(j)) + \lambda_M(\Lambda(j))], \text{ for all } j \in I(1, s). \tag{7.3.28}$$

Remark 7.3.2. We remark that the mean-square convergence and stability, respectively, imply convergence and stability in the probability.

7.3.3 *Conclusions*

In this section, we present the scope and significance of the use of an auxiliary dynamic, however trivial they may be. Further, we relate at least three differential or difference systems of equations. The first is referred to as (i) a nominal or simple and smooth system; the second one can be considered (ii) a simplified perturbed system (P) associated with a nominal or unperturbed system (U); and the third one is considered to be (iii) a complex highly nonstationary, nonlinear and interconnected large-scale system. The convergence and stability in the context of the variational systems are analyzed in a systematic and unified way in the framework of the existing work in the literature. In particular, a few silent features are presented as conclusions in the following.

The convergence and stability conditions are explicitly presented in terms of the rate coefficients of complex dynamic system parameters. Furthermore, this exhibits the following simple conclusions:

C_1. The weighted self-inhibitory effects are greater than the cross-coupling structural perturbation effects.
C_2. The conditions are algebraically simple and easy to compute.
C_3. The conditions are sufficient and conservative. However, they are reliable and robust.
C_4. They are robust with respect to parametric changes of the system.
C_5. It shows that structural perturbations can be considered as stabilizing agents.
C_6. It sheds light on the fundamental problem of "complexity versus stability."
C_7. It also sheds light on the problem of "stochastic versus deterministic."

7.4 Prototype Model of Short-Run Market Equilibrium under Discrete-Time Wiener Process Disturbance and Qualitative Analysis — II

This section is devoted to studying a short-run dynamic market in the area of economics under Ito-type stochastic perturbations. In this work, external perturbations are introduced via the Wiener process with $E[w(t)|w(s+1+k)] = 0$, for $t \in [k, k+1)$ and given $s \in [k-1, k)$

due to the fact that the Wiener process is an independent process with mean zero and variance $\sigma^2 t$, for $t > 0$. Hence, the sequence $\{\xi_k\}_{k=0}^{k}$, where $\xi_k = w(k^-)$, $k \in N$, is a discrete-time Wiener process. Moreover, for each term of the sequence, ξ_k, there is a unique $k, k \in N$, and an open random interval for each $[\xi_k, \xi_{k+1^-})$. In short, the sequences $\{[\xi_k, \xi_{k+1^-})\}_{k=1}^{\infty}$ and $\{[k, k+1)\}_{k=1}^{\infty}$ are one-to-one and onto. Further, ξ_k is a sample continuous Wiener process on the state of each interval $[\xi_k, \xi_{k+1^-})$ as state of the process, and also ξ_k is sample continuous on the intervals $[k, k+1) \equiv [\xi_k, \xi_{k+1^-})$ as the domain/state of the discretized Wiener process with endpoints as time and also states. The stochastic stability analysis of the short-run market system under white noise process perturbation is outlined. Furthermore, the roles and scopes of a smooth auxiliary system, the method of variation of constants and the comparison method in the 21st century are illustrated in a unified and systematic way. Stability analysis of the short-run market equilibrium process under white noise stochastic perturbations:

$$p_k = (\beta/\alpha)p_k^*, \tag{7.4.1}$$

where p_k = price, p_k^* = anticipated or expected price and ξ_k is a white noise disturbance. Here, β and α are the rate coefficients for the supply and demand functions, respectively. We assume that the random disturbances are serially uncorrelated and have equal variance. The variables p_k and p_k^* are deviations from the equilibrium price. A variety of extrapolative expectations were considered for p_k^*, all of which were shown to be special cases of the following distributed-lag model:

$$p_k = a_1 p_{k-1} + a_2 p_{k-2} + \cdots + a_n p_{k-n} + \xi_k, \tag{7.4.2}$$

for which $n \geq 2$. A short-run market equilibrium process under external random structural perturbations is formulated. The discrete-time dynamic model (7.4.2) can be rewritten as a system of difference equations:

$$\Delta y(k+1) = Ay(k) + \epsilon(k), y(k_0) = y_0, \tag{7.4.3}$$

where

$$y(k) = [p_{k-n}, p_{k-n+1}, \ldots, p_{k-1}]^T, \tag{7.4.4}$$

$$\epsilon(k) = [0, \ldots, 0, \xi_k]^T \tag{7.4.5}$$

and

$$A_{n\times n} = \begin{bmatrix} -1 & 1 & 0 & \dots & 0 & 0 \\ 0 & -1 & 1 & \dots & 0 & 0 \\ 0 & 0 & -1 & \dots & 0 & 0 \\ \vdots & \vdots & \vdots & \ddots & \vdots & \vdots \\ 0 & 0 & 0 & \dots & -1 & 1 \\ a_n & a_{n-1} & a_{n-2} & \dots & a_2 & a_1 - 1 \end{bmatrix}, \tag{7.4.6}$$

where ϵk is a discrete-time Gaussian process; in particular, the Wiener process is defined on the complete probability space $(\Omega, \mathfrak{F}, P)$ with mean and variance:

$$E[e(k)^T(k)\epsilon(k)y(k)] = y^T(k)Cy(k) + \sigma^2(k). \tag{7.4.7}$$

We utilize the role and scope of the auxiliary discrete-time process associated with (7.4.3) (7.3.7) in Section 7.3 and (1.1.1), (1.1.2) as

$$\Delta x(k) = F(k, x(k)), x(k_0) = x_0, \tag{7.4.8}$$

where $F(k, x(k)) = -x(k)$.

We consider the system (7.4.8) as a perturbed system of

$$\Delta m(k) = 0, m(k_0) = m_0, \tag{7.4.9}$$

where, without loss of generality, $m_0 > 0$. From the features of the auxiliary IVP (7.4.8) as outlines before, for instance, the close solution is represented as:

$$\begin{aligned} m(k, k_0, m_0) &= \Phi(k, k_0)m_0 = (1)^k I m_0 = m_0, \text{ for } k \geq k_0, \\ m(k, k_0 - 1, m_0) &= \Phi(k, k_0 - 1)m_0 = I(1)^{k-1}, \text{ for } k_0 - 1, \end{aligned} \tag{7.4.10}$$

and

$$\begin{aligned} &\frac{\partial}{\partial m_0} m(k, k_0, m_0) = I(1)^k = 1I \quad \text{and} \\ &\frac{\partial^2}{\partial m_0 \partial m_0} m(k, k_0, m_0) = 0, \quad \text{for } k \geq k_0, \end{aligned} \tag{7.4.11}$$

where 0 is an nxn matrix.

Employing the method of variation of constants in Theorem 1.1.3, in particular, Corollary 1.1.5, we obtain a solution process for (7.4.8) via the solution process of (7.4.9) as

$$\frac{d}{ds}(m(k,k_0,sp(k)+(1-s)p(k-1)) = p(k)-p(k-1) \quad \text{iff}$$

$$\int_0^1 \frac{d}{ds}m(k,k_0,sp(k)+(1-s)p(k-1))ds$$
$$= m(k,k_0,p(k)) - m(k,k_0,p(k-1))$$

$$x(k,k_0,m_0) - m(k,k_0,m_0) = \sum_{s=k_0+1}^{k}[-x_p(k,k_0,p(k-1))] \quad \text{iff}$$

$$x(k,k_0,m_0) = (1)^k m_0 + [(-1)^1 m_0 + (-1)^2 m_0$$
$$+ \cdots (-1)^q m_0 + \cdots (-1)^k m_0].$$

This implies that for the initial state at $k_0, x(k_0) = x_0$,

$$x(k_0+1,k_0,m_0) = [(1)^1 + (-1)^1]m_0 = 0m_0,$$
$$x(k_0+2,k_0,m_0) = [(1)^2 + (-1)^1 + (-1)^2]m_0 = 1m_0,$$
$$x(k_0+3,k_0,m_0) = [(1)^3 + (-1)^1 + (-1)^2 + (-1)^3]m_0 = 0m_0,$$

....

$$x(k+q,k_0,m_0)$$
$$= [(1)^q + (-1)^1 + (-1)^2 + \cdots + (-1)^i + \cdots + (-1)^q]m_0$$

...,

$$x(k+q+1,k_0,m_0)$$
$$= [(1)^{q+1} + (-1)^1 + (-1)^2 + \cdots + (-1)^i + \cdots + (-1)^{q+1}]m_0,$$

...,

$$x(k,k_0,m_0) = [(1)^k + (-1)^1 + (-1)^2 + \cdots + (-1)^q + \cdots + (-1)^k]m_0,$$

where the initial time interval is $[k_0-1,k_0]$ and also the operating interval is $k \in [k_0, k_0+k]$, for $k \in N$. Thus, in IVP (7.4.8), the forward solution process of the IVP is uniquely defined by the functional or delay initial value for the discrete-time dynamic process in a systematic and unified way.

Remark 7.4.1. A remark similar to Remark 7.3.1 can be made and is stated here for the sake of completeness. The smooth auxiliary systems (7.4.7) and (7.4.8) reduce to the classical case of investigating a very trivial system. This exhibits the role and scope of the auxiliary system, the method of variation of constants and the variational comparison method in the 21st century in a unified and systematic way.

Procedure for development of variational comparison approach: We consider the following type of energy or Lyapunov function in the context of the auxiliary system (7.4.9) for the short-run market equilibrium dynamic system (7.4.3) as

$$V(k-1,y) = y^T D(k-1)y, \tag{7.4.12}$$

where $k \geq k_0$ and D is a diagonal matrix with positive entries along the diagonal and zero as the off-diagonal entries, and applying Corollaries 1.1.2 and 1.1.3 in the context of (7.4.4), we have

$$D(k) = \operatorname{diag}\{d_1(k), \ldots, d_i(k), \ldots, d_n(k)\}, \tag{7.4.13}$$

where $d_i(k) > 0$, for $1 \leq i \leq n$. The L-operator for (7.4.3) in the context of auxiliary system discrete-time process of (7.4.7) and (7.4.8) is given by, for $p = k-1$ and $p+1 = k$, $m(k,k,m_0) = m_0$ and is a solution process of (7.4.3) and (7.4.7), respectively. The details are exhibited in the following remark.

Remark 7.4.2. Following the proof of Theorem 1.2.3, in particular, Corollary 1.2.5,

$$V(p+1, m(k,p+1,y(p+1))) = V(p+1, m(k,p+1,y(p)+Ay(p))),$$

and also noting

$$V(p, m(k,p,y(p))) = V(p, m(k,p+1,y(p)+F(p,y(p)))),$$

we have

$$\begin{aligned}
&E[V(p+1, m(k,p+1,y(p+1))) - V(p, m(k,p,y)|\mathcal{F}_p)] \\
&\quad = \Delta_{(7.4.3)} V(p, m(k,p,y)) = \mathcal{L}_a m(k,p,y) \\
&\qquad + \mathcal{L}_e m(k,p,y) + \mathcal{L}_O m(k,p,y)
\end{aligned}$$

$$
\begin{aligned}
&= m^T(k,p,y)A^T D(p)m(k,p,y) + m^T(k,p,y)D(p)Am(k,p,y)\\
&\quad + m^T(k,p,y)A^T D(p)Am(k,p,y)\\
&\quad + m^T(k,p,y)CD(p)m(k,p,y) + \sigma^2(p)\\
&= y^T A^T D(p)y + y^T D(p)Ay + y^T A^T D(p)Ay\\
&\quad + y^T CD(p)Ay + \sigma^2(p)\\
&= y^T(A^T D(p) + D(p)A + A^T D(p)A + CD(p) - D(p))y + \sigma^2(p).
\end{aligned}
\tag{7.4.14}
$$

Hence, we have

$$
\begin{aligned}
\Delta_{(7.4.3)}V(k-1) = y^T(&A^T D(k-1) + D(k-1)A + A^T D(k-1)A\\
&+ CD(k-1) - D(k-1))y + \sigma^2(k-1),
\end{aligned}
\tag{7.4.15}
$$

$$
\Delta_{(7.4.3)}V(k-1) \leq ((\lambda_M(\Gamma) + \lambda_M(\Sigma)) - 1)V(k-1,y) + \beta(k-1),
\tag{7.4.16}
$$

where Γ and Σ are $n \times n$ deterministic and diffusion rate matrices defined as follows:

$$
\begin{aligned}
\Gamma = &\left(\sqrt{D(k-1)}A\left(\sqrt{D(k-1)}\right)^{-1}\right)\\
&+ \left(\sqrt{D}(k-1)A\left(\sqrt{D(k-1)}\right)^{-1}\right) \quad \text{and}\\
\Sigma = &\left(\sqrt{D(k-1)}\right)^{-1} d_n(k-1)C\left(\sqrt{D(k-1)}\right)^{-1}.
\end{aligned}
$$

Further, λ_M stands for the largest eigenvalue of a matrix. Furthermore,

$$
\beta(k-1) = d_n(k-1)\sigma^2(k-1).
$$

Hence,

$$
\Delta_{(7.4.3)}V(k-1,y) \leq GV(k-1,y) + \beta(k-1),
\tag{7.4.17}
$$

where G and β are defined and

$$
\beta(k-1) = d_n(k-1)\sigma^2(k-1).
\tag{7.4.18}
$$

Under these conditions, the comparison equation is

$$\Delta u(k) = Gu(k-1) + \beta(k), \tag{7.4.19}$$

where

$$G = \lambda_M(\Gamma) + \lambda_M(\Sigma) - 1 < \alpha,$$
$$\lim_{k\to\infty} \beta(k-1) = 0 \tag{7.4.20}$$

and

$$0 < \alpha < 1. \tag{7.4.21}$$

In the following, we provide the sufficient conditions for the short-run market equilibrium dynamic system (7.4.3) via the method of energy or Lyapunov function in the context of the variational comparison approach.

7.4.1 *Convergence analysis*

In this section, applying the procedure of convergence analysis of (7.4.3) is outlined. To illustrate the scope and significance of the variational comparison method, in this work, we utilize the different forms of Lyapunov-like functions and derive the different sets of sufficient conditions for the mean-square convergence of the short-run market equilibrium dynamic system (7.4.3) under random environmental perturbations.

Mean-Square Convergence — 1: The solution process of (7.4.3) converges in the mean square if the conditions (7.4.17), (7.4.18), (7.4.19) and (7.4.20) are satisfied.

Remark 7.4.3. A more relaxed condition for convergence and stability can be reformulated as follows. Beginning with equation (7.4.15), one can write

$$\Delta_{(7.4.3)}V(k-1,y) = -y^T D(k-1)y + \left(\sqrt{D(k-1)}\right)^T \times y^T \Lambda \left(\sqrt{D(k-1)}\right) + \beta(k-1), \tag{7.4.22}$$

where Λ is an $n \times n$ aggregate rate matrix defined as follows:

$$\begin{aligned}\Lambda = \Bigg[&\sqrt{D(k-1)}A\left(\sqrt{D(k-1)}\right)^{-1} \\ &+ \left(\sqrt{D(k-1)}\right)^{-1} A^T \sqrt{D(k-1)} \\ &+ \left(\sqrt{D(k-1)}\right)^{-1} A^T D(k-1) A \left(\sqrt{D(j)}\right)^{-1} \\ &+ \left(\sqrt{D(k-1)}\right)^{-1} d_n(k-1) C \left(\sqrt{D(k-1)}\right)^{-1}\Bigg], \end{aligned} \quad (7.4.23)$$

where its largest eigenvalue is $\lambda_M(\Lambda)$, and further,

$$\beta(k-1) = d_n(k-1)\sigma^2(k-1). \quad (7.4.24)$$

Then,

$$\Delta_{(7.4.3)} V(k-1, y) \leq \lambda_M(\Lambda) V(k-1, y) - V(k-1, y), \quad (7.4.25)$$

and thus

$$\Delta_{(7.4.3)} V(k-1, y) \leq G V(k-1, y) + \beta(k-1), \quad (7.4.26)$$

where G is defined by

$$Gu = (\lambda_M(\Lambda) - 1)u. \quad (7.4.27)$$

$\lambda_M(\Lambda)$ and $\beta(k-1)$ are defined as in (7.4.23) and (7.4.24), respectively. The corresponding comparison difference equation is defined similarly as in (7.4.19):

$$\Delta u(k) = Gu(k-1) + \beta(k-1) \quad (7.4.28)$$

in the context of (7.4.22)–(7.4.27) and (7.4.24).

Mean-Square Convergence — 2: The mean-square convergence conditions for the solution process of (7.4.3) in the context of the Lyapunov function are as follows:

$$(\lambda_M(\Lambda) - 1) < 0, \quad (7.4.29)$$

$$\lim_{k \to \infty} \beta(k-1) = 0. \quad (7.4.30)$$

If one also assumes that there exist positive numbers $\Lambda > 0$ and $\beta > 0$ such that conditions (7.4.18) and (7.4.19) are met, then process (7.4.3) converges almost surely in this context.

7.4.2 Stability analysis

In this section, sufficient conditions for the stability analysis of the short-run market equilibrium dynamic system (7.4.3) via the method of energy or Lyapunov function in the context of variational comparison approach is illustrated. It is assumed that $x \equiv 0$ is the unique equilibrium state of the short-run market equilibrium process. For this purpose, it is also assumed that $\sigma^2(k) \equiv 0$ in (7.4.3), for all $k \in \mathbb{I}(k_0)$.

Remark 7.4.4. We observe the following under the assumptions of $y(k) \equiv 0$ and $\sigma^2(k) \equiv 0$:

(a) The set of sufficient conditions for Mean-Square Convergence — 1 of (7.4.3) given in (7.4.17), (7.4.18), (7.4.19) and (7.4.20) also forms the sufficient conditions for the mean-square stability of the equilibrium of state $y(k) \equiv 0$ of (7.4.3).
(b) The sets of sufficient conditions for convergence in the probability of (7.4.3) provided in (a) also form the sufficient conditions for stability in the probability of the equilibrium state $y(k) \equiv 0$ of (7.4.3).
(c) The sets of sufficient conditions for the almost-sure convergence of (7.4.3) provided in (a) are also sufficient conditions for the almost-sure stability of equilibrium state $y(k) \equiv 0$ of (7.4.3).

Remark 7.4.5. The local solution process of the auxiliary system (in general, $k = p + 1$ and $p = k - 1$, or, in particular, $p = k_0 - 1$ and $p + 1 = k_0$) plays a role in the development of variational comparison theorems. Since the inception of the ideas of "variational comparison theorem and constant methods" in 1975, these conceptual algorithms have been playing a significant role not only in the theory of differential or difference equations but also in the study of classical qualitative analysis of solution processes of complex systems of differential or difference equations, dynamic error analysis and dynamic reliability theory. In the absence of an auxiliary system, the smooth Lyapunov function (quadratic form $-y^T Py$, where P is an $n \times n$ definite matrix) and its variants were used in the literature:

$$D = \text{diag}\{\alpha, \dots, \alpha\} = \alpha I, \tag{7.4.31}$$

with $\alpha > 0$, for $k \geq k_0$, where I is an $n \times n$ identity matrix. In this case, (7.4.14) is reduced to

$$LV(k-1, y) = 2\alpha y^T Ay + \alpha y^T A^T Ay + \alpha^T yCy + \alpha\sigma^2(k). \quad (7.4.32)$$

In addition, the Lyapunov function is the square of the Euclidean norm $y^T y$. In this context, (7.4.14) reduces to

$$LV(k-1, y) = y^T(A^T + A + A^T A + C)y. \quad (7.4.33)$$

We also compare as noted in the literature, $A = F - I_{n\times n}$ is substituted into (7.4.33) to yield

$$LV(k-1, y) = y^T(F^T F + 2(F - F^T) + C - I_{n\times n})y. \quad (7.4.34)$$

In the quadratic form, F and F^T are equivalent. That is, $y^T F^T y = y^T Fy$. Thus, the terms $2F$ and $-2F^T$ vanish, leaving

$$LV(k-1, y) = y^T(F^T F + C - I_{n\times n})y. \quad (7.4.35)$$

From the above presented results, we can draw a few conclusions similar to the conclusion in Section 7.3.3.

7.5 Statistical Data Analysis: Discrete-Time Dynamic Model for Local Sample Mean and Covariance Processes

7.5.0 *Introduction*

In real-world dynamic modeling problems, the future states of continuous-time dynamic processes are influenced by the past state history and response/reaction time delay processes influencing the present states. In this section, the influence of state history, the concept of a lagged adaptive expectation process and the idea of a moving average lead to the development of a general interconnected discrete-time dynamic model of local sample mean and covariance statistic processes. A few byproducts of the discrete-time sample mean and variance statistic process are to (a) initiate ideas for the usage of a discrete-time interconnected dynamic approach parallel to the continuous-time dynamic process, (b) shorten computation time and (c) significantly reduce state error estimates. Moreover, the dynamic model is the generalization of the statistic of a random sample drawn from the "static" population.

7.5.1 *Data organization*

The theoretical foundations and underlying conceptual computations are briefly outlined. Based on the underlying quantitative variables, we first restate the variables in terms of algebraic mathematical symbols. For this purpose, we need to specify an obvious assumption, as described in the following.

Let us assume that the data collection is based on a data collection scheduled worksheet. This is denoted by $I_1(N_0)$. The work sheet is an ordered arrangement of data collection process. On the basis of this and the above data arrival/collection arrangement, we associate a data process. We recognize that there is an obvious one-to-one relation between the data collection worksheet, $I_1(N_0)$ and a data arrangement process according to the worksheet. In fact, data collection worksheet, $I_1(N_0)$ and arranged data, $x_1, x_2, x_3, \ldots, x_i, \ldots, x_{N_0}$. In short the sequence, $\{x_i\}_{i=1}N_0$, is defined on $I_1(N_0)$. Finally, in the above description, we define order and size preserving concept coupled with statistical data.

7.5.2 *Discrete-time dynamic model for local sample mean and covariance process*

In this section, we use the idea of moving average to derive a discrete-time dynamic model for the mean and covariance processes, motivated by the state and parameter estimation problems of the continuous-time nonlinear stochastic dynamic model of the energy commodity market network. For this purpose, we need to introduce a few definitions and notations.

Let τ be a finite constant time delay. Here, τ is an unknown random variable that characterize the influence of the past performance history of the state of the dynamic process. It plays an important role in the development of mathematical models of continuous- and discrete-time dynamic processes. Based on the practical nature of the data collection process, it is essential to either transform this time delay into positive integers or design a suitable data collection schedule or discretization process. We define the discretized version of τ as

$$r = \left[\left|\frac{\tau}{\Delta t}\right|\right] + 1. \tag{7.5.1}$$

Definition 7.5.1. Let $x = [x_1, x_2, \ldots, x_n]^T$ be a continuous-time multivariate stochastic process defined on an interval $[t_0 - \tau, T]$ into $\mathfrak{R}^n$, for some $T > 0$. For $t \in [t_0 - \tau, T]$, let $\mathcal{F}_t$ be an increasing sub-sigma algebra of a complete probability space $(\Omega, \mathcal{F}, \mathcal{P})$ for which $x(t)$ is $\mathcal{F}_t$ measurable. Define

$$I(a, b) - \{i \in \mathbb{Z} : a \leq i \leq b, a, b \in \mathbb{Z}\}. \tag{7.5.2}$$

The partition $\mathbb{P}$ is defined as follows:

$$\mathbb{P} = \{t_k : t_k = t_0 + k\Delta t, k \in I(-r, N)\}. \tag{7.5.3}$$

Let $\{x(i)\}_{i=1}^{N_0}$ be a finite sequence corresponding to the stochastic process x and partition $\mathbb{P}$ defined in (7.5.3). We further recall that $x(t_k)$ is $\mathcal{F}_{t_k}$ measurable for $k \in I(-r, N)$. For convenience, we write $x(t_k) \equiv x(k)$.

Without loss of generality, we assume that the real data observation/collection partition schedule $\mathbb{P}$ is defined in (7.5.3). Now, we present the definitions of iterative process and simulation time schedules.

Definition 7.5.2. The iterative process time schedule in relation with the real data collection schedule is defined by

$$IP = \{t_{i-r} : \text{ for } t_i \in \mathbb{P}\}. \tag{7.5.4}$$

The simulation time is based on the order p of the time series model of m_k-local conditional sample mean and covariance processes.

Definition 7.5.3. The simulation process time schedule in relation with the real data observation schedule is defined by

$$SP = \begin{cases} \{t_{i+r} : \text{ for } t_i \in \mathbb{P}\} & \text{if } p \leq r, \\ \{t_{i+p} : \text{ for } t_i \in \mathbb{P}\} & \text{if } p > r. \end{cases} \tag{7.5.5}$$

Definition 7.5.4. For $r \geq 1, k \in I(0, N)$ and each $m_k \in I(2, r + k - 1)$, a partition P_k of the closed interval $[t_{k-m_k}, t_{k-1}]$ is called local at time t_k, and it is defined by

$$P_k := t_{k-m_k} < t_{k-m_k+1} < \cdots < t_{k-1}. \tag{7.5.6}$$

Moreover, P_k is referred to as the m_k-point sub-partition $\mathbb{P}$ in (7.5.3) of the closed sub-interval $[t_{k-m_k}, t_{k-1}]$.

Definition 7.5.5. For each $k \in I(0, N)$ and $m_k \in I(2, r+k-1)$, a local finite sequence at t_k of the size m_k is a restriction of $\{x(t_k)\}_{k=-r}^{N}$ to P_k in (7.5.6). This restriction sequence is defined by

$$S_{m_k,k} := \{x(k-1+l)\}_{l=-m_k+1}^{0}. \tag{7.5.7}$$

As m_k varies from 2 to $r+k-1$, the corresponding respective local sequence $S_{m_k,k}$ at t_k varies from $\{x(t_l)\}_{l=k-2}^{k-1}$ to $\{x(t_l)\}_{l=r+1}^{k-1}$. As a result of this, the sequence defined in (7.5.7) is also called a m_k-local moving sequence.

Remark 7.5.1. Note that for the simulation time schedule $k = q$, the observation size m_k ranges from 2 to $q-1$.

Furthermore, the average corresponding to the local sequence $S_{m_k,k}$ in (7.5.7) is defined by

$$\bar{S}_{m_k,k} = \begin{pmatrix} \bar{S}^1_{m_k,k} \\ \bar{S}^2_{m_k,k} \\ \vdots \\ \bar{S}^n_{m_k,k} \end{pmatrix} = \frac{1}{m_k} \sum_{l=-m_k+1}^{0} x(k-1+l). \tag{7.5.8}$$

The average/mean defined in (7.5.8) is also called the m_k-local average/mean.

For $k \in I(0, N)$, the m_k-local covariance matrix corresponding to the local sequence $S_{m_k,k}$ in (7.5.7) is defined by

$$\sum_{m_k,k} = \begin{pmatrix} S^{1,1}_{m_k,k} & S^{1,2}_{m_k,k} & S^{1,3}_{m_k,k} & \cdots & S^{1,n}_{m_k,k} \\ S^{2,1}_{m_k,k} & S^{2,2}_{m_k,k} & S^{2,3}_{m_k,k} & \cdots & S^{2,n}_{m_k,k} \\ \vdots & \ddots & \ddots & \ddots & \vdots \\ S^{n,1}_{m_k,k} & S^{n,2}_{m_k,k} & S^{n,3}_{m_k,k} & \cdots & S^{n,n}_{m_k,k} \end{pmatrix}, \tag{7.5.9}$$

where $s^{j,l}_{m_k,k} \equiv s^{j,l}_{m_k,k}(x), j, l \in I(1, n)$, is the local sample covariance statistic between x_j and x_l at t_k described by

$$s^{j,l}_{m_k,k} := \begin{cases} \frac{1}{m_k} \sum_{a=-m_k+1}^{0} (x_j(k-1+a) - \bar{S}^j_{m_k,k}) & \\ \quad \times (x_l(k-1+a) - \bar{S}^j_{m_k,k}), & \text{for small } m_k, \\ \frac{1}{m_k-1} \sum_{a=-m_k+1}^{0} (x_j(k-1+a) - \bar{S}^j_{m_k,k}) & \\ \quad \times (x_l(k-1+a) - \bar{S}^j_{m_k,k}), & \text{for large } m_k. \end{cases} \tag{7.5.10}$$

In the following, we derive an interconnected discrete-time local conditional sample average/mean and covariance dynamic process. Denoting $x(k) \equiv x(t_k)$, for $k \in I(1, N)$, we state and prove the following theorem.

Theorem 7.5.1. *Let $\{x(t_k)\}_{k=-r+1}^{N}$ be a conditional random sample of continuous-time stochastic dynamic process with respect to sub-σ $\mathcal{F}_{t_k}$, where t_k belong to partition $\mathbb{P}$ in (7.5.3). Let $\bar{S}_{m_k,k}$ and $\sum_{m_k,k}$ be m_k-local conditional sample average and local conditional sample covariance at t_k, for each k in $I(0, N)$, respectively. Then, an interconnected discrete-time dynamic model of local conditional sample mean and sample covariance statistic is described by*

$$\begin{cases} \bar{S}_{m_{k-p+1},k-p+1} = \frac{m_{k-p}}{m_{k-p+1}} \bar{S}_{m_{k-p},k-p}, \quad \bar{S}_{m_0,0} = \bar{S}_0 \\ \Sigma_{m_k,k} = \begin{cases} \frac{m_{k-1}}{m_k} \left[\sum_{j=1}^{p} \left[\frac{m_{k-j}}{\prod_{l=0}^{j=l} m_{k-l}} \right] \Sigma_{m_{k-j},k-j} \right. \\ \qquad \left. + \frac{m_{k-p}}{\prod_{l=0}^{p=l} m_{k=l}} \bar{S}_{m_{k-p},k-p} \bar{S}^{T}_{m_{k-p},k-p} \right] + \mathcal{E}_{m_{k-l},k-1}, \\ \qquad \text{for small } m_k, m_{k-1} < m_k; \\ \sum_{j=1}^{p} \left[\frac{m_{k-j}-1}{\prod_{l=0}^{j=l} m_{k-l}} \right] \Sigma_{m_{k-j},k-j} \\ \qquad + \frac{m_{k-p}}{\prod_{l=0}^{p=l} m_{k=l}} \bar{S}_{m_{k-p},k-p} \bar{S}^{T}_{m_{k-p},k-p} + \epsilon_{m_{k-l},k-1}, \\ \qquad \text{for large } m_k, m_{k-1} < m_k; \\ \Sigma_{m_j,j} = \Sigma_j, \quad j \in I(-p, 0), \textit{ initial conditions}, \end{cases} \end{cases} \tag{7.5.11}$$

where p is the order of the system and

$$\eta = \begin{pmatrix} \eta^1 \\ \eta^2 \\ \vdots \\ \eta^n \end{pmatrix}, \quad \mathcal{E}_{m_k,k} = \begin{pmatrix} \mathcal{E}^{1,1}_{m_k,k} & \mathcal{E}^{1,2}_{m_k,k} & \mathcal{E}^{1,3}_{m_k,k} & \cdots & \mathcal{E}^{1,n}_{m_k,k} \\ \mathcal{E}^{2,1}_{m_k,k} & \mathcal{E}^{2,2}_{m_k,k} & \mathcal{E}^{2,3}_{m_k,k} & \cdots & \mathcal{E}^{2,n}_{m_k,k} \\ \vdots & \ddots & \ddots & \ddots & \vdots \\ \mathcal{E}^{n,1}_{m_k,k} & \mathcal{E}^{n,1}_{m_k,k} & \mathcal{E}^{n,1}_{m_k,k} & \cdots & \mathcal{E}^{n,n}_{m_k,k} \end{pmatrix},$$

$$\epsilon_{m_k,k} = \begin{pmatrix} \epsilon^{1,1}_{m_k,k} & \epsilon^{1,2}_{m_k,k} & \epsilon^{1,3}_{m_k,k} & \cdots & \epsilon^{1,n}_{m_k,k} \\ \epsilon^{2,1}_{m_k,k} & \epsilon^{2,2}_{m_k,k} & \epsilon^{2,3}_{m_k,k} & \cdots & \epsilon^{2,n}_{m_k,k} \\ \vdots & \ddots & \ddots & \ddots & \vdots \\ \epsilon^{n,1}_{m_k,k} & \epsilon^{n,1}_{m_k,k} & \epsilon^{n,1}_{m_k,k} & \cdots & \epsilon^{n,n}_{m_k,k} \end{pmatrix},$$

$$\begin{cases} \eta_{m_{k-p},k-p} = \frac{1}{m_{k-p+1}}\Big[\sum_{\mathfrak{l}=-m_{k-p+1}+1}^{-m_{k-p}+1} x(k-p+\mathfrak{l}) \\ \quad - x(k-p-m_{k-p}+1) \\ \quad - x(k-p-mk-p) + x(k-p)\Big], \\ \mathcal{E}_{m_{k-l},k-l} = \frac{m_k-1}{m_k}\Big[\sum_{\mathfrak{l}=1}^{p} \frac{x(k-\mathfrak{l})x^T(k-\mathfrak{l})}{\prod_{a=0}^{\mathfrak{l}=1} m_{k-a}} \\ \quad - \sum_{\mathfrak{l}=1}^{p} \frac{x(k-\mathfrak{l}-m_{k-\mathfrak{l}})x^T(k-\mathfrak{l}-m_{k-\mathfrak{l}})}{\prod_{a=0}^{\mathfrak{l}=1} m_{k-a}} \\ \quad - \sum_{\mathfrak{l}=1}^{p} \frac{x(k+1-\mathfrak{l}-m_{k-\mathfrak{l}})x^T(k+1-\mathfrak{l}-m_{k-\mathfrak{l}})}{\prod_{a=0}^{\mathfrak{l}=1} m_{k-a}}\Big] \\ \quad + \frac{m_k-1}{m_k}\left[\sum_{\mathfrak{l}=1}^{p}\left[\frac{\sum_{v=-\mathfrak{l}+2-m_{k-\mathfrak{l}+1}}^{-\mathfrak{l}+2-m_{k-\mathfrak{l}}} x(k-1+v)x^T(k-1+v)}{\prod_{a=0}^{\mathfrak{l}-1} m_{k-a}}\right]\right. \\ \quad \left. + \sum_{\mathfrak{l}=1}^{p}\left[\frac{\sum_{\substack{v,s=-\mathfrak{l}+2-m_{k-\mathfrak{l}+1} \\ v\neq s}}^{-\mathfrak{l}+1} x(k-1+v)x^T(k-1+s)}{\prod_{a=0}^{\mathfrak{l}-1} m_{k-a}}\right]\right] \\ \quad - \frac{1}{m_k}\sum_{\substack{v,s=-m_k+1 \\ v\neq s}}^{0} x(k-1+v)x^T(k-1+s), \\ \epsilon_{m_{k-l},k-l} = \sum_{\mathfrak{l}=1}^{p} \frac{x(k-\mathfrak{l})x^T(k-\mathfrak{l})}{\prod_{a=0}^{\mathfrak{l}-1} m_{k-a}} - \sum_{\mathfrak{l}=1}^{p} \frac{x(k-\mathfrak{l}-m_{k-\mathfrak{l}})x^T(k-\mathfrak{l}-m_{k-\mathfrak{l}})}{\prod_{a=0}^{\mathfrak{l}-1} m_{k-a}} \\ \quad - \sum_{\mathfrak{l}=1}^{p} \frac{x(k-\mathfrak{l}+1-m_{k-\mathfrak{l}})x^T(k-\mathfrak{l}+1-m_{k-\mathfrak{l}})}{\prod_{a=0}^{\mathfrak{l}-1} m_{k-a}} \\ \quad + \sum_{\mathfrak{l}=1}^{p}\left[\frac{\sum_{v=-\mathfrak{l}+2-m_{k-\mathfrak{l}+1}}^{-\mathfrak{l}+2-m_{k-\mathfrak{l}}} x(k-1+v)x^T(k-1+v)}{\prod_{a=0}^{\mathfrak{l}-1} m_{k-a}}\right] \\ \quad + \sum_{\mathfrak{l}=1}^{p}\left[\frac{\sum_{\substack{v,s=-\mathfrak{l}+2-m_{k-\mathfrak{l}+1} \\ v\neq s}}^{-\mathfrak{l}+1} x(k-1+v)x^T(k-1+s)}{\prod_{a=0}^{\mathfrak{l}-1} m_{k-a}}\right] \\ \quad - \frac{1}{m_k-1}\sum_{\substack{v,s=-m_k+1 \\ v\neq s}}^{0} x(k-1+v)x^T(k-1+s). \end{cases} \tag{7.5.12}$$

Proof. First, we prove Theorem 7.5.1 for small m_k.

For $m_{k-1} < m_k$,

$$\bar{S}^j_{m_k,k} = \frac{1}{m_k}\sum_{\mathfrak{l}=1-m_k}^{0} x_j(k-1+\mathfrak{l})$$

$$= \frac{1}{m_k}\left[\sum_{\mathfrak{l}=1-m_k}^{1-m_{k-1}} x_j(k-1+\mathfrak{l}) - x_j(k-m_{k-1})\right.$$

$$\left. - x_j(k-1-m_{k-1}) + \sum_{\mathfrak{l}=-m_{k-1}}^{-\mathfrak{l}} x_j(k-1+\mathfrak{l}) + x_j(k-1)\right]$$

$$= \frac{m_{k-1}}{m_k}\bar{S}^j_{m_{k-1},k-1} + \frac{1}{m_k}\left[\sum_{\mathfrak{l}=1-m_k}^{1-m_{k-1}} x_j(k-1+\mathfrak{l})\right.$$

$$\left. - x_j(k-m_{k-1}) - x_j(k-1-m_{k-1}) + x_j(k-1)\right],$$

$$S^{i,j}_{m_k,k} = \frac{1}{m_k}\left[\sum_{\mathfrak{l}=-m_k+1}^{0} (x_i(k-1+\mathfrak{l}))(x_j(k-1+\mathfrak{l}))\right.$$

$$\left. - \frac{1}{m_k}\left(\sum_{a=-m_k+1}^{0} x_i(k-1+a)\right)\left(\sum_{a=-m_k+1}^{0} x_j(k-1+a)\right)\right]$$

$$= \frac{1}{m_k}\left[\sum_{\mathfrak{l}=-m_k+1}^{-m_{k-1}-1} (x_i(k-1+\mathfrak{l}))(x_j(k-1+\mathfrak{l}))\right.$$

$$+ \sum_{\mathfrak{l}=-m_{k-1}}^{-\mathfrak{l}} (x_i(k-1+\mathfrak{l}))(x_j(k-1+\mathfrak{l}))$$

$$\left. + (x_i(k-1))(x_j(k-1))\right]$$

$$= \frac{1}{m_k}\left[\sum_{\mathfrak{l}=-m_{k-1}}^{-1} (x_i(k-1+\mathfrak{l}))(x_j(k-1+\mathfrak{l}))\right.$$

$$- \frac{1}{m_{k-1}}\left(\sum_{a=-m_{k-1}}^{-1} x_i(k-1+a)\right)\left(\sum_{a=-m_{k-1}}^{-1} x_j(k-1+a)\right)$$

$$+ \frac{1}{m_{k-1}} \left(\sum_{a=-m_{k-1}}^{-1} x_i(k-1+a) \right) \left(\sum_{a=-m_{k-1}}^{-1} x_j(k-1+a) \right)$$

$$+ \sum_{\mathfrak{l}=-m_k+1}^{-m_{k-1}+1} (x_i(k-1+\mathfrak{l}))(x_j(k-1+\mathfrak{l}))$$

$$- x_i(k-1-m_{k-1})x_j(k-1-m_{k-1})$$

$$\left. - x_i(k-m_{k-1})x_j(k-m_{k-1}) + x_i(k-1)x_j(k-1) \right]$$

$$- \frac{1}{m_k^2} \sum_{a=-m_k+1}^{0} x_i(k-1+a) \sum_{a=-m_k+1}^{0} x_j(k-1+a)$$

$$= \frac{m_{k-1}}{m_k} s^{i,j}_{m_{k-1},k-1} + \frac{m_{k-1}}{m_k} \bar{S}^i_{m_{k-1},k-1} \bar{S}^j_{m_{k-1},k-1}$$

$$+ \frac{\sum_{\mathfrak{l}=-m_k+1}^{-m_{k-1}+1} x_i(k-1+\mathfrak{l})x_j(k-1+\mathfrak{l})}{m_k}$$

$$+ \frac{\begin{array}{c} x_i(k-1)x_j(k-1) - x_i(k-1-m_{k-1})x_j(k-1-m_{k-1}) \\ - x_i(k-m_{k-1})x_j(k-m_{k-1}) \end{array}}{m_k}$$

$$- \bar{S}^i_{m_k,k} \bar{S}^j_{m_k,k}.$$

Hence,

$$s^{i,j}_{m_k,k} = \frac{m_{k-1}}{m_k} s^{i,j}_{m_{k-1},k-1} + \frac{m_{k-1}}{m_k} \bar{S}^i_{m_{k-1},k-1} \bar{S}^j_{m_{k-1},k-1} - \bar{S}^i_{m_k,k} \bar{S}^j_{m_k,k}$$

$$+ \frac{\sum_{\mathfrak{l}=-m_k+1}^{-m_{k-1}+1} x_i(k-1+\mathfrak{l})x_j(k-1+\mathfrak{l})}{m_k}$$

$$+ \frac{\begin{array}{c} x_i(k-1)x_j(k-1) - x_i(k-1-m_{k-1})x_j(k-1-m_{k-1}) \\ - x_i(k-m_{k-1})x_j(k-m_{k-1}) \end{array}}{m_k}. \tag{7.5.13}$$

Next, we find an expression connecting $s^{i,j}_{m_{k-1},k-1}, \bar{S}^i_{m_{k-1},k-1}$ $\bar{S}^j_{m_{k-1},k-1}$ and $\bar{S}^i_{m_k,k}\bar{S}^j_{m_k,k}$. By definition and simplification,

$$
\begin{aligned}
m_k^2\bar{S}^i_{m_k,k}\bar{S}^j_{m_k,k} &= \sum_{\mathfrak{l}=-m_k+1}^{0} x_i(k-1+\mathfrak{l}) \sum_{\mathfrak{l}=-m_k+1}^{0} x_j(k-1+\mathfrak{l}) \\
&= \sum_{\mathfrak{l}=-m_k+1}^{0} x_i(k-1+\mathfrak{l})x_j(k-1+\mathfrak{l}) \\
&\quad + \sum_{\substack{\mathfrak{l},s=-m_k+1 \\ \mathfrak{l}\neq s}}^{0} x_i(k-1+\mathfrak{l})x_j(k-1+s) \\
&= (m_{k-1})s^{i,j}_{m_{k-1},k-1} + m_{k-1}\bar{S}^i_{m_{k-1},k-1}\bar{S}^j_{m_{k-1},k-1} \\
&\quad + x_i(k-1)x_j(k-1) \\
&\quad - x_i(k-1-m_{k-1})x_j(k-1-m_{k-1}) \\
&\quad - x_i(k-m_{k-1})x_j(k-m_{k-1}) \\
&\quad + \sum_{\mathfrak{l}=-m_k+1}^{-m_{k-1}+1} x_i(k-1+\mathfrak{l})x_j(k-1+\mathfrak{l}) \\
&\quad + \sum_{\substack{\mathfrak{l},s=-m_k+1 \\ \mathfrak{l}\neq s}}^{0} x_i(k-1+\mathfrak{l})x_j(k-1+s). \qquad (7.5.14)
\end{aligned}
$$

Substituting (7.5.14) into (7.5.13), we have

$$
\begin{aligned}
s^{i,j}_{m_k,k} &= \frac{m_{k-1}}{m_k}\left[\frac{m_{k-1}}{m_k}s^{i,j}_{m_{k-1},k-1} + \frac{m_{k-1}}{m_k}\bar{S}^i_{m_{k-1},k-1}\bar{S}^j_{m_{k-1},k-1}\right] \\
&\quad + \frac{m_{k-1}}{m_k}\left[\frac{\begin{array}{l} x_i(k-1)x_j(k-1) - x_i(k-1-m_{k-1}) \\ \quad \times x_j(k-1-m_{k-1}) - x_i(k-m_{k-1})x_j(k-m_{k-1}) \end{array}}{m_k}\right.
\end{aligned}
$$

$$
\left. + \frac{\sum_{l=-m_k+1}^{-m_{k-1}+1} x_i(k-1+l)x_j(k-1+l)}{m_k} \right]
$$

$$
- \frac{\sum_{\substack{l,s=-m_k+1 \\ l \neq s}}^{0} x_i(k-1+l)x_j(k-1+s)}{m_k^2}. \tag{7.5.15}
$$

Likewise, using (7.5.14),

$$
\begin{aligned}
& m_{k-1}^2 \bar{S}^i_{m_{k-1},k-1} \bar{S}^j_{m_{k-1},k-1} \\
&\quad = (m_{k-2}) s^{i,j}_{m_{k-2},k-2} + (m_{k-2}) \bar{S}^i_{m_{k-2},k-2} \bar{S}^j_{m_{k-2},k-2} \\
&\qquad + x_i(k-2)x_j(k-2) \\
&\qquad - x_i(k-2-m_{k-2})x_j(k-2-m_{k-2}) \\
&\qquad - x_i(k-1-m_{k-2})x_j(k-1-m_{k-2}) \\
&\qquad + \sum_{l=-m_{k-1}}^{-m_{k-2}} x_i(k-1+l)x_j(k-1+l) \\
&\qquad + \sum_{\substack{l,s=-m_{k-1} \\ l \neq s}}^{-1} x_i(k-1+l)x_j(k-1+s).
\end{aligned}
$$

Also,

$$
\begin{aligned}
& m_{k-2}^2 \bar{S}^i_{m_{k-2},k-2} \bar{S}^j_{m_{k-2},k-2} \\
&\quad = (m_{k-3}) s^{i,j}_{m_{k-3},k-3} + (m_{k-3}) \bar{S}^i_{m_{k-3},k-3} \bar{S}^j_{m_{k-3},k-3} \\
&\qquad + x_i(k-3)x_j(k-3) \\
&\qquad - x_i(k-3-m_{k-3})x_j(k-3-m_{k-3}) \\
&\qquad - x_i(k-2-m_{k-3})x_j(k-2-m_{k-3})
\end{aligned}
$$

$$+\sum_{\mathfrak{l}=-m_{k-2}}^{-m_{k-3}} x_i(k-1+\mathfrak{l})x_j(k-1+\mathfrak{l})$$

$$+\sum_{\substack{l,s=-m_{k-2}\\ l\neq s}}^{-2} x_i(k-1+l)x_j(k-1+s).$$

Continuing in this sense and substituting $\bar{S}^i_{m_{k-i},k-i}, \bar{S}^j_{m_{k-i},k-i}$, $i = 2,\ldots,p-1$, into $\bar{S}^i_{m_{k-1},k-1}, \bar{S}^j_{m_{k-j},k-1}$, we have

$$\begin{aligned}
&(m_{k-1})\bar{S}^i_{m_{k-1},k-1}\bar{S}^j_{m_{k-1},k-1}\\
&\quad=\sum_{i=2}^{p}\left[\frac{m_{k-i}}{\prod_{a=1}^{\mathfrak{l}-1} m_{k-a}}\right] s^{i,j}_{m_{k-\mathfrak{l}},k-\mathfrak{l}} + \frac{m_{k-p}}{\prod_{a=1}^{p-1} m_{k-a}}\bar{S}^i_{m_{k-p},k-p}\bar{S}^j_{m_{k-p},k-p}\\
&\qquad+\sum_{\mathfrak{l}=2}^{p}\frac{x_i(k-\mathfrak{l})x_j(k-\mathfrak{l})}{\prod_{a=1}^{\mathfrak{l}-1} m_{k-a}}\\
&\qquad-\sum_{\mathfrak{l}=2}^{p}\frac{x_i(k-\mathfrak{l}-m_{k-1})x_j(k-\mathfrak{l}-m_{k-\mathfrak{l}})}{\prod_{a=1}^{p-1} m_{k-a}}\\
&\qquad-\sum_{\mathfrak{l}=2}^{p}\frac{x_i(k+1-\mathfrak{l}-m_{k-1})x_j(k+1-\mathfrak{l}-m_{k-\mathfrak{l}})}{\prod_{a=1}^{p-1} m_{k-a}}\\
&\qquad+\sum_{\mathfrak{l}=2}^{p}\left[\frac{\sum_{l=-\mathfrak{l}+2-m_{k-\mathfrak{l}}+1}^{-\mathfrak{l}+2-m_{k-\mathfrak{l}}} x_i(k-1+\mathfrak{l})x_j(k-1+l)}{\prod_{a=1}^{\mathfrak{l}-1} m_{k-a}}\right]\\
&\qquad+\sum_{\mathfrak{l}=2}^{p}\left[\frac{\sum_{\substack{l,s=-\mathfrak{l}+2-m_{k-\mathfrak{l}}+1\\ l\neq s}}^{-\mathfrak{l}+1} x_i(k-1+\mathfrak{l})x_j(k-1+s)}{\prod_{a=1}^{\mathfrak{l}-1} m_{k-a}}\right].
\end{aligned}\tag{7.5.16}$$

Finally, the results follow by substituting (7.5.16) into (7.5.15). This completes the proof.

Proof. Next, we prove Theorem 7.5.1 for large m_k.
For large m_k,

$$
\begin{aligned}
s^{i,j}_{m_k,k} &= \frac{1}{m_k - 1}\left[\sum_{\mathfrak{l}=-m_k+1}^{0}(x_i(k-1+\mathfrak{l}))(x_j(k-1+\mathfrak{l}))\right.\\
&\quad \left. - \frac{1}{m_k}\left(\sum_{a=-m_k+1}^{0} x_i(k-1+a)\right)\left(\sum_{a=-m_k+1}^{0} x_j(k-1+a)\right)\right]\\
&= \frac{1}{m_k - 1}\left[\sum_{\mathfrak{l}=-m_k+1}^{-1}(x_i(k-1+\mathfrak{l}))(x_j(k-1+\mathfrak{l}))\right.\\
&\quad - \frac{1}{m_{k-1}}\sum_{\mathfrak{l}=-m_{k-1}}^{-1} x_i(k-1+\mathfrak{l})\sum_{\mathfrak{l}=-m_{k-1}}^{-1} x_j(k-1+\mathfrak{l})\\
&\quad \left. + \frac{1}{m_{k-1}}\sum_{\mathfrak{l}=-m_{k-1}}^{-1} x_i(k-1+\mathfrak{l})\sum_{\mathfrak{l}=-m_{k-1}}^{-1} x_j(k-1+\mathfrak{l})\right]\\
&\quad + \frac{1}{m_k - 1}[x_i(k-1)x_j(k-1) - x_i(k-1-m_{k-1})\\
&\quad \times x_j(k-1m_{k-1}) - x_i(k-m_{k-1})x_j(k-m_{k-1})]\\
&\quad + \frac{1}{m_k - 1}\left[\sum_{\mathfrak{l}=-m_k+1}^{-m_{k-1}+1} x_i(k-1+\mathfrak{l})\sum_{\mathfrak{l}=-m_k+1}^{-m_{k-1}+1} x_j(k-1+\mathfrak{l})\right.\\
&\quad \left. - \frac{1}{m_k}\sum_{\mathfrak{l}=-m_k+1}^{0} x_i(k-1+\mathfrak{l})\sum_{\mathfrak{l}=-m_k+1}^{0} x_j(k-1+\mathfrak{l})\right]\\
&= \frac{m_{k-1}-1}{m_k - 1}s^{i,j}_{m_{k-1},k-1} + \frac{m_{k-1}-1}{m_k - 1}\bar{S}^i_{m_{k-1},k-1}\bar{S}^j_{m_{k-1},k-1}\\
&\quad - \frac{m_k}{m_k - 1}\bar{S}^i_{m_k,k}\bar{S}^j_{m_k,k}
\end{aligned}
$$

$$+\frac{\begin{array}{c}x_i(k-1)x_j(k-1)-x_i(k-1-m_{k-1})\\ \times x_j(k-1-m_{k-1})-x_i(k-m_{k-1})x_j(k-m_{k-1})\end{array}}{m_k-1}$$

$$+\frac{\sum_{\mathfrak{l}=-m_k+1}^{-m_{k-1}+1}x_i(k-1+\mathfrak{l})x_j(k-1+\mathfrak{l})}{m_k-1}.$$

Hence,

$$\begin{aligned}s_{m,k}^{i,j}&=\frac{m_{k-1}-1}{m_k-1}s_{m_{k-1},k-1}^{i,j}+\frac{m_{k-1}}{m_k-1}\bar{S}_{m_{k-1},k-1}^{i}\bar{S}_{m_{k-1},k-1}^{j}\\&\quad\times\frac{m_k}{m_k-1}\bar{S}_{m_k,k}^{i}\bar{S}_{m_k,k}^{j}\\&\quad+\frac{\begin{array}{c}x_i(k-1)x_j(k-1)-x_i(k-1-m_{k-1})\\ \times xj(k-1-m_{k-1})-x_i(k-m_{k-1})x_j(k-m_{k-1})\end{array}}{m_k-1}\\&\quad+\frac{\sum_{\mathfrak{l}=-m_k+1}^{-m_{k-1}+1}x_i(k-1+\mathfrak{l})x_j(k-1+\mathfrak{l})}{m_{k-1}}.\end{aligned}\tag{7.5.17}$$

Next, we find an expression connecting $\bar{S}_{m_k,k}^{i}\bar{S}_{m_k,k}^{j}$, $\bar{S}_{m_{k-1},k-1}^{i}$, $\bar{S}_{m_{k-1},k-1}^{j}$ and $s_{m_k-1,k-1}^{i,j}$. By definition and simplification,

$$\begin{aligned}m_k^2\bar{S}_{m_k,k}^{i}\bar{S}_{m_k,k}^{j}&=\sum_{\mathfrak{l}=-m_k+1}^{0}x_i(k-1+\mathfrak{l})\sum_{\mathfrak{l}=-m_k+1}^{0}x_j(k-1+\mathfrak{l})\\&=\sum_{\mathfrak{l}=-m_k+1}^{0}x_i(k-1+\mathfrak{l})x_j(k-1+\mathfrak{l})\\&\quad+\sum_{\substack{l,s=-m_k+1\\ l\neq s}}^{0}x_i(k-1+\mathfrak{l})x_j(k-1+s)\\&=(m_{k-1}-1)s_{m_{k-1},k-1}^{i,j}+m_{k-1}\bar{S}_{m_{k-1},k-1}^{i}\bar{S}_{m_{k-1},k-1}^{j}\\&\quad+x_i(k-1)x_j(k-1)\\&\quad-x_i(k-1-m_{k-1})x_j(k-1-m_{k-1})\\&\quad-x_i(k-m_{k-1}x_j(k-1))\end{aligned}$$

$$+ \sum_{\mathfrak{l}=-m_k+1}^{-m_{k-1}+1} x_i(k-1+\mathfrak{l})x_j(k-1+\mathfrak{l})$$

$$+ \sum_{\substack{l,s=-m_k+1 \\ l \neq s}}^{0} x_i(k-1+\mathfrak{l})x_j(k-1+s). \qquad (7.5.18)$$

Substituting (7.5.18) into (7.5.17), we have

$$s_{m,k}^{i,j} = \frac{m_{k-1}-1}{m_k} s_{m_{k-1},k-1}^{i,j} + \frac{m_{k-1}}{m_k} \bar{S}_{m_{k-1},k-1}^{i} \bar{S}_{m_{k-1},k-1}^{j}$$

$$+ \frac{\begin{array}{c} x_i(k-1)x_j(k-1) - x_i(k-1-m_{k-1}) \\ \times x_j(k-1-m_{k-1}) - x_i(k-m_{k-1})x_j(k-m_{k-1}) \end{array}}{m_k}$$

$$+ \frac{\sum_{\mathfrak{l}=-m_k+1}^{-m_{k-1}+1} x_i(k-1+\mathfrak{l})x_j(k-1+\mathfrak{l})}{m_k}$$

$$- \frac{\sum_{\substack{l,s=-m_k+1 \\ l \neq s}}^{0} x_i(k-1+\mathfrak{l})x_j(k-1+s)}{m_k^2}. \qquad (7.5.19)$$

Likewise,

$$m_{k-1}^2 \bar{S}_{m_{k-1},k-1}^{i} \bar{S}_{m_{k-1},k-1}^{j}$$

$$= (m_{k-2}) s_{m_{k-2},k-2}^{i,j} + (m_{k-2}) \bar{S}_{m_{k-2},k-2}^{i} \bar{S}_{m_{k-2},k-2}^{j}$$

$$+ F^{-1} x_i(k-1) F^{-1} x_j(k-1)$$

$$- x_i(k-2-m_{k-2})x_j(k-2-m_{k-2})$$

$$- x_i(k-1-m_{k-2})x_j(k-1-m_{k-2})$$

$$+ \sum_{\mathfrak{l}=-m_{k-1}}^{-m_{k-2}} x_i(k-1+\mathfrak{l})x_j(k-1+\mathfrak{l})$$

$$+ \sum_{\substack{l,s=-m_{k-1} \\ l \neq s}}^{-1} x_i(k-1+l)x_j(k-1+s),$$

$$
\begin{aligned}
&m_{k-2}^2 \bar{S}^i_{m_{k-2},k-2} \bar{S}^j_{m_{k-2},k-2} \\
&\quad = (m_{k-3}) s^{i,j}_{m_{k-3},k-3} + (m_{k-3}) \bar{S}^i_{m_{k-3},k-3} \bar{S}^j_{m_{k-3},k-3} \\
&\qquad - x_i(k-3-m_{k-3}) x_j(k-3-m_{k-3}) \\
&\qquad - x_i(k-2-m_{k-3}) x_j(k-2-m_{k-3}) \\
&\qquad + \sum_{l=-m_{k-2}-1}^{-m_{k-3}-1} x_i(k-1+l) x_j(k-1+l) \\
&\qquad + \sum_{\substack{l,s=-m_{k-2} \\ l \neq s}}^{-2} x_i(k-1+l) x_j(k-1+s).
\end{aligned}
$$

Continuing in this sense and substituting $\bar{S}_{m_{k-i},k-i}$, $i = 2, \ldots, p-1$ into $\bar{S}_{m_{k-1},k-1}$, we have

$$
\begin{aligned}
&(m_{k-1}) \bar{S}^{i,j}_{m_{k-1},k-1} \\
&= \sum_{l=2}^{p} \left[\frac{m_{k-i}-1}{\prod_{m_{k-a}}^{l-1}} \right] s^{i,j}_{m_{k-i},k-i} + \frac{m_{k-p}}{\prod_{m_{k-a}}^{p-1}} \bar{S}^i_{m_{k-p},k-p} \bar{S}^j_{m_{k-p},k-p} \\
&\quad + \sum_{l=2}^{p} \frac{F^{-i+1} x_i(k-1) F^{-i+1} x_j(k-1)}{\prod_{a=1}^{l-1} m_{k-a}} \\
&\quad + \sum_{l=2}^{p} \frac{F^{-i+1-m_{k-i}} x_i(k-1) F^{-i+1-m_{k-i}} x_j(k-1)}{\prod_{a=1}^{l-1} m_{k-a}} \\
&\quad - \sum_{l=2}^{p} \frac{F^{-i+2-m_{k-i}} x_i(k-1) F^{-i+2-m_{k-i}} x_j(k-1)}{\prod_{a=1}^{l-1} m_{k-a}} \\
&\quad + \sum_{l=2}^{p} \left[\frac{\sum_{l=-i+2-m_{k-i+1}}^{-i+2-m_{k-i}} x_i(k-1+l) F^l x_j(k-1)}{\prod_{a=1}^{l-1} m_{k-a}} \right] \\
&\quad + \sum_{l=2}^{p} \left[\frac{\sum_{\substack{l,s=-i+2-m_{k-i+1} \\ l \neq s}}^{-i+1} x_i(k-1+l) x_j(k-1+s)}{\prod_{a=1}^{l-1} m_{k-a}} \right]. \qquad (7.5.20)
\end{aligned}
$$

Finally, the result follows by substituting (7.5.20) into (7.5.19). This completes the proof.

Remark 7.5.2. For each $k \in I_0(N), n = 1, p = 2$ and small m_k, the interconnected system (7.5.11) reduces to the following special case:

$$X(k;2) = A(k, X(k-1;2);2)X(k-1;2) + e(k;2), \tag{7.5.21}$$

where $X(k;2), A(k;2)$ and $e(k;2)$ are defined by

$$X(k;2) = \begin{pmatrix} X_1(k;2) \\ X_2(k;2) \end{pmatrix}, \quad X_1(k;2) = \bar{S}_{m_{k-1},k-1},$$

$$X_2(k;2) = \begin{pmatrix} s^2_{m_{k-1},k-1} \\ s^2_{m_k,k} \end{pmatrix}, \quad A(k;2) = \begin{pmatrix} A_{11}(k;2) & A_{12}(k;2) \\ A_{21}(k;2) & A_{22}(k;2) \end{pmatrix},$$

$$A_{11}(k;2) = \frac{m_{k-2}}{m_k - 1}, \quad A_{12}(k;2) = (0\ 0),$$

$$A_{21}(k;2) = \begin{pmatrix} 0 \\ \frac{(m_k-1)m_{k-2}}{m_k^2 m_{k-1} \bar{S}_{m_{k-2},k-2}} \end{pmatrix},$$

$$A_{22}(k;2) = \begin{pmatrix} 0 & 1 \\ \frac{(m_k-1)m_{k-2}}{m_k^2 m_{k-1}} & \frac{(m_k-1)m_{k-2}}{m_k^2 m_{k-1}} \end{pmatrix}, \quad e(k;2) = \begin{pmatrix} e_1(k;2) \\ e_2(k;2) \end{pmatrix},$$

$$e_1(k;2) = \eta_{m_{k-2},k-2}, \quad e_2(k;2) = \begin{pmatrix} 0 \\ \mathcal{E}_{m_{k-1},k-1} \end{pmatrix} \quad \text{and}$$

$$\begin{cases} \eta_{m_{k-2},k-2} = \frac{1}{m_k}\Big[\sum_{i=-m_{k-1}+1}^{-m_{k-2}+1} F^i x_{k-2} F^{-m_{k-2}+1} x_{k-2} \\ \qquad\qquad - F^{-m_{k-2}} x_{k-2} + F^0 x_{k-2}\Big], \\ \mathcal{E}_{m_{k-1},k-1} = \frac{m_k-1}{m_k}\Big[\frac{(F^0 x_{k-1})^2 - (F^{-m_{k-1}} x_{k-1})^2 - (F^{1-m_{k-1}} x_{k-1})^2}{m_k} \\ \qquad\qquad + \frac{(F^{-1} x_{k-1})^2 - (F^{1-m_{k-2}} x_{k-1})^2 - (F^{-m_{k-2}} x_{k-1})^2}{m_k m_{k-1}}\Big] \\ \qquad\qquad + \frac{m_k-1}{m_k}\Bigg[\frac{\sum_{i=-m_{k-1}}^{-m_{k-2}} (F^i x_{k-1})^2}{m_k m_{k-1}} + \frac{\sum_{\substack{i,j=-m_{k-1} \\ i\neq j}}^{-1} F^i x_{k-1} F^j x_{k-1}}{m_k m_{k-1}} \\ \qquad\qquad + \frac{\sum_{i=1-m_k}^{1-m_{k-1}} (F^i x_{k-1})^2}{m_k}\Bigg] \\ \qquad\qquad - \frac{\sum_{\substack{i,j=1-m_k \\ i\neq j}}^{0} F^i x_{k-1} F^j x_{k-1}}{m_k^2}. \end{cases}$$

7.5.3 *Existence of solution process*

To prove the existence of a solution for system (7.5.11), we need the following lemma.

Lemma 7.5.1. *Define* $D(n) = \sum_{j=1}^{n} jj!$. *For* $n \in \mathbb{Z}, (\mathbb{Z}^+ = \{1, 2, 3, \ldots\}), D(n) = (n+1)! - 1$.

Define $\phi_{j,k} = \frac{m_k - 1}{m_k} \frac{m_{k-j}}{\prod_{l=0}^{j-1} m_{k-1}}$, *for* $j = I(1,p), k \in I(0,N)$. *We show that for some positive* p, *the sum* $\sum_{j=1}^{p} \phi_{j,k} < 1$, *for all* k, *such that* $m_{k-1} < m_k$.

Lemma 7.5.2. *Suppose* $m_{k-1} \le m_k$. *For all* $k \in I(0,N)$, *we have* $\phi_{1,k} + \phi_{2,k} < 1$.

Proof. Suppose $m_{k-1} \le m_k$. For all $k \in I(0,N)$, we have

$$\phi_{1,k} + \phi_{2,k} = \frac{m_k - 1}{m_k^2} \left[m_{k-1} + \frac{m_{k-2}}{m_{k-1}} \right]$$

$$\le \frac{m_k - 1}{m_k^2} [m_{k-1} + 1] \le \frac{m_k^2 - 1}{m_k^2} < 1.$$

Lemma 7.5.3. *Suppose* $m_{k-1} \le m_k$ *such that* $m_k = m_{k-1} + 1$. *For all* $k \in I(0,N)$, *we have* $\sum_{j=1}^{p} \phi_{j,k} < 1$.

Proof. Suppose $m_k = m_{k-1} + 1$. We can write $m_{k-p} = m_{k-1} - (p-1)$. For all $k \in I(0,N)$, we have

$$\sum_{j=1}^{p} \phi_{j,k} = \frac{m_k - 1}{m_k} \sum_{j=1}^{p} \frac{m_{k-j}}{\prod_{l=0}^{j-1} m_{k-l}}$$

$$= \frac{m_{k-1}}{(m_{k-1}+1)^2} \left[m_{k-1} + \frac{m_{k-1} - 1}{m_{k-1}} \right.$$

$$\left. + \frac{m_{k-1} - 2}{m_{k-1}(m_{k-1} - 1)} + \cdots + \frac{m_{k-1} - (p-1)}{\prod_{l=0}^{p-2} (m_{k-1} - l)} \right]$$

$$= \frac{m_{k-1}}{(m_{k-1}+1)^2} \frac{1}{m_{k-1}!} \left[\sum_{j=m_{k-1}-(p-1)}^{m_{k-1}} jj! \right]$$

$$= \frac{m_{k-1}}{(m_{k-1}+1)^2} \frac{1}{m_{k-1}!} [D(m_{k-1}) - D(m_{k-1} - p)]$$

$$\leq \frac{m_{k-1}}{(m_{k-1}+1)^2} \frac{(m_{k-1}+1)!}{m_{k-1}!} = \frac{m_{k-1}}{m_{k-1}+1} < 1.$$

The following theorem shows the existence of a solution for the system of equations (7.5.11).

Theorem 7.5.2. *Suppose $m_k = m_{k-1} + 1$ and $\mathbb{E}[|x|] < \infty$. Then, $\{\bar{S}_{m_k,k}\}$ and $\{\Sigma_{m_k,k}\}$ described in (7.5.11) have a unique solution.*

Proof. The proof follows from Lemmas 7.5.1 and 7.5.3.

7.5.4 *Numerical application*

In this section, we briefly compare the applications of dynamic model (7.5.11) and GARCH in the context of four energy commodities: daily Henry Hub natural gas, daily crude oil, daily coal and weekly ethanol data. We compare the estimates $s^2_{\hat{m}_k,k}$ with those derived from the usage of a GARCH(1, 1) model which is defined by

$$z_t|\mathcal{F}_{t-1} \sim \mathcal{N}(0, h_t),$$
$$h_t = \alpha_0 + \alpha_1 h_{t-1} + \beta_1 z^2_{t-1}, \quad \alpha_0 > 0, \alpha_1, \beta_1 \geq 0. \tag{7.5.22}$$

The parameters α_0, α_1 and β_1 of the GARCH(1, 1) conditional variance model (7.5.22) for the four commodities of natural gas, crude oil, coal and ethanol are estimated. The estimates of the parameters are given in Table 7.2.

Table 7.2: Parameter estimates for GARCH(1, 1) model (7.5.22).

Data set	α_0	α_1	β_1
Natural gas	6.863×10^{-5}	0.853	0.112
Crude oil	9.622×10^{-5}	0.917	0.069
Coal	3.023×10^{-5}	0.903	0.081
Ethanol	4.152×10^{-5}	0.815	0.019

7.6 Role and Scope of Hybrid Dynamic Processes in Conceptual Computational Large-Scale Data Analysis

We associate systems (6.1.1), (6.1.2) and (6.1.3) with the following hybrid systems, as described subsequently. A dynamic hybrid system corresponding to (6.1.1) is described by

$$\begin{aligned} dx &= \mathbf{A}(t,x)\cdot d\boldsymbol{\xi}, t \neq t_k, x(t_k) = z, \\ \Delta y(t_k) &= \mathbf{B}(t_k,y)\cdot I(\boldsymbol{\xi})(t_k,t_{k+1}), y(t_k) = z, \\ y(t_0) &= x(t_0) = x_0, k \in I(0,N), \end{aligned} \tag{7.6.1}$$

where $\{t_{N0}, t_{N1}, \ldots, t_{Nk}, \ldots, t_{NN}\}$ is a partition of $J = [t_0, t_0 + \alpha]$ such that $t_0 = t_{N0} < t_{N1} < \cdots < t_{NN} = t_{NN} = t_0 + \alpha; t_k = t_{Nk}$ and $h^k = h_N^k = t_{k+1} - t_k = \Delta t_k$ for $k \in I(0,N)$; $z \in R^n$. Let us denote $J_k = [t_k, t_{k+1}]$.

The stochastic large-scale hybrid system (7.6.1) is decomposed into interconnected subsystems of the following form:

$$\begin{aligned} dx_i &= [\mathbf{a}^i(t,x_i) + \mathbf{c}^i(t,x)]\cdot d\boldsymbol{\xi}, t \neq t_k, x_i(t_k) = z_i, \\ \Delta y_i(t_k) &= [\mathbf{b}^i(t_k,y_i) + \mathbf{d}^i(t_k,y)]\cdot I(\boldsymbol{\xi})(t_k,t_{k+1}), y_i(t_k) = z_i, \\ y_i(t_0) &= x_i(t_0) = x_i^0, k \in I(0,N) \quad \text{and} \quad i \in I(1,p). \end{aligned} \tag{7.6.2}$$

The isolated hybrid subsystems corresponding to (7.6.2) are described by

$$\begin{aligned} dx_i &= \mathbf{a}^i(t,x_i)\cdot d\boldsymbol{\xi}, t \neq t_k, x_i(t_k) = z_i, \\ \Delta y_i(t_k) &= \mathbf{b}^i(t_k,y_i)\cdot I(\boldsymbol{\xi})(t_k,t_{k+1}), y_i(t_k) = z_i, \\ y_i(t_0) &= x_i(t_0) = x_i^0, k \in I(0,N) \quad \text{and} \quad i \in I(1,p). \end{aligned} \tag{7.6.3}$$

Let $x(t) = (x_1^T(t), x_2^T(t), \ldots, x_i^T(t), \ldots, x_p^T(t))^T = x(t, t_0, x_0)$ and $y(t_k) = (y_1^T(t_k), y_2^T(t_k), \ldots, y_i^T(t_k), \ldots, y_p^T(t_k))^T$ be the solution processes of the continuous-time and discrete parts of the large-scale system (6.1.2) through (t_0, x^0), respectively. Let $\hat{x}_{ix_i^0} = \hat{x}_i(t, t_0, x_i^0)$ and $\hat{y}_i(t_k) = \hat{y}_i(t_k, t_0, x_i^0)$ be the solution processes of the continuous-time and discrete parts of the ith isolated system (6.1.3) through (t_0, x_i^0), respectively. Furthermore, for $k \in I(1,N)$, let $x^k(t) = x(t, t_k, z)$ and $y^k(t_k) = y(t_k, t_{k-1}, z)$ be the solution processes of the large-scale hybrid system (7.6.2) through $(t_{k-1}, z), t \in J_{k-1}$, and let $\{y^k\}_0^N$ be an R^n-valued random process such that for $0 \leq j \leq k-1, y^j$ is F_{t_k}-measurable and $y^k = (y_1^{kT}, y_2^{kT}, \ldots, y_i^{kT}, \ldots, y_p^{kT})^T$, where $y_i^k \in R[(\Omega, F, P), R^{n_i}]$. We denote the solution processes of the large-scale and ith isolated hybrid systems, (7.6.2) and (7.6.3), by $x^{k-1}(t) = x(t, t_{k-1}, y^{k-1}), y^{k-1}(t_k) = y(t_k, t_{k-1}, y^{k-1})$ and $\hat{x}_i^{k-1}(t) = \hat{x}_i(t, t_{k-1}, y_i^{k-1}), \hat{y}_i^{k-1}(t_k) = \hat{y}_i(t_k, t_{k-1}, y_i^{k-1})$ through (t_{k-1}, y^{k-1}) and (t_{k-1}, y_i^{k-1}), respectively.

Similarly, other algorithms, namely, the improved Euler and Runge–Kutta methods or the Simpson rule, the trapezoidal rules and other rules, can be introduced in place of $\Delta y_i(k, \omega)$ in (5.1.7).

The hybrid system corresponding to (5.1.1) is described by

$$\begin{aligned} dx &= f(t, x, \omega), \quad x(t_k, \omega) = z, \quad t \neq tk, \\ \Delta y(k, \omega) &= B(t_k, y(k, \omega), \omega), \quad y(k, \omega) = z, \\ y(t_0, \omega) &= y_0(\omega) = x(t_0, \omega) = x_0(\omega), \quad k \in I(0, N), \end{aligned} \tag{7.6.4}$$

where $\{t_{N0}, t_{N1}, \ldots, t_{Nk}, \ldots, t_{NN}\}$ is a partition of $J = [t_0, t_0 + \alpha]$ such that $t_0 = t_{N0} < t_{N1} < \cdots < t_{NN} = t_0 + \alpha$, $t_k = t_{Nk}$, and $h^k = h_N^k = t_{k+1} - t_k = \Delta t_k$, for $k \in I(0, N)$; $z \in \mathbb{R}^n$, and $J_k = [t_k, t_{k+1}]$.

The stochastic large-scale hybrid system (7.6.4) is decomposed into p interconnected subsystems as follows:

$$\begin{aligned} dx_i &= f^i(t, x_i, \omega) + R^i(t, x, \omega), \quad x_i(t_k, \omega) = z_i, \quad t \neq t_k, \\ \Delta y_i(k, \omega) &= b^i(t_k, y_i(k, \omega), \omega) + d^i(t_k, y(k, \omega), \omega), \quad y_i(k, \omega) = z_i, \\ y_i(t_0, \omega) &= x_i(t_0, \omega) = x_i^0(\omega), \quad k \in I(0, N), \quad i \in I(1, p). \end{aligned} \tag{7.6.5}$$

The isolated hybrid subsystems corresponding to (7.6.5) are described as follows:

$$\begin{aligned} dx_i &= f^i(t, x, \omega), \quad x_i(t_k, \omega) = z_i, \quad t \neq t_k, \\ \Delta y_i(k, \omega) &= b^i(t_k, y_i(k, \omega), \omega), \quad y_i(t_k, \omega) = z_i, \\ y_i(t_0, \omega) &= x_i(t_0, \omega) = x_i^0(\omega), \quad k \in [0, N], \quad i \in I(1, p). \end{aligned} \tag{7.6.6}$$

Let $x(t, \omega) = x(t, t_0, x_0, \omega) = (x_1^T, x_2^T, \ldots, x_i^T, \ldots, x_p^T)^T$ and $y(k, \omega) = y(k, k_0, x_0, \omega) = (y_1^T(k), y_2^T(k), \ldots, y_i^T(k), \ldots, y_p^T))^T$ be the solution processes of the continuous-time and discrete parts of the large-scale system (5.1.2) through (t_0, x^0) and (k_0, x^0), respectively. Let

$$\hat{x}_{ix_i^0}(t) = \hat{x}(t, t_0, x_i^0)$$

and

$$\hat{y}_i(k) = \hat{y}_i(k, k_0, x_i^0)$$

be the solution processes of the continuous-time and discrete-time parts of the ith isolated systems (5.1.3) through (t_0, x_i^0) and (k_0, x^0), respectively. Further, for $k \in I(1, N)$, let $x^k(t, \omega) = x(t, t_k, z, \omega)$ and $y^k(t_k, \omega) = y(t_k, t_{k-1}, z)$ be the solution processes of the large-scale hybrid systems (7.6.5) through (t_{k-1}, z), $t \in J_{k-1}$, and let $\{y^k\}_0^N$ be an R^n-valued random process, $y^k = (y_1^{kT}, y_2^{kT}, \ldots, y_i^{kT}, \ldots, y_p^{kT})^T$, where $y_i^k \in R[(\Omega, F, P), R^n]$.

We denote the solution processes of the ith isolated hybrid system, (7.6.5) and (7.6.6), by

$$x^{k-1}(t) = x(t, t_{k-1}, y^{k-1}),$$
$$y^{k-1}(t) = y(t, t_{k-1}, y^{k-1})$$

and

$$\hat{x}^{k-1}(t) = \hat{x}_i(t, t_{k-1}, y_i^{k-1}),$$
$$\hat{y}^{k-1}(t_k) = \hat{y}_i(t, t_{k-1}, y_i^{k-1})$$

through (t_{k-1}, y^k) and (t_{k-1}, y_i^{k-1}), respectively.

7.7 Notes and Comments

The content of Section 7.1 is adapted from Ladde and Lawrence [27] and the related work of Harlow and Delph [12]. The material in Section 7.2 is new and based on the ongoing work of Aleman and Ladde [1], Ladde and Ladde [18, 19], and Ladde, Sathananthan and Pirapakaran [35] and related to the work of Bellomo and Pistone [4]. The prototype models of short-run markets under Markovian random structural and Gaussian process perturbations in Sections 7.3 and 7.4 are based on the works of Griffin and Ladde [10, 11] and Kulkarni and Ladde [17], respectively. See also Griffin [10], Griffin and Ladde [11], Revankar [45] and Turnovsky [48]. Section 7.5 is adapted from Otunuga, Ladde and Ladde [43] and related to the works of Casella and Berger [6] and Otunuga, Ladde and Ladde [44]. Section 7.6 covers a reformulation of a discrete-time iterative process in the framework of a hybrid dynamic process. It is based on work of Ladde [24].

Appendix

A.1 Stochastic

Let us consider the nonlinear system of difference equations with random parameters

$$\begin{aligned} x_k &= h(k-1, x_{k-1}, A(\omega)), \quad k \geq k_0 + 1, \\ x_{k_0} &= x_0, \end{aligned}$$

where $x_0 \in \mathbb{R}^n$ and $A \in \mathbb{R}^m$ are random vectors defined on a complete probability space, (Ω, F, P). It is also assumed that $h(k, x, a)$ is continuously differentiable with respect to $(x, a)^T$. We denote by $x_k = x(k, k_0, x_0, A)$ a solution process of (A.1.1). The system can be rewritten as follows:

$$\begin{aligned} z_k &= H(k-1, z_{k-1}), \quad k \geq k_0 + 1, \\ z_{k_0} &= z_0, \end{aligned} \tag{A.1.1}$$

where $z_k \in \mathbb{R}^{n+m}, z_k = [x_k^T, A^T]^T, H = [h^T, A^T]^T$ and $z_0 = [x_0^T, A^T]^T$. We establish the differentiability of the solution process z_k of (A.1.1) with respect to the initial data z_0. This theorem stochasticizes (Ladde/Lak, 1980) and also provides a rigorous and systematic proof.

Theorem A.1.1. *Let h be a continuously differentiable function with respect to $z = (x^T, A^T)$, satisfying* (A.1.1). *Then, for* $1 \leq i \leq$

$m+n$, $\frac{\partial z}{\partial z_{o_i}}$ exists; moreover, $\frac{\partial z}{\partial z_{o_i}}$ is a solution of the following linear difference equation:

$$y_k = H_z(k-1, z_{k-1})y_{k-1}, \quad y_{k_0} = y_0 = e_i, \tag{A.1.2}$$

where $z_k = z(k, k_0, z_0)$,

$$H_z = \frac{\partial H}{\partial z}, \quad e_i \in \mathbb{R}^{n+m},$$

$z_k = z(k, k_0, z_0)$ is the solution process of (A.1.1) and e_i is the vector with the ith component as 1 and 0 elsewhere.

Proof. Let $z_k = z(k, k_0, z_0)$ and $\bar{z} = z(k, k_0, \bar{z}_0)$ be solutions of (A.1.1) through (k_0, z_0) and $(k_0, \bar{z}_0)$, respectively, where

$$\bar{z}_0 = z_0 + \alpha e_i, \quad \alpha \in \mathbb{R}'\{0\}. \tag{A.1.3}$$

Consider the difference

$$\bar{z}_k - z_k = H(k-1, \bar{z}_{k-1}) - H(k-1, z_{k-1}). \tag{A.1.4}$$

To show that $\frac{\partial z}{\partial z_{o_i}}, 1 \leq i \leq n+m$, exists and is a solution of (A.1.2), it suffices to show that

$$\lim_{\alpha \to 0} \frac{H(k-1, \bar{z}_{k-1}) - H(k-1, z_{k-1})}{\alpha} \tag{A.1.5}$$

exists and is a solution of (A.1.2).

For this purpose, define a function $G : [0,1] \to \mathbb{R}^{n+m}$,

$$G(r) = H(k-1, z_{k-1} + r(\bar{z}_{k-1} - z_{k-1})).$$

Then,

$$\begin{aligned} G'(r) &= \frac{d}{dr}(H(k-1, z_{k-1} + r(\bar{z}_{k-1} - z_{k-1}))) \\ &= H_z(k-1, z_{k-1} + r(\bar{z}_{k-1} - z_{k-1}))(\bar{z}_{k-1} - z_{k-1}). \end{aligned} \tag{A.1.6}$$

Integrating (A.1.5) with respect to r from 0 to 1 yields

$$\int_0^1 G'(r)dr = \int_0^1 H_z(k-1, z_{k-1} + r(\bar{z}_{k-1} - z_{k-1}))(\bar{z}_{k-1} - z_{k-1})dr,$$

or

$$H(k-1,\bar{z}_{k-1}) - H(k-1,z_{k-1})$$
$$= \int_0^1 H_z(k-1, z_{k-1} + r(\bar{z}_{k-1} - z_{k-1}))dr(\bar{z}_{k-1} - z_{k-1}).$$

From this and (A.1.4), one can obtain

$$\begin{aligned}\frac{\bar{z}_k - z_k}{\alpha} &= H_z(k-1, z_{k-1})\frac{\bar{z}_{k-1} - z_{k-1}}{\alpha} \\ &\quad + \int_0^1 [H_z(k-1, z_{k-1} + r(\bar{z}_{k-1} - z_{k-1})) \\ &\quad - H_z(k-1, z_{k-1})]\left(\frac{\bar{z}_{k-1} - z_{k-1}}{\alpha}\right)dr \\ &= H_z(k-1, z_{k-1})\left(\frac{\bar{z}_{k-1} - z_{k-1}}{\alpha}\right) \\ &\quad + \mu_k(\alpha)\left(\frac{\bar{z}_{k-1} - z_{k-1}}{\alpha}\right), \qquad \text{(A.1.7)}\end{aligned}$$

where

$$\mu_k(\alpha) = \int_0^1 [H_z(k-1, z_{k-1} + r(\bar{z}_{k-1} - z_{k-1})) - H_z(k-1, z_{k-1})]dr. \qquad \text{(A.1.8)}$$

For each $k \geq k_0 + 1$, we note that $\bar{z}_0 = z_0 + \alpha e_i$; therefore, $\mu_k(\alpha) \to 0$ as $\alpha \to 0$. Hence, we have the existence of $\frac{\partial z}{\partial z_{0_i}}, 1 \leq i \leq n+m$.

Let

$$y_k = y_k(k_0, z_0, e_i) \qquad \text{(A.1.9)}$$

be a solution of (A.1.2) through (k_0, e_i), and let

$$\Phi(k, k_0, z_0), \text{ with } \Phi(k_0, k_0, z_0) = I, \qquad \text{(A.1.10)}$$

denote the fundamental matrix solution of (A.1.2). Then, the solution process (A.1.9) can be written as

$$y_k = \Phi(k, k_0, z_0)e_i. \qquad \text{(A.1.11)}$$

For $k \geq k_0$, define

$$m_k = \frac{\bar{z}_k - z_k}{\alpha} - y_k,$$
$$m_{k_0} = 0. \tag{A.1.12}$$

From this definition, (A.1.2) and (A.1.8), we obtain

$$\begin{aligned} m_k &= H_z(k-1, z_{k-1}) \left(\frac{\bar{z}_{k-1} - z_{k-1}}{\alpha} \right) \\ &\quad + \mu_k(\alpha) \left(\frac{\bar{z}_{k-1} - z_{k-1}}{\alpha} \right) - H_z(k-1, z_{k-1}) y_{k-1} \\ &= (H_z(k-1, z_{k-1}) + \mu_k(\alpha)) m_{k-1} + \mu_k(\alpha) y_{k-1}. \end{aligned}$$

Therefore, m_k satisfies the following difference equation:

$$u_k = (H_z(k-1, z_{k-1}) + \mu_k(\alpha)) u_{k-1},$$
$$u_{k_0} = 0, \tag{A.1.13}$$

where y_k is the solution process of (A.1.2) and $\mu_k(\alpha)$ is defined in (A.1.8).

Next, we consider the difference equation

$$w_k = (H_z(k-1, z_{k-1}) + \mu_k(\alpha)) w_{k-1},$$
$$w_{k_0} = w_0 = 0. \tag{A.1.14}$$

The fundamental matrix solution of (A.1.14) is given by

$$\Phi_u(k, k_0, z_0), \text{ where } \Phi_u(k_0, k_0, z_0) = I, \tag{A.1.15}$$

and thus the solution process (A.1.14) is represented by

$$w_k = w(k, k_0, w_0) = \Phi_u(k, k_0, z_0) w_0. \tag{A.1.16}$$

Using the variation of constants formula of (1.1.18), the solution process of (A.1.13), $u(k, k_0, w_0)$, with $m_{k_0} = w_0$, is given by

$$\begin{aligned} u(k, k_0, w_0) &= w(k, k_0, w_0) \\ &\quad + \sum_{s=k_0+1}^{k} \Phi_u(k, k_0, z_0) \Phi_u^{-1}(s, k_0, z_0) \mu_s(\alpha) y_{s-1}. \end{aligned}$$

From (A.1.11), (A.1.16) and the fact that $w_0 = 0$, the above equation reduces to

$$u(k, k_0, 0) = \sum_{s=k_0+1}^{k} \Phi_u(k, k_0, z_0)\Phi_u^{-1}(s, k_0, z_0)\mu_s(\alpha)\Phi_u(s-1, k_0, z_0)e_i.$$

For a fixed value of k, the right-hand side of the above equality tends to the zero vector as $\alpha \to 0$. This fact, together with the uniqueness of the solution process of (A.1.13), establishes the convergence of $\frac{\partial z}{\partial z_{0_i}}$ into y_k as $\alpha \to 0$. Therefore, $\frac{\partial z}{\partial z_{0_i}}$ is a solution process of (A.1.2). This completes the proof.

Lemma A.1.1. *Assume that*

$$F \in M[I(k_0 + 1) \times R^n, R[\Omega, R^n]]$$

and its sample derivative

$$\frac{\partial F}{\partial y}(k, y, \omega) = F_y(k, y, \omega)$$

exists and is sample continuous in y, for each $k \in I_0(k_0 + 1)$. Then,

$$F(k, y, \omega) - F(k, x, \omega) = \int_0^1 \frac{\partial F}{\partial y}(k, sy + (1 - s)x, \omega)(y - x)ds. \tag{A.1.17}$$

In particular, if $x = 0$ and $F(k, 0, \omega) \equiv 0$, then

$$F(k, y, \omega) = \int_0^1 \frac{\partial F}{\partial y}(k, 0, \omega)yds. \tag{A.1.18}$$

Theorem A.1.2. *Let W be an n-dimensional random vector with probability density function $f_W(w)$. Let G be a transformation defined on $\mathcal{R}^n$ into itself such that $G(w)$ is continuously differentiable in w and, moreover, G is invertible. Then, the probability density function of*

$$U = G(W) \tag{A.1.19}$$

is given by

$$f_U(u) = f_W(G^{-1}(u)) \left| \frac{\partial G^{-1}(u)}{\partial u} \right|, \tag{A.1.20}$$

where $|\frac{G^{-1}(u)}{\partial u}|$ is the determinant of the Jacobian matrix of G^{-1}.

Lemma A.1.2. *Let u_0 be a random variable and $m(k,\omega)$ and $\nu(k,s,\omega)$ be sequences of random variables. Assume that*

$$m(k,\omega) \le u_0 + \sum_{s=k_0+1}^{k} \nu(k,s,\omega)m(s-1,\omega). \tag{A.1.21}$$

Then,

$$m(k,\omega) \le u_0 \left[\exp\left(\sum_{s=k_0+1}^{k} \nu(k,s,\omega)\right)\right]. \tag{A.1.22}$$

Lemma A.1.3. *Let $m(k,\omega)$, $n(k,\omega)$, and $\nu(k,s,\omega)$ be sequences of random variables. Assume that*

$$m(k,\omega) \le n(k,\omega) + \sum_{s=k_0+1}^{k} \nu(k,s,\omega)m(s-1,\omega). \tag{A.1.23}$$

Then,

$$m(k,\omega) \le n(k,\omega) + \sum_{s=k_0+1}^{k} n(s,\omega)\nu(k,s,\omega)\left[\exp\left(\sum_{\tau=s+1}^{k} \nu(k,\tau,\omega)\right)\right]. \tag{A.1.24}$$

A.2 Ordinary Differential Equations with Random Parameters

In this section, we provide a few results concerning ordinary differential systems with random parameters and initial value problems.

Lemma A.2.1. *Assume that $F \in M[J \times D, R[\Omega, R^n]]$ and its sample derivative $\frac{\partial F}{\partial y}(t,y,\omega)$ exists and is sample continuous in y, for each $t \in J$, where D is an open convex set in R^n. Then,*

$$\begin{aligned} &F(t,y_1,\omega) - F(t,y_2,\omega) \\ &\quad = \int_0^1 \left[\frac{\partial}{\partial y} F(t, sy_1 + (1-s)y_2, \omega)\right](y_1 - y_2)ds. \end{aligned} \tag{A.2.1}$$

Lemma A.2.2. *Assume that the hypothesis of Lemma* A.2.1 *holds. Further, assume that* $x(t, \omega) = x(t, t_0, y_0, \omega)$ *and* $\bar{x}(t, \omega) = x(t, t_0, \bar{y}_0, \omega)$ *are solution processes of* (2.1.1) *through* (t_0, x_0) *and* $(t_0, \bar{y}_0)$, *respectively, existing for* $t \geq t_0$, *such that* $y_0, \bar{y}_0$ *belong to* D. *Then, for* $t \geq t_0$,

$$x(t, t_0, y_0, \omega) - x(t, t_0, \bar{y}_0, \omega) = \int_0^1 \Phi(t, t_0, \bar{y}_0 + s(y_0 - \bar{y}_0), \omega) ds (x_0 - \bar{x}_0), \qquad \text{(A.2.2)}$$

where $\Phi(t, t_0, x_0, \omega)$ *is the fundamental matrix solution process of the variational system*

$$z' = F_y(t, y(t, \omega), \omega) z, \quad z(t_0) = z_0. \qquad \text{(A.2.3)}$$

Theorem A.2.1. *Assume that the hypothesis of Lemma* A.2.1 *holds. Further, assume that* $R \in M[J_+ \times D, R[\Omega, R^n]]$ *and* $R(t, y, \omega)$ *is a.s. sample continuous in* y, *for fixed* $t \in J$. *Let* $y(t, \omega) = y(t, t_0, y_0, \omega)$ *be a solution process of*

$$y' = F(t, y, \omega) + R(t, y, \omega), \quad y(t_0) = y_0, \qquad \text{(A.2.4)}$$

existing for $t \geq t_0$, *and let* $x \in (t, t_0, y_0, \omega) = x(t, \omega)$ *be a solution process of* (2.1.1) *through* (t_0, y_0). *Then, for* $t_0 \leq s \leq t$,

$$\frac{d}{ds} x(t, s, y(s, \omega)) = \Phi(t, s, y(s, \omega), \omega) R(s, y(s, \omega), \omega), \qquad \text{(A.2.5)}$$

where $x(t, s, y(s, \omega), \omega)$ *is the solution process of* (2.1.1) *through* $(s, y(s, \omega))$. *Moreover,*

$$y(t, t_0, y_0, \omega) = x(t, t_0, y_0, \omega) + \int_{t_0}^t \Phi(t, s, y(s, \omega), \omega) R(s, y(s, \omega), \omega) ds. \qquad \text{(A.2.6)}$$

Lemma A.2.3. *Let* C *be a given constant and* k *a given nonnegative sample continuous process on an interval* J. *Let* m *be a sample continuous process defined on* J *into* R_+ *and satisfy*

$$m(t,\omega) \le C + \int_{t_0}^{t} k(s,\omega)m(s,\omega)ds, \quad t \ge t_0 \text{ and } t \in J. \tag{A.2.7}$$

Then,

$$m(t,\omega) \le C \exp\left[\int_{t_0}^{t} k(s,\omega)ds\right], \quad t \in J. \tag{A.2.8}$$

Theorem A.2.2. *Let us assume that m, n and ν are nonnegative processes. Further, assume that*

$$m(t,\omega) \le n(t,\omega) + \int_{t_0}^{t} \nu(t,s,\omega)m(s,\mu)ds. \tag{A.2.9}$$

Then,

$$m(t,\omega) \le n(t,\omega) + \int_{t_0}^{t} n(s,\omega)\nu(t,s,\omega)\exp\left[\int_{s}^{t} \nu(t,u,\omega)du\right] ds. \tag{A.2.10}$$

A.3 Boundary Value Problems

In the following, we present a fixed point theorem and a result concerning a measurable selection of multivalued maps.

Theorem A.3.1 (Schauder's Fixed Point Theorem). *Let X be a real Banach space and C be a closed, bounded and convex subset of X. Let F be a compact map from C into itself. Then, F has a fixed point.*

Theorem A.3.2. *Let $(\Omega, \mathcal{F})$ be a measurable space. Let (X, d) be a separable metric space, $\mathcal{L}$ be the Borel σ-algebra of (X, d) and S be a multivalued map defined on Ω into $2^x \backslash \{\phi\}$ such that:*

(i) *$S(\omega)$ is complete for all $\omega \in \Omega$;*
(ii) *$\rho(x, S(\cdot))$ is measurable for every $x \in X$, where $\rho(x, S)$ is the distance from x to S.*

Then, S admits an $(\mathcal{F}, \mathcal{L})$-measurable selection if there exists a map g from Ω into X such that $g^{-1}(B) \in \mathcal{F}$, for $B \in \mathcal{L}$ and $g(\omega) \in S(\omega)$.

Under the Carathéodory type of condition, we present a comparison theorem.

Theorem A.3.3. *Assume that*

(i) $g \in M[E, R[\Omega, R^n]]$ *and* $g(t, u, \omega)$ *is sample continuous in* u *for fixed* t*, where* $E = [t_0, t_0 + a) \times D$ *and* D *is an open set in* R^n*;*
(ii) $g(t, u, \omega)$ *is almost surely sample quasimonotone nondecreasing in* u*, for each* t*;*
(iii) $r(t, \omega)$ *is the sample maximal solution of the random differential system*

$$u' = g(t, u, \omega), \quad u(t_0, \omega) = u_0(\omega), \tag{A.3.1}$$

existing on $[t_0, t_0 + a)$*;*
(iv)

$$z \in C[[t_0, t_0 + a), R[\Omega, R^n]], t(r(t)) \in E \ w.p.\ 1,$$
$$z(t_0, \omega) \leq u_0(\omega) \ w.p.\ 1$$

and

$$D^+ z(t, \omega) \leq g(t, z(t, \omega), \omega) \qquad a.e.\ t \in [t_0, t_0 + a).$$

Then,

$$z(t, \omega) \leq r(t, \omega) \quad \text{for } t \in [t_0, t_0 + a). \tag{A.3.2}$$

Corollary A.3.1. *If in Theorem* A.3.3, *the inequalities are reversed, then the conclusion of Theorem* A.3.3 *is replaced by*

$$z(t, \omega) \geq \rho(t, \omega), \quad t \in [t_0, t_0 + a), \tag{A.3.3}$$

where $\rho(t, \omega)$ *is the sample minimal solution process of* (A.3.1).

References

[1] R. Aleman and G. S. Ladde, Development of distribution function corresponding to solution of difference or differential equations, Unpublished Manuscript, 2023.

[2] K. E. Atkinson, *An Introduction to Numerical Analysis*. John Wiley, New York, 1989.

[3] I. Babuska, M. Prager, and E. Vitasek, *Numerical Processes in Differential Equations, I. Equations*. Parha, John Wiley & Sons Ltd, London, UK, 1966.

[4] N. Bellomo and G. Pistone, Time evolution of problems, *Mechanical Research Communications*, **6**, 75–80 (1979).

[5] A. T. Bharucha-Reid and M. Sambandham, *Random Polynomials*. Academic Press, New York, 1986.

[6] G. Casella and R. L. Berger, *Statistical Inference*, 2nd edn. Duxbury Advanced Series. Duxbury, Pacific Grove, CA, 2002.

[7] J. F. Chamayou and J. L. Dunau, Random difference equations with logarithmic distribution and the triggered shot noise, *Advances in Applied Mathematics*, **29**(3), 454–470 (2002).

[8] J. F. Chamayou and J. L. Dunau, Random difference equations: An asymptotical result, *Journal of Computational and Applied Mathematics*, **154**(1), 183–193 (2003).

[9] J. Golec and G. S. Ladde, Euler-type approximation for systems of stochastic differential equations, *Journal of Applied Mathematics and Simulation*, **28**, 357–385 (1989).

[10] B. L. Griffin and G. S. Ladde, Qualitative properties of stochastic iterative processes under random structural perturbations, *Abstracts of the American Mathematical Society*, **18**(1), 144 (1997).

[11] B. L. Griffin and G. S. Ladde, Qualitative properties of stochastic iterative processes under random structural perturbations, *Mathematics and Computers in Simulation*, **67**, 181–200 (2004).

[12] D. G. Harlow and T. J. Delph, Numerical solutions of random differential equations, *Simulation*, **33**(3), 243–258 (1991).

[13] K. Hayashi, On stability of numerical solutions of ordinary systems by one-step methods, *TRU Mathematics*, **5**, 67–83 (1969).

[14] P. Henerici, *Discrete Variable Methods in Ordinary Differential Equations*. John Wiley & Sons, New York, 1968.

[15] T. Jankowski, Some remarks on numerical solution of initial problems for systems of differential equations, *Aplikace Matematiky*, **24**(6), 421–426 (1979).

[16] P. E. Kloeden and E. Platen, *Numerical Solution of Stochastic Differential Equations*. Springer-Verlag, New York, 1992.

[17] R. M. Kulkarni and G. S. Ladde, Stochastic stability of short-run market equilibrium: A comment, *The Quarterly Journal of Economics*, **91**, 731–735 (1979).

[18] A. G. Ladde and G. S. Ladde, *An Introduction to Differential Equations: Deterministic Modeling, Methods and Analysis*, Vol. 1. World Scientific Publisher, New Jersey, 2012.

[19] A. G. Ladde and G. S. Ladde, *An Introduction to Differential Equations: Stochastic Modeling, Methods and Analysis*, Vol. 2. World Scientific Publisher, New Jersey, 2013.

[20] G. S. Ladde, Variational comparison theorem and perturbations of nonlinear systems, *Proceedings of the American Mathematical Society*, **52**, 181–187 (1975).

[21] G. S. Ladde, Stability and oscillations in single species process with past memory, *International Journal of Systems Science*, **10**(6), 639–647 (1979).

[22] G. S. Ladde, A few recent advancements in the study of hybrid systems. In Seenita Sivasundaram (ed.), *Proceedings of ICNPAA 2002: IVth International Conference on Nonlinear Problems in Aviation and Aerospace* (European Conference Publications, Cambridge, United Kingdom), pp. 289–296, 2003.

[23] G. S. Ladde, Hybrid systems: Convergence and stability analysis of stochastic large-scale approximation schemes, *Dynamic Systems and Applications*, **13**, 487–512 (2004).
[24] G. S. Ladde, Modern hybrid dynamic inequalities: Role and scope in 21st century, Under Preparation, 2024.
[25] G. S. Ladde and V. Lakshmikantham, *Random Differential Inequalities*. Academic Press, New York, 1980.
[26] G. S. Ladde, V. Lakshmikantham, and S. Leela, A technique in perturbation theory, *Rocky Mountain Journal of Mathematics*, **6**, 133–140 (1976).
[27] G. S. Ladde and B. A. Lawrence, On joint probability density functions of discrete time processes, *Mathematics and Computer in Simulation*, **63**, 629–650 (2003).
[28] G. S. Ladde and M. Sambandham, Stochastic versus deterministic, *Mathematics and Computers in Simulation*, **24**(6), 507–514 (1982).
[29] G. S. Ladde and M. Sambandham, Random difference inequalities. In Lakshmikantham, V. (ed.), *Trends in the Theory and Practice of Nonlinear Analysis*. North Holland, Amsterdam, Netherlands, pp. 231–240, 1985.
[30] G. S. Ladde and M. Sambandham, Numerical solution to random difference equation. In Lakshmikantham V. (ed.), *Nonlinear Analysis and Application*. Marcel Dekker, New York, pp. 279–288, 1987.
[31] G. S. Ladde and M. Sambandham, Variations of constant formula and error estimate to stochastic difference systems, *Journal of Mathematical Physics*, **22**, 557–584 (1988).
[32] G. S. Ladde and M. Sambandham, Numerical treatment of random polynomials, *Applied Mathematics and Computations*, **55**, 13–30 (1993).
[33] G. S. Ladde and M. Sambandham, *Stochastic versus Deterministic Systems of Differential Equations*. Marcel Dekker, Inc., New York, 2004.
[34] G. S. Ladde and M. Sambandham, Generalized variational comparison theorems and nonlinear iterative process under random parametric perturbations, *Communications in Applied Analysis*, **14**, 273–300 (2010).
[35] G. S. Ladde, S. Sathyanathan, and R. Peranakan, *Proceedings of Neural, Parallel and Scientific Computation*, Vol. 1, pp. 257–260, 1995.

[36] G. S. Ladde and D. D. Siljak, Connective stability of large-scale stochastic systems, *International Journal of Systems Science*, **6**, 713–721 (1975).

[37] G. S. Ladde and D. D. Siljak, Convergence and stability of distributed stochastic iterative processes, *IEEE Transactions on Automatic Control*, **31**, 665–672 (1990).

[38] V. Lakshmikantham and S. Leela, *Differential and Integral Inequalities — Theory and Applications*. Academic Press, New York, 1969.

[39] V. Lakshmikantham and D. Trigiante, *Theory of Difference Equations: Numerical Methods and Applications*. Marcel Dekker, Inc., New York, 2002.

[40] M. D. Lax, Numerical solutions of random nonlinear equations, **2**, 163–169 (1985).

[41] M. B. Nevelson and R. Z. Hasminski, *Stochastic Approximation and Recursive Estimation*. American Mathematical Society, Providence, RI, 1973.

[42] T. Ohashi, On conditions for convergence of one-step methods for ordinary differential equations, *TRU Mathematics*, **6**, 59–62 (1970).

[43] O. M. Otunuga, G. S. Ladde, and N. G. Ladde, Discrete time dynamic model for local sample mean and covariance. Preprint (2016).

[44] O. M. Otunuga, G. S. Ladde, and N. G. Ladde, Local lagged adapted generalized method of moments: An innovative estimation and forecasting approach and its applications, *Journal of Time Series Econometrics* (2019). doi:10.1515/jtse-2016-0024.

[45] N. S. Revankar, Stochastic stability of short-run market equilibrium: A comment, *The Quarterly Journal of Economics*, **91**, 724–727 (1971).

[46] D. D. Siljak, *Large-Scale Dynamic Systems-Stability and Structures*. North Holland, Amsterdam, Netherlands, 1978.

[47] D. P. Squier, One-step methods for ordinary differential equations, *Numerische Mathematik*, **13**, 176–179 (1969).

[48] S. J. Turnovsky, Stochastic stability of short-run market equilibrium under variations in supply, *The Quarterly Journal of Economics*, **82**, 666–681 (1968).

[49] Y. Wallach, *Alternating Sequential/Parallel Processing.* Springer, Heidelberg, 1962.

[50] M. T. Wasan, *Stochastic Approximation*, Vol. 58. Cambridge University Press, London, UK, 1969.

[51] A. E. Woul, *New Computing Environments: Parallel, Vector, and Systolic.* SIAM, Philadelphia, PA, 1986.

Index

E

F

G

H

I

J

L

T

V

W

Printed in the United States
by Baker & Taylor Publisher Services